The Evolution of Darwinism

Selection, Adaptation, and Progress in Evolutionary Biology

"How extremely stupid of me not to have thought of that!"
Thomas Henry Huxley, upon first encountering
Dar[illegible] theory of evolution by natural selection

Alas, th[illegible] Darwin's theory is deceptive. From the v[illegible] differing [illegible]pretations, and [illegible] led on a range of fundam[illegible] the nature of selection, the scope of adaptation, and the question of evolutionary progress. This book traces these issues from Darwin's own evolving quest for understanding to ongoing contemporary debates, and explores their implications for the greatest questions of all: where we came from, who we are, and where we might be heading.

Written in a clear and nontechnical style, this book will be of interest to students, scholars, and anyone wishing to understand the development of evolutionary theory.

Timothy Shanahan is Professor of Philosophy at Loyola Marymount University.

The Evolution of Darwinism

Selection, Adaptation, and Progress in Evolutionary Biology

TIMOTHY SHANAHAN
Loyola Marymount University

PUBLISHED BY THE PRESS SYNDICATE OF THE UNIVERSITY OF CAMBRIDGE
The Pitt Building, Trumpington Street, Cambridge, United Kingdom

CAMBRIDGE UNIVERSITY PRESS
The Edinburgh Building, Cambridge CB2 2RU, UK
40 West 20th Street, New York, NY 10011-4211, USA
477 Williamstown Road, Port Melbourne, VIC 3207, Australia
Ruiz de Alarcón 13, 28014 Madrid, Spain
Dock House, The Waterfront, Cape Town 8001, South Africa

http://www.cambridge.org

First published 2004

Printed in the United States of America

Typeface ITC New Baskerville 10/13 pt. *System* LaTeX 2ε [TB]

A catalog record for this book is available from the British Library.

Library of Congress Cataloging in Publication Data

Shanahan, Timothy, 1960–
The evolution of Darwinism : selection, adaptation, and progress in evolutionary
biology / Timothy Shanahan.
p. cm.
Includes bibliographical references and index.
ISBN 0-521-83413-9 (hardback) – ISBN 0-521-54198-0 (pbk.)
1. Evolution (Biology) – History. 2. Natural selection – History. 3. Adaptation
(Biology) – History. 4. Evolution (Biology) – Philosophy. 5. Natural selection –
Philosophy. 6. Adaptation (Biology) – Philosophy. I. Title.
QH361.S43 2004
576.8′2–dc22 2003058439

ISBN 0 521 83413 9 hardback
ISBN 0 521 54198 0 paperback

"What a magnificent view one can take of the world: Astronomical causes, modified by unknown ones, cause changes in geography & changes of climate superadded to change of climate from physical causes – these superinduce changes of form in the organic world, as adaptation & these changing affect each other, & their bodies, by certain laws of harmony keep perfect in these themselves – instincts alter, reason is formed, & the world peopled with Myriads of distinct forms from a period short of eternity to the present time, to the future – How far grander than idea from cramped imagination that God created.... How beneath dignity of him, who is supposed to have said let there be light & there is light."

– Charles Darwin, D Notebook, pp. 36–37 [6 August 1838]

Contents

Introduction

> Let me lay my cards on the table. If I were to give an award for the single best idea anyone has ever had, I'd give it to Darwin, ahead of Newton and Einstein and everyone else.
>
> (Dennett 1995, p. 21)

Listen to Your Mother

In later life the eminent physiologist Sir Charles Sherrington recalled that, as a young man in 1873, as he was departing his home for a summer holiday, his mother persuaded him to take along a copy of the *Origin of Species*, saying "It sets the door of the universe ajar!" (quoted in Young 1992, p. 138). Sherrington's mother was right. No other scientific theory has had such a tremendous impact on our understanding of the world and of ourselves as has the theory Charles Darwin presented in that book.

This claim will undoubtedly sound absurd to some familiar with the history of science. Surely the achievements of Copernicus, Galileo, Newton, Einstein, Bohr, and other scientists who developed revolutionary views of the world are of at least equal, if not greater, significance. Aren't they? Not really. Although it is true that such scientific luminaries made fundamentally important contributions to our understanding of the physical structure of the world, in the final analysis their theories are about *that* world, whether or not it includes life, sentience, and consciousness. Darwin's theory, by contrast, although it encompasses the entire world of living things, the vast majority of which are not human, has always been understood to have deep implications for our understanding of *ourselves*. Look at it this way: Part of what makes human beings distinct from other living things is our impressive cognitive abilities. Unlike other species that simply manage to make a living in the world, we strive – and sometimes succeed – in *understanding* the world as well. It is partly in virtue of our ability to understand key aspects of the world that we have been so successful as a species. Our best means of understanding the natural world in a genuinely deep sense is through the scientific theories we create. But note: These scientific theories are the products of brains, which are themselves the products of natural processes. Darwin's theory provided the framework for the first credible naturalistic explanation for human existence, including the origin, function, and nature of those capacities that enable us to ponder why we have the characteristics we do. In other words, there is an important *asymmetry* between Darwin's and all other scientific theories. No other scientific theory purports to explain the capacities that permit us to devise and contemplate scientific theories, but Darwin's theory – precisely because the correct explanation for the evolution of human cognitive abilities lies within its domain – provides just such a framework. There is simply no other scientific theory that even comes close to playing this central role in our quest for self-understanding. The importance of understanding Darwin's theory cannot be overestimated.

"How Extremely Stupid Not to Have Thought of That!"

> If superior creatures from space ever visit earth, the first question they will ask, in order to assess the level of our civilization, is: 'Have they discovered evolution yet?'
>
> (Dawkins 1989a, p. 1)

In one sense, of course, Darwin's theory of evolution by natural selection is among the simplest scientific theories ever advanced. Living things vary among themselves. These variations arise randomly, that is, without regard to whether a given variation would be beneficial or not. Those living things with advantageous variations tend to stick around a bit longer than others, and give rise to more like themselves. Hence their numbers increase. That's the essence of Darwin's theory. What could be simpler? As Darwin's friend and scientific advocate Thomas Henry Huxley (1825–95) is reported to have exclaimed after first encountering the idea of natural selection, "How extremely stupid of me not to have thought of that!"

Alas, the apparent simplicity of Darwin's theory is deceptive. From the very beginning Darwin's great idea has been subject to differing interpretations, and even now professional opinion is sharply divided on a range of fundamental issues. These are not challenges to Darwinism from without (such as "Scientific Creationism") that question the entire project of giving naturalistic explanations of living things but, rather, debates *within* Darwinism about the most basic causes, processes, and expected outcomes of natural selection. Central among these are debates about the nature and operation of natural selection, the scope and limits of adaptation, and the question of evolutionary progress.

Selection, Perfection, Direction

> As natural selection works solely by and for the good of each being, all corporeal and mental endowments will tend to progress towards perfection.
>
> (Darwin 1859, p. 489; 1959, p. 758)

So wrote Charles Darwin in all six editions of the *Origin of Species.*[1] What he meant by this claim, how later biologists have treated the issues it addresses, and whether (or in what sense) this claim might be true, are the subjects of this book.

Part I focuses on natural selection, the central theoretical principle of Darwinism. Selection explains why living things display complex adaptations, giving them the appearance of having been intelligently designed. But life exists on many "levels," with biological systems organized hierarchically from genes and cells up through species and ecosystems. Selection is usually thought of as acting upon organisms. But does selection act at other levels as well? How did Darwin think about the level(s) at which selection operates and forges adaptations (Chapter 1)? Does selection operate at levels "above" individual organisms, e.g., at the level of groups (Chapter 2)? What has led biologists to argue about the correct "unit of selection," and how are such disputes best resolved (Chapter 3)?

Part II examines the issue of biological "perfection." The two most striking general facts about the living world that require explanation are the sheer diversity of forms of life, and the incredible adaptive fit between living things and their environments. It has sometimes even been claimed that organisms are *perfectly* adapted to their ways of life. But is the idea of perfect adaptation even coherent? How did Darwin view the issue of biological perfection (Chapter 4)? How have biologists after Darwin understood the relationship between natural selection and adaptation (Chapter 5)? What degree of biological perfection does the theory of natural selection predict, and what factors prevent living things from achieving perfect adaptation (Chapter 6)?

Part III examines the controversial issue of "evolutionary progress." It has seemed obvious to many biologists that there has been an overall direction in the evolution of life toward more complex, sophisticated organisms. Once there were only the simplest sorts of living things – replicating molecules, perhaps. Now the world burgeons with innumerable species displaying amazing adaptations fitting them for every conceivable niche in the economy of nature. How could anyone who accepts an evolutionary view of life deny that progress has occurred? Yet perhaps no other issue in evolutionary biology has inspired such passionate controversy. How did Darwin approach the issue of evolutionary progress (Chapter 7; additional discussion of this highly contested issue appears in the Appendix)? How have later biologists addressed this issue (Chapter 8)? Does talk of "higher" and "lower" organisms make sense? Are some organisms more "advanced" than others? Is there an overall direction to evolution? In the final analysis, does it make any sense at all to describe evolution as "progressive" (Chapter 9)?

Although different parts of the book focus on each of the three issues of "selection," "perfection" (adaptedness), and "direction" (progress), they are closely related to one another, and the interconnections between them are as interesting as the details of each one taken separately. As noted above, Darwinism is uniquely important as a scientific theory in large part because it bears directly on the origin, nature, and destiny of the human species, including explanations for both our "corporeal and mental endowments," as Darwin called them. The final chapter explores these issues as they relate to our self-understanding as a species. Can selection account for the most distinctive human characteristics? How well adapted, in body and mind, are human beings? Was there anything inevitable about the evolution of *Homo sapiens*? Finally, given our best current understanding of evolution, what sort of fate might our species anticipate? Such questions are addressed by reviewing the results of earlier chapters with an eye to understanding their significance for human evolution. They form the bulk of Chapter 10.

Science and Religion

> [W]e are not here concerned with our hopes or fears, only with the truth as far as our reason allows us to discover it.
>
> (Darwin 1871, vol. 2, p. 405)

Having said this, one might naturally expect to find an extended discussion of the implications of evolutionary ideas for traditional religious conceptions of humankind. After all, for many nonbiologists (and even for some biologists), "Darwinism" is inextricably linked to theological issues.[2] This is understandable. In the public mind, Darwinism and "creationism" are often seen as locked in a battle for the hearts and minds (and souls) of men. From the very beginning, friends and foes alike have seen in Darwin's theory profound implications for religious beliefs about the origin, nature, and destiny of human beings. Are we the special creations of a loving Deity, made in His image, or the accidental by-products of a blind, purposeless process which never had us (or anything, for that matter) in mind in the first place? Do we have immaterial souls which distinguish us from all other living things, making possible self-consciousness, a conscience attuned to the dictates of morality, and the hope for immortality, or are we simply bipedal primates whose peculiar adaptation consists in a hypertrophied neocortex, enabling us to ponder questions whose answers lie forever beyond the range of our impressive (but bounded) cognitive abilities? Do each of us as individuals have a glorious (or hellific) future to anticipate, or will each of us at the moment of death simply cease to exist, the personal analog of the extinction that has determined the destiny of 99.99 percent of all species that have ever existed?

It would be tempting to try to draw definitive conclusions about such matters from a survey of Darwinian ideas. Many have succumbed to this temptation, often cloaking deeply entrenched personal opinions in the thinnest of scientific attire (e.g., Provine 1988). Matters are rarely so simple, and the implications of Darwinism for perennial questions such as "the meaning of life" are not straightforward (Miller 1999; Ruse 2000; Stenmark 2001). The reader will look in vain for such a discussion in the present book, which focuses on Darwinism *per se*, rather than on its relationship to other (nonscientific) issues. I want to leave entirely open the question of whether a Darwinian view of life is compatible with a religious view of life. (This is, incidentally, the very same approach that Darwin took in the *Origin of Species.*) The reasons for this exclusion are both practical and philosophical. Practically, this would be a much different, and much longer, book were it to address such issues. Philosophically, the relationship between evolutionary ideas and religious beliefs is far more subtle and complex than is often supposed. Besides, any *serious* discussion of the relationship between Darwinism and religious belief presupposes an historically informed and philosophically critical understanding of evolution – just what this book attempts to provide. Readers are invited to follow out the implications for religious belief of the various evolutionary ideas discussed in this book, if they wish, but they will receive no direct assistance from this book itself. Its central concerns lie elsewhere.

Methodological Confessions

> [One] does not know a science completely as long as one does not know its history.
>
> (Auguste Comte; quoted in Kragh 1987, p. 12)

Like life itself, scientific theories are historical entities whose present forms are products of the past, and are thus fully comprehensible only when understood against this background. This is perhaps especially true for ideas concerning evolution, since controversy has accompanied evolutionary thought from the very beginning. Consequently, the discussions that follow approach each of the main topics of the book (selection, adaptation, progress) historically by looking first at early views (especially those of Darwin), then moving forward as the ideas were further developed and modified in the twentieth century, and finally ending with contemporary views and debates. There is plenty of history in the pages that follow. Nonetheless, one thing the reader will not find in this book is history for history's sake. I have enormous respect for historians and for the work they do. The fruits of their researches inform many of the discussions that follow. But the history presented here always has one eye on the present, in the sense that contemporary debates determine which aspects of the history of evolutionary thought merit detailed discussion. In this sense the history discussed here is "presentist" – a serious sin from the perspective of some historians, but one which is necessary to accomplish the task at hand.[3]

The historical treatments that follow are therefore necessarily selective. When a cartographer surveys a tract of land, certain features stand out as peaks and high points, while others drop below the line of sight. Both are important, but every feature of the landscape cannot be included in the final map. Likewise, in surveying the scientific landscape of the development of evolutionary biology, certain episodes stand out as deserving of special treatment. This study is organized around these high points.[4]

Darwin's Long Shadow

> No other field of science is as burdened by its past as is evolutionary biology. . . . The discipline of evolutionary biology can be defined to a large degree as the ongoing attempt of Darwin's intellectual descendants to come to terms with his overwhelming influence.
>
> (Horgan 1996, p. 114)

Our examination of the three major topics of this book – selection, perfection, and direction – begins with an examination of Darwin's views on each of these topics. Understanding Darwin's views is fundamental. Darwinism begins with Darwin, and if we wish to understand how Darwinism has changed – the "evolution of Darwinism" – then we will need to know what Darwinism was in its original formulation(s). Such understanding can then serve to anchor our examinations of later developments. Getting clear about Darwin's own view is important for an additional reason. More than any other figure, Darwin continues to function as the patron saint of evolutionary biology. Showing that one's own view is the same as Darwin's can serve as a powerful rhetorical device in legitimating one's view. It therefore becomes important to have an accurate account of Darwin's views on these topics.

Given the number of years that have passed between the publication of Darwin's works and the present, it would be natural to suppose that all is now well understood about how he conceived of the fundamental nature of the evolutionary process. But this would be mistaken. Although he generally wrote with admirable clarity, the exact nature of Darwin's views on a number of basic issues remains a matter of scholarly dispute. Understanding precisely what he had in mind raises difficult interpretive problems which, given his critical historical role in the development of evolutionary biology, are worth examining and attempting to resolve.

The title of this book reflects the *dual* goals it aims to achieve: First, to convey an understanding of the *sort of evolution* that forms the basis for contemporary Darwinism (i.e., evolution and its products as understood from a Darwinian perspective); second, to understand how *Darwinism itself* has evolved (i.e., developed historically) in its understanding of the living world. Accomplishing both of these aims requires tackling a range of difficult historical, scientific, and philosophical issues. Let's get to it.

PART I

SELECTION

1

Darwin and Natural Selection

> Natural Selection, as we shall hereafter see, is a power incessantly ready for action, and is as immeasurably superior to man's feeble efforts, as the works of Nature are to those of Art.
>
> (Darwin 1859, p. 61)

Introduction

"After having been twice driven back by heavy south-western gales, Her Majesty's ship *Beagle*, a ten-gun brig, under the command of Captain Fitz Roy, R. N., sailed from Devonport on the 27th of December, 1831" (Darwin 1839, p. 1). So begins Darwin's travel journal, *The Voyage of the Beagle*, published in 1839. The purpose of the expedition was to survey the South American coast and to make chronometrical measurements. The twenty-two-year-old Darwin had signed on as (unofficial) ship naturalist and (official) "gentleman dining companion" for the captain. The expedition was planned as a two-year voyage. In fact, it would be nearly five years before the *Beagle* returned to England (29 October 1836). Its voyage proved to be the seminal experience in Darwin's life.

A Theory by Which to Work

The story of Darwin's discovery of "evolution by means of natural selection" has been told many times (e.g., Bowler; 1989; Young 1992). Although scholars continue to debate the relative importance of one or another element in this story, there is nonetheless widespread agreement on the basic factors that led Darwin to his theory. Prior to his voyage on the *Beagle*, Darwin had spent three years at Cambridge University, training to be a country parson, and before that had studied medicine at the University of Edinburgh. Having discovered that he was more interested in beetle collecting and "geologizing" than either medicine or theology, Darwin abandoned his course of studies and eagerly sought and (with the help of some well-placed connections) secured a place aboard the H.M.S. *Beagle* for its voyage around the world. At each place the ship docked, Darwin made arduous trips inland to collect plants, animals, fossils, and rocks. Despite being seasick for much of the voyage, he took extensive notes on the geology and biology of each area. On his return to England in October 1836, thanks to the correspondence he maintained with scientists at home, Darwin was welcomed as a respected and accomplished naturalist. He immediately set to work sorting out the material and observations he had collected on the voyage.

Darwin opened his first private notebook recording his evolutionary speculations in July 1837.[1] In it he considered how the "transmutation" of one species into another could account for some of the observations made during his voyage. For example, finches on the Galapagos Archipelago (six hundred miles due west of Ecuador) differed dramatically from one island to another, yet all resembled finches on the South American mainland in their basic structure, despite the fact that the volcanic islands represented a quite different environment. The resemblance could be explained, Darwin realized, by supposing that a few individuals from the mainland were carried by storms out to the islands, where their descendants then became modified to each different island environment. Over sufficient time, each form had evolved into a new species. Darwin also realized that this explanation could be generalized. In a world characterized by environmental change, some individuals will vary in a way that better fits them to the new circumstances. With sufficient change, the descendants of these individuals will form new species. Others will fail to adapt and will go extinct, leaving gaps between those forms remaining. This would account for the large differences between some species but not between others. Darwin became convinced that this account was true, and by the end of 1837 was in search of a cause of this species formation.

Famously, it was Darwin's reading ("for amusement") of the Reverend Thomas Malthus's *Essay on the Principle of Population* (1798) in September 1838 that, he said, provided the crucial insight he needed (Darwin 1958, pp. 119–20). Malthus had noted that populations tend to increase faster than their food supply, leading to a struggle for existence amongst their members. Darwin realized that any variations among individuals providing an advantage over others would help those individuals to survive, and disadvantageous variations would tend to be eliminated from the population. If the beneficial variations were passed on to offspring, there would be a gradual change as successive individuals became better adapted to their environments. As Darwin later wrote: "Here, then, I had at last a theory by which to work" (Darwin 1958, p. 120). Having the theory in hand, he began collecting additional evidence to show that it would explain a wide range of otherwise puzzling phenomena.

The theory was sketched out briefly for the first time in an essay in 1842, and then enlarged further in an essay of 1844 (F. Darwin 1909). It is significant that in the latter work Darwin was putting his ideas on paper in the same year that a book espousing a very different account of the evolution of life appeared. Although it enjoyed a degree of popular success, *Vestiges of the Natural History of Creation* (1844), written by Robert Chambers but, wisely, published anonymously, was generally scorned by the scientific community as embodying the worst sort of unfounded evolutionary speculation. Chambers's suggestion, for example, that mammals had evolved from birds via platypuses as an intermediary, received the ridicule it deserved. Darwin had no intention of subjecting his own ideas to the same hostile reception. He decided to amass much more evidence to support his theory before going public with it.

As it turned out, it would be another fifteen years before Darwin would be ready to present his theory to the world, during which time he continued to work on various biological problems.[2] The crucial event that forced his hand was the arrival in the post in June 1858 of a paper by another English naturalist, Alfred Russel Wallace (1823–1913), which sketched out a theory so similar to Darwin's own that Darwin wrote to his friend and confidant the geologist Charles Lyell, "If Wallace had my MS [manuscript] sketch written out in 1842, he could not have made a better short abstract!" (F. Darwin 1887, vol. 1, p. 473). Darwin immediately set to work on composing an "abstract" of his theory. The result was *On the Origin of Species,* published in November 1859.

The *Origin* was an instant bestseller, quickly selling out its entire first printing of fifteen hundred copies on the day it was published (24 November 1859). In Darwin's lifetime it sold over twenty-seven thousand copies in Britain alone. Much of its success can be attributed to the fact that Darwin wrote it as a summary of his theory rather than as the more extensively documented tome he had originally intended, thus making it accessible to a much wider audience. Others had proposed evolutionary views before. What was novel in Darwin's theory was the central role given to what he called "natural selection," a seemingly simple idea with profound implications. In the "Introduction" Darwin provides the best concise statement of evolution by natural selection anyone has ever given:

> As many more individuals of each species are born than can possibly survive; and as, consequently, there is a frequently recurring struggle for existence, it follows that any being, if it vary however slightly in any manner profitable to itself, under the complex and sometimes varying conditions of life, will have a better chance of surviving, and thus be *naturally selected.* From the strong principle of inheritance, any selected variety will tend to propagate its new and modified form. (Darwin 1859, p. 5; emphasis in original)

Later we will examine various aspects of Darwin's theory in detail, but at the outset it is important to understand what was different – and to many of his contemporaries, objectionable – about this theory. As a number of writers have pointed out, it wasn't so much Darwin's advocacy of evolution that was novel or disturbing. By 1859 evolutionary ideas had become almost commonplace. Rather, what was disconcerting was the idea that natural selection operating on chance variations produced the diversity and apparent design in nature. Darwin's theory seemed to make evolution more blind and haphazard than anyone had imagined. One way to appreciate the novelty of these aspects of Darwin's theory is to contrast it with an account of evolution in which chance variation and natural selection are *not* key explanatory elements. We can then return to examine specific aspects of Darwin's theory more closely.

"Nature's Plan of Campaign"

Jean Baptiste Pierre Antoine de Monet, Chevalier de Lamarck (1744–1829) stands out as the most important evolutionary theorist before Darwin. Some previous thinkers, for example, George Louis Leclerc, Comte de Buffon (1707–88), had toyed with the idea of limited species change based on different environments, but no fully developed evolutionary theory appeared before Lamarck's at the beginning of the nineteenth century. His evolutionary speculations appear in three works: In the introduction to his *System of Invertebrate Animals* (1801); more fully in his most famous work, *Zoological Philosophy* (1809); and finally, in the introduction to his *Natural History of Invertebrates* (1815).[3]

In keeping with the natural history tradition since Aristotle, Lamarck accepted the idea that the major classes of organisms can be arranged in a linear series of increasing complexity. But, whereas Aristotle was content simply to describe this series, Lamarck wanted to explain it as a true historical sequence produced by a gradual evolutionary process taking place over an immense period of time. According to Lamarck, "Nature, in successively producing all species of animals, beginning with the most imperfect or the simplest, and ending her work with the most perfect, has caused their organization gradually to become more complex" (Lamarck 1809, p. 60). The various classes of organisms we observe today (e.g., insects, fishes, amphibians, reptiles, birds, mammals) were explained as the result of this primary complexifying process.

To explain this process Lamarck postulated an "endowment" (or "law"), according to which animal life has the inherent power of acquiring progressively more complicated organization. As organisms move up this ladder of organization, vacant morphological space at the bottom is continually being replenished with lower forms (e.g., worms) arising from spontaneous generation from inanimate matter. In Lamarck's view, biogenesis (the origination of life from nonlife) was not a singular unique event in the history of the earth, but rather a continuous and ongoing process. It follows that different lineages begin their ascent up the ladder of complexity at different times. Thus part of the diversity we observe is simply the result of different lineages having begun at different times, with the secondary result that each has so far progressed to a different stage in its upward ascent. The lineage that includes *Homo sapiens* is the oldest, because it alone has reached the highest stage of development. Given the movement involved in this picture, an escalator rather than a ladder is perhaps a better representation.

This complexifying process is the primary cause of organic diversity. As Lamarck realized, however, another force must also be at work: "If the cause which is always tending to make organization more complex were the only one affecting the form and the organs of animals, the increasing complexity of organization would everywhere follow an extremely regular progression. But this is not the case" (Lamarck 1809, p. 130). That is, were the intrinsic tendency toward increasing complexity the *only* cause of evolutionary change, then one might expect to observe a single linear sequence of forms, grading smoothly from the simplest to the most complex. In fact, however, the living world is characterized by tremendous diversity in which it is difficult to locate every species on a simple scale of increasing complexity. A second biological datum requiring explanation is the diversity of forms *within* each major class of organism. "Mammals" comprise many different kinds of animals, for example, rodents, canines, felines, etc. Likewise, "felines" are represented by leopards, lions, jaguars, tigers, ocelots, and so on. As Lamarck noted, "The organization of animals, in its growing complexity, from the least to the most perfect, presents only an *irregular gradation* of which the whole extent displays a large number of anomalies or deviations which have no apparent order in their diversity" (Lamarck 1809, p. 221). In order to account for this diversity of forms, Lamarck realized, there must be other forces at work besides the intrinsic drive toward perfection.

To explain this level of diversity Lamarck posited a secondary process of adaptation to environmental conditions. To survive, organisms must be able to interact successfully with their environments which are always changing. As environments change, new needs (*besoin*) are induced within organisms. These needs result in changes in the animal's "efforts" or "habits," with a corresponding increased use of relevant parts of the body. Lamarck postulated "vital fluids" that are forced into specific parts of the body, causing these body parts to hypertrophy, thus helping the organism to meet its needs more effectively. Likewise, if an organ or part is no longer needed, it falls into disuse and gradually atrophies, eventually disappearing altogether. Structural changes thus induced are then passed on to offspring. The cumulative effect of this process is the appearance of different kinds of organisms, and eventually entirely different species. This is the infamous "inheritance of acquired characteristics" doctrine usually associated with Lamarck, but that he neither originated nor was specifically criticized by his contemporaries for holding, since it was widely accepted in his day. The controversial part of Lamarck's theory for his contemporaries was the way he incorporated the idea of "vital fluids" as responsible for naturally-occurring structural changes. Lamarck's view was radical because, rather than being fitted by God or nature with a constant structure for specific environments, he saw organisms as undergoing changes simply as a result of natural processes operating within and upon organisms in the particular environments in which they found themselves. It was the speculative *naturalism* of Lamarck's account, rather than its evolutionary character *per se*, that so many of his contemporaries found objectionable.

In summary, Lamarck viewed the production of living things as the result of two different kinds of forces. On the one hand, there are forces that underlie the natural tendency of living things to complexify according to a preordained scale of perfection, an inherent power of acquiring progressively more complicated organization that tends toward the production of a *regular gradation* of living things from simple to complex. On the other hand, interfering forces orthogonal to these prevent living things from arriving at their idealized natural state. Adaptation to different environmental conditions disrupts the smooth progression in complexity, resulting in diversity. As Lamarck described this twofold process: "The progression in the complexity of organization suffers, here and there, in the general series of animals, from anomalies produced by the influence of the circumstances of the environment, and by those of the habits contracted" (Lamarck 1809, p. 133). Consequently, only the main types of organization (families or classes) could be arranged in a single series of increasing complexity. Because of adaptation to changing environments, species cannot be arranged in a simple series of higher or lower. Thus the central upward tendency of nature "only appears in a general way, and not in the details" (Lamarck 1815, p. 52). The following quote nicely captures Lamarck's overall view:

> Nature's plan of campaign in the production of animals is clearly marked out by [a] primal and predominant cause, which endows animal life with the ability to complicate organization progressively, and to complicate and perfect gradually, not only the total organization, but also each system of organs in particular.... But a quite separate cause, an accidental and consequently variable one, has here and there cut across the execution of this plan, without however destroying it.... This cause... has given rise to whatever real discontinuities there may be in the series, and to the terminated branches which depart from it, at various points, and diminish its simplicity, and finally to the anomalies to be seen in the various organ-systems of the different organizations. (Lamarck 1815, p. 133)

Despite the fact that Lamarck is now considered to have gotten it almost completely wrong, his theory was nonetheless a serious effort to explain certain accepted but problematic facts about nature. First, many forms uncovered in the fossil record are no longer extant. Likewise, there is no evidence in the fossil record of many of the forms we see today. Clearly there has been a tremendous replacement of organic forms over time. Second, an inspection of extant animals shows that they form a graded series of increasing complexity. Organisms can be more or less arranged along a *scala naturae* ranging from bacteria to *Homo sapiens*, with each step along the ladder exhibiting greater complexity. Third, organisms display amazing diversity, which must be explained in some way. Finally, organisms seem exquisitely well-suited for their particular environments. Organic replacement, increasing complexity, diversity, and fitness are four primary biological phenomena Lamarck correctly recognized as in special need of explanation. Providing correct explanations of each is, of course, important, but the importance of correctly identifying and taking seriously the problems to be solved should not be underestimated. Lamarck's contributions in this regard were seminal.

Ideals of Natural Order

Clearly there are fundamental differences between Lamarck's and Darwin's theories, differences that are critical for understanding both the nature of Darwinian evolution and the nature of Darwinian explanations. Stephen Toulmin (1961) suggests that every scientific explanation presupposes an "ideal of natural order" that permits the inquirer to distinguish between what is the "natural," normal state of a thing, to be taken for granted and used in framing explanations, and which phenomena depart from this natural state and therefore require explanation. A short digression into the history of physics will help to bring Toulmin's central idea into focus.

Consider the very different starting points for Aristotelian and Galilean dynamics. Aristotle formulated his physics of motion by generalizing from a commonsense explanation of a moving object: A cart being pulled by a horse. The cart continues to move just insofar as the horse continues to pull it along. Two factors are at work: The external agency (the horse) keeping the body in motion, and resistance (the weight of the cart) tending to bring the motion to a stop. Aristotle realized that this explanation could be generalized for any moving body. Explaining the motion of any body means recognizing that a body moves at the rate appropriate to an object of its weight, when subjected to just that particular balance of force and resistance. In order for an object to remain in motion, a force must be continually exerted. Relax the force being exerted, and the object in motion will eventually come to rest. Being "at rest" is the natural state of any natural substance, and requires no special explanation. Being "in motion" requires special explanation. Complete rest, or steady motion under a balance of actions and resistances, is the natural motion of an object. Anything that can be shown to exemplify this balance will thereby be explained.

As is well known, the science of motion underwent a dramatic revolution in the seventeenth century in which the ideal of natural order at the heart of Aristotelian physics was abandoned and replaced by another, quite different conception. The most radical single step was taken by Galileo, who argued that rest and uniform motion are equally "natural" for bodies, with neither in need of explanation. Only *changes* in motion, for example, acceleration, require special explanation. This looks, at first glance, very like our modern "law of inertia." Yet Galileo's conception of motion is no more identical with our own than is Aristotle's. In some important respects it is closer to Aristotle's conception than it is to ours. Whereas Aristotle's model of motion was a cart being pulled by a horse, Galileo's model was that of a ship moving steadily across the ocean and disappearing over the horizon, its motion describing a curve. Only some active force could deflect the ship from its circular path. He thus took *circular* motion to be entirely natural and therefore not in need of explanation. Such a conception proved extremely useful for "explaining" the motions of the heavenly bodies. Because they move with uniform speed in perfect circles (Galileo believed), their motion is entirely natural and therefore in need of no special explanation.

When we turn to Newton we find that the ideal of natural motion has changed once again. The paradigm example of motion is now a body moving at uniform speed in a Euclidean straight line, completely unaffected by any external forces. A body's motion is treated as natural and not in need of explanation only when it is unaffected by all forces, including its own weight – a situation that is never observed in the real world. But the ideal example doesn't need to be observed because it provides a standard against which a body's actual motion requires explanation. Newton's first law of motion, the principle of inertia, represents an ideal of natural order supplying a standard of rationality and intelligibility for understanding and explaining natural phenomena. Once this new theoretical ideal was accepted, and with a little help from the hypothesis of universal gravitation, dozens of previously puzzling phenomena fell into an intelligible pattern. Newton's ideal of natural order structured physical explanations right up to the twentieth century, when Einstein's development of relativity theory fundamentally altered our conceptions of the physical world once again.

Stepping back now from the details of the different models just described, it becomes clear that what counts as a successful explanation in physics, and indeed even of what natural phenomena require explanation, is intimately related to ideas about the fundamental order of nature. Any dynamical theory involves some explicit or implicit reference to a paradigm example which specifies the manner in which, in the natural (or ideal) course of events, bodies may be expected to move. By comparing the motion of any actual body with this paradigm example, that which requires explanation can be determined. Every step of the explanatory project is governed and directed by the fundamental conceptions of the theory.

What is true in physics is equally true in biology. As Ernst Mayr (1988) notes, others before Darwin had attempted to explain the diversity of living things, but Darwin provided a new *kind* of theory by reversing what could be taken for granted, and what required special explanation. Once again, Aristotle made the significant original contribution to explaining biological phenomena. As in his physics, so, too, in his biology, Aristotle employed a natural state model according to which the forces acting on an entity or set of entities can be partitioned into two kinds: Forces that ground the natural tendency of the kind of entity being considered, and interfering forces which may prevent the entities in question from arriving at their natural state. A familiar nonbiological example is water, which has a natural tendency to flow from higher to lower elevation, but whose actual movement in that direction can be obstructed by interfering forces such as a dam, becoming frozen, and so on. Aristotle's favored biological example was an acorn whose natural tendency to develop into an oak tree can be thwarted by any of a number of interfering forces, such as drought, consumption by a squirrel, and so on. In Aristotle's model, individual organisms (oak trees, squirrels, etc.) are specimens of types each of whose essence is fixed and immutable. Individual variability is real but represents departure from the ideal type defining each species. Departure from this ideal type therefore requires explanation. Aristotle devotes considerable attention in the *Generation of Animals* to accounting for "monsters" and other less dramatic deviations from the ideal species type. As in his physics, so, too, in his biology, a natural-state model determines which natural phenomena are and are not in need of explanation.

From our current perspective, Lamarck's theory can be seen as transitional between Aristotle's and Darwin's. Like Aristotle, Lamarck held that species themselves are fixed and immutable. Species themselves do not change.[4] Unlike Aristotle, Lamarck believed that the living world is characterized by significant change and replacement over time. Organisms come to occupy different rungs (i.e., instantiate different species) as they progress up the ladder of phylogenetic development. On this scheme, individual organisms evolve; species do not. The apparent replacement of some species by others is to be explained by individuals of the former being gradually transformed, that is, evolving, into individuals of the latter – something that is unintelligible in Aristotle's biological theory.

Darwin's approach differs fundamentally from that of both Aristotle and Lamarck. According to Aristotle's essentialist, "typological" approach to variability, the type (species) is fixed and primary, and individual variability is derived and in need of explanation in terms of interfering forces. Likewise, for Lamarck, "All of the races of living bodies continue to exist *in spite of* their variations" (Lamarck 1809, p. 55). Individual variations are viewed as a kind of "noise"' disrupting the directional process from the simple to the complex. They are evolutionary dead-ends, not leading to new lines of development. Darwin's "populational" theory entails a complete reversal of these approaches (Mayr 1963, 1976; Sober 1980, 1985). According to this view, individual variability is fundamental (and largely unexplained), and the existence of types (e.g., species) requires special explanation. Species exist precisely *because* of naturally occurring variations. Organic variation is the natural result of the absence of interfering forces. Uniformity (species) results from interfering forces (e.g., geographical isolation, which prevents individuals from interbreeding). For example, whereas Aristotle and Lamarck would explain variations in the height of oak trees as due to interfering forces affecting the oak's natural tendency, Darwin would treat the variation as natural (as reflecting a "norm of reaction" in contemporary parlance), with the fact that the trees instantiate the restricted height distribution they do as in need of explanation (e.g., in terms of selection against individuals that depart significantly from the mean). In Darwin's hands the explanandum (that which requires explanation) and the explanans (that which does the explaining) are reversed. Aristotle and Lamarck each treat variations as somewhat unfortunate consequences of imperfections in the process; Darwin treats variations as the indispensable precondition of continuing evolutionary development.

The contrast between Lamarck's and Darwin's theories of organic change can be understood in another way as well (Sober 1984, 1994). Lamarck's theory is premised on a "developmental stage" (or ontogenetic) conception in which phylogeny (the series of changes characterizing a lineage through time) is modeled upon ontogeny (the series of developmental changes undergone by an individual), in two distinct ways. First, the overall process of evolution is modeled on the development of individual organisms. Just as individual organisms develop according to a preset plan (laid down in their hereditary material), so, too, evolution as a whole is viewed as a directional unfolding from lesser to greater complexity according to "nature's plan of campaign," as Lamarck called it. Second, in Lamarck's scheme evolution is driven by changes in individuals, not in species. The explanation for why giraffes have long necks is that in the past individual giraffes stretched their necks to reach the higher foliage, this altered feature was passed on to offspring, and the process was repeated until the long-necked creatures of today appeared. Changes among individual organisms drive the process.

The contrasts with Darwin's theory are striking. Whereas Lamarck's theory treats evolution as preprogrammed, in Darwin's theory whatever direction there is in the process is dictated by changing environments and the ability of populations to respond, both of which are highly contingent. Darwin also separates ontogenetic and phylogenetic explanations, restricting each to just one stage in the overall evolutionary process. Developmental ontogeny explains individual characteristics, while selection explains populational characteristics, and hence phylogeny. Why does *this* giraffe have a long neck? Because it inherited long-neck genes whose instructions were expressed in an appropriate environment. Why do giraffes (as a species) have long necks? Because in the past individuals with long necks enjoyed greater survival and reproductive success than those with shorter necks, and these more successful individuals differentially passed on their characteristics to offspring. So far as evolution is concerned, organisms are essentially fixed in their attributes, while species evolve.

Natural Selection

In its essentialism regarding species, Lamarck's theory harkens back to Aristotle; in its transformism concerning life as a whole, it anticipates Darwin. As a conceptual bridge between pre-evolutionary biology and contemporary evolutionary biology it thus occupies a historically crucial position. Yet as daring and novel as it was, Lamarck's theory was a dead-end in the history of evolutionary theorizing, whereas Darwin's theory has given rise to a vigorous research program extending far beyond anything that even Darwin could have imagined. But it has also given rise to numerous controversies, many of which center on the operation of natural selection. Given the centrality of natural selection in Darwin's theory, it is of fundamental importance to understand how selection operates. This turns out to be considerably more difficult than it first seems. In a seminal article, Richard Lewontin noted that "The generality of the principles of natural selection means that any entities in nature that have variation, reproduction, and heritability may evolve" (Lewontin 1970, p. 1). That seems clear enough. However, this seemingly straightforward observation, as Lewontin was well aware, harbors difficult problems. Selection is often thought of as operating on individual organisms. In principle, at least, it could operate on other sorts of biological entities as well. But which ones? What kinds of characteristics must a biological entity have in order for it to be subject to selection? How would selection of these other entities relate to selection of individual organisms? Even if selection could operate on these other biological entities, which ones does it in fact operate on, and what are the consequences for understanding the evolution of life on Earth?

Consideration of these problems has led to one of the most vigorous controversies in contemporary evolutionary biology: the "units of selection" debate (Brandon and Burian 1984; Sober and Wilson 1994). Whereas some biologists have asserted that selection operates exclusively on individual organisms, others have advocated models according to which selection operates on other biological entities as well. The issues dividing these biologists are complex and multifaceted, and will be the subject of later chapters, but they were prefigured in Darwin's writings. Understanding his view on this issue is thus essential to making sense out of subsequent debates. Consequently, it is worthwhile to examine Darwin's views on such issues in some detail, with the aim of finding out precisely how he conceived of the operation of natural selection in the evolutionary process. Did he have a settled view about the entities upon which selection can or does operate? If so, what was it?

Darwin and Organism Selection

According to a contemporary slogan intended to unambiguously identify biological entities with their respective evolutionary roles, "genes mutate, organisms are selected, and species evolve" (Hull 1988). Darwin knew nothing about genes, of course, but it seems obvious that he would have accepted the claim that organisms are selected, and that species evolve. After all, those are two of the key ideas constituting his theory. It is also easy to show that Darwin *generally* viewed selection as operating amongst individual organisms rather than on biological entities at some higher (or lower) level of organization. Despite the subtitle of the *Origin of Species*, the "preservation of favoured races" is construed as an *effect* of the struggle for survival at the level of individual organisms. It seems quite clear that when Darwin writes "Hence, as more individuals are produced than can possibly survive, there must in every case be a struggle for existence, either one individual with another of the same species, or with the individuals of distinct species, or with the physical conditions of life" (Darwin 1859, p. 63), the struggle being described is between individual organisms. He was even willing to be more precise. Most often, Darwin thought, the struggle will be intraspecific: "[T]he struggle almost invariably will be most severe between individuals of the same species" (Darwin 1859, p. 75).

Just as the struggle for existence is primarily between individual organisms, so, too, is selection primarily for or against the individual. In a pack of wolves, the swiftest and slimmest will be more effective predators, and hence there will be selection for wolves possessing such characteristics (Darwin 1859, p. 90). Sexual selection, too, in which possessing some feature attractive to the opposite sex gives one an edge in the competition for mates, is presented in such a way that individuals are selected because they have some advantage over other individuals within their immediate group. By definition, sexual selection takes place within a species, pitting conspecific against conspecific, and thus represents individual selection in the clearest sense (Darwin 1859, pp. 87–90). Similar examples of Darwin's preference for explanations in terms of individual selection are easy to produce. Clearly, whenever a biological phenomenon required a selectionist explanation, Darwin preferred to construe selection as operating amongst individual organisms. This point is simply not controversial.

"One Special Difficulty"

This tidy picture is complicated when one considers Darwin's treatment of certain "special difficulties." Special difficulties require special explanations including, in this case, consideration of selection operating on biological entities other than (or in addition to) individual organisms. For example, in Chapter VII of the *Origin*, Darwin considers "one special difficulty, which at first appeared to me insuperable, and actually fatal to my whole theory. I allude to the neuters or sterile females in insect-communities" (Darwin 1859, p. 236). Later in the same chapter he declared that castes of sterile workers in the social insects pose "by far the most serious special difficulty, which my theory has encountered" (Darwin 1859, p. 242). The "special difficulty" for Darwin was not (as it became for later Darwinians) to explain sterility and extreme altruistic behavior (although, as we shall see, Darwin did offer an explanation for these puzzles) but, rather, to explain how natural selection could produce a neuter caste whose members were so structurally different from their parents and from one another: "[F]or these neuters often differ widely in instinct and in structure from both the males and fertile females, and yet, being sterile, they cannot propagate their kind" (Darwin 1859, p. 236). "[T]he difficulty," Darwin wrote, "lies in understanding how such correlated modifications of structure could have been slowly accumulated by natural selection" (Darwin 1859, p. 237).

More precisely, Darwin recognized and attempted to resolve two distinct problems concerning sterile castes of workers in social insects. The first problem concerned the origin and maintenance of sterile castes. Why this should be a problem for Darwin's theory is clear. Sterile individuals, by definition, do not reproduce. Instead, they appear to sacrifice their reproductive interests for the benefit of the rest of the hive or colony. If natural selection favors those individuals more proficient at reproducing themselves, then sterile individuals are obviously at a distinct disadvantage relative to their more prolific conspecifics, and should be eliminated from the struggle for existence in short order. Yet social insects, with their sterile castes, are among the most widespread and successful living systems on earth. The existence of sterile castes among social insects seems inexplicable on the assumption that all selection is for individually advantageous characteristics. What possible individual advantage can accrue to being sterile? There appears to be none. How, then, is the presence of sterile castes to be explained?

Despite the serious threat it posed to his theory, Darwin apparently thought that this problem could be handled rather easily, and so his discussion of it is surprisingly brief:

> How the workers have been rendered sterile is a difficulty; but not much greater than that of any other striking modification of structure; for it can be shown that some insects and other articulate animals in a state of nature occasionally become sterile; and if such insects had been social, and it had been *profitable to the community* that a number should have been annually born capable of work, but incapable of procreation, I can see no very great difficulty in this being effected by natural selection. (Darwin 1859, p. 236; emphasis added)

The key idea in this passage is that in addition to operating on individually advantageous characteristics, selection can also operate on characteristics "profitable to the community." Apparently, Darwin was willing to entertain the idea that there could be selection for characteristics beneficial to the community, even though they were of no use (and actually detrimental) to the individuals possessing those characteristics.[5]

But did Darwin really entertain the idea of selection operating on more inclusive entities than individual organisms? Michael Ruse (1980) offers a spirited defense of the claim that, contrary to appearances, Darwin never departed from a strict individual selectionist perspective. According to Ruse, by the end of the 1860s "there was nothing implicit about Darwin's commitment to individual selection. He had looked long and hard at group selection and rejected it" (Ruse 1980, p. 620).[6] Again: "In the nonhuman world Darwin was a firm, even aggressive, individual selectionist . . . [who], for organisms other than man . . . unequivocally invoked individual selection" (Ruse 1980, p. 629). On this view, when Darwin does seem clearly to come out in support of some sort of higher-level selection process, such lapses constitute a "quaver in his commitment to individual selection" when he "for once did lose sight of the individual and allow that possibly the unit of selection may have been the group" (Ruse 1980, pp. 626–7).

How, then, should Darwin's apparent group selectionist explanation of sterile neuters be understood? According to Ruse, there is no appeal to higher-level selection here. Rather, the key to understanding Darwin's argument is to note that the sterile altruists are genetically related to the fertile members of the colony. Although they are themselves reproductively disadvantaged by being sterile, nonetheless by helping their relatives to survive and reproduce they are assisting in the propagation of copies of their genes, many of which are shared with close relatives. Instead of passing on their genes directly through producing offspring, sterile individuals do so indirectly through the offspring of their fertile relatives. Such a process (later named "kin selection") cannot be considered higher-level (i.e., community-level) selection, Ruse argues, because selection is not preserving characteristics exclusively of value to nonrelatives. Consequently, "Darwin was certainly an individual selectionist at this point" (Ruse 1980, p. 619).

Despite the attractions of this interpretation in simplifying our image of Darwin considerably and even allowing him to anticipate important developments in twentieth-century evolutionary biology, it suffers from two serious difficulties. First, it depends on the assumption that Darwin could not have been proposing a higher-level selection process if the individuals in question are genetically related. In other words, it assumes that higher-level selection requires that individuals sacrifice themselves for nonrelatives. The rationale for this assumption is far from clear. Second, and more directly relevant in the present context, it is unclear that *Darwin* made any such assumption. Ruse's interpretation depends on familiarity with a solution to the problem that was not clearly understood until well over a century after the publication of the *Origin*. It is true that recent explanations (from the mid-1960s on) of sterile castes among social insects have focused on explanations in terms of benefits conferred on genetic relatives by sterile individuals (e.g., Hamilton 1963, 1964). But clearly such explanations cannot be simply read back into Darwin's account if we wish to understand how *he* approached the problem.[7] Our best guide to what Darwin thought is what he actually said, interpreted in the context of his other remarks on similar issues. Interpreting Darwin as offering a "kin selection" solution to the problem of sterile castes runs the risk of reading back into Darwin's writings what we, now, believe to be the correct explanation of the problem at hand, rather than considering Darwin's solution on its own terms.

Fortunately, there is plenty of material to help us bring Darwin's views about the operation of selection into sharper focus. His answer to the second problem concerning sterile neuters, in particular, provides important further clues to his thinking. Recall the essential difficulty: "[W]ith the working ant we have an insect differing greatly from its parents, yet absolutely sterile; so that it could never have been transmitted successively acquired modifications of structure or instinct to its progeny. It may well be asked how is it possible to reconcile this case with the theory of natural selection?" (Darwin 1859, p. 237). Darwin thought that the problem of explaining how natural selection could produce a neuter caste differing widely in instinct and in structure from both the males and fertile females was much greater than the problem of explaining how natural selection could have rendered the workers sterile in the first place. But he thought that the problem was solvable:

> I can see no real difficulty in any character having become correlated with the sterile condition of certain members of insect-communities . . . when it is remembered that selection may be applied to the family, as well as to the individual, and may thus gain the desired end. . . . Thus I believe it has been with social insects: a slight modification of structure, or instinct, correlated with the sterile condition of certain members of the community, has been advantageous to the community: consequently the fertile males and females of the same community flourished, and transmitted to their fertile offspring a tendency to produce sterile members having the same modification. (Darwin 1859, pp. 237–8)

Here Darwin seems to draw an explicit contrast between "selection applied to the family" and selection applied "to the individual," suggesting that he was well aware of the distinction between the two processes. The explanation offered for the existence of sterile castes is the fact that such a condition "has been advantageous to the community" in relation to *other communities* lacking this feature. This suggests that he was thinking of a selective advantage accruing to the community that is distinct from, and in this case contrary to, benefits for at least some of the individual members of that community. In this way, individually deleterious traits might nonetheless be selected for if such traits are linked to some advantage for the community as a whole.

This interpretation has the virtue of taking Darwin's own statement of his view seriously. Unfortunately, this straightforward interpretation is complicated by the fact that Darwin's remarks still contain some ambiguities. When he cautions that we should not forget that "selection may be applied to the family, as well as to the individual," does he intend to remind his readers that there is another kind of selection *in addition to* individual selection? Or does the "as well as" clause in his remark indicate that he thinks that selection at the level of the family acts *in concert with* selection at the level of individuals, with both processes conjointly producing the phenomenon to be explained? His remark later in this passage that some "slight modification of structure, or instinct, correlated with the sterile condition of certain members of the community, has been advantageous to the community" implies that he is thinking of selection at the level of the family or community *rather than* selection at the level of individuals as the preferred explanation of sterility. Yet elsewhere he writes: "In social animals [natural selection] will adapt the structure of each individual for the benefit of the community; if each in consequence profits by the selected change" (Darwin 1859, p. 87). This supposes that selection will adapt the structure of each individual to the benefit of the community *only if* such adaptation *also* benefits the individual. So in this case benefit to the individual is primary. However, in the sixth edition of the *Origin* (1872) the passage is changed to read as follows: "In social animals [natural selection] will adapt the structure of each individual for the benefit of the whole community; if the community profits by the selected change" (Darwin 1959, p. 172). The change of emphasis has now been reversed! Darwin's view is not altogether as clear as we might like, so we are left with some uncertainty in representing his thought.

To make matters worse, another interpretive problem arises when we consider the remarks elided from the long quote above. There Darwin uses the following comparisons with sterile insect castes to make his point: "Thus, a well-flavoured vegetable is cooked, and the individual is destroyed; but the horticulturist sows seeds of the same stock, and confidently expects to get nearly the same variety . . ." (Darwin 1859, pp. 237–8). In this example, it is not the family as a discrete unit that is the object of selection but, rather, the *characteristics* of the family that are carried in the seeds. Neuter insects are presumably meant to be analogous to the "well-flavoured vegetable" that is cooked, in that neither is individually reproductively successful, yet the characteristics of each are preserved in other members of their family. There are, of course, disanalogies as well. In the case of the vegetables, "tastiness" is a characteristic of both parents and offspring, whereas sterility is a characteristic of certain individuals only (i.e., a certain subset of the offspring of fertile parents). Whereas the characteristics correlated with the sterility of neuter insects are supposed to be of benefit to their community, the tastiness of certain individual vegetables is not obviously of benefit to the "community" of which they are a part. Finally, what is missing from this example is some characteristic correlated with tastiness whose existence is to be explained in terms of selection for being "well-flavoured" in the way in which sterility is supposed to be explained by being correlated with (for example) large mandibles in the soldier caste of some ant species.

Darwin's second example is somewhat more helpful, inasmuch as it introduces the issue of sterile offspring: "I have such faith in the powers of selection, that I do not doubt that a breed of cattle, always yielding oxen with extraordinarily long horns, could be slowly formed by carefully watching which individual bulls and cows, when matched, produced oxen with the longest horns; and yet no one ox could ever have propagated its kind" (Darwin 1859, p. 238). Here the analogy with neuter insects is closer. A particular characteristic had by sterile offspring but not by their parents (e.g., long horns in oxen, large mandibles in soldier ants) can become correlated with the sterile offspring, even though (by definition) such individuals cannot pass on this characteristic to their offspring. Where the analogy breaks down, however, is in the causes responsible for the correlations in question. In the case of the long-horned oxen, the cause is artificial selection operating on their parents. Having extraordinarily long horns is presumably of no benefit to the parents nor to the herd, although it may be valued by the breeder. In the case of the neuter insects some structure correlated with sterility proved to be advantageous to the community, including their parents. As a result, the fertile individuals who produced such useful offspring flourished and continued to produce sterile offspring having the same modification.

As Darwin concludes a bit later, "With these facts before me, I believe that natural selection, by acting on the fertile parents, could form a species which should regularly produce neuters..." (Darwin 1859, p. 241). Darwin's talk of selection acting on the fertile parents might lead one to conclude that they, rather than the community, are the beneficiaries of the presence of neuters. As he proceeds to note, however, he conceives of neuter castes as benefiting the entire community of which they are a part: "We can see how useful their production may have been to a social community of insects, on the same principle that the division of labour is useful to civilized man" (Darwin 1859, pp. 241–2). Who, then, is the primary beneficiary of the division of labor – the community or individual men? Darwin doesn't say. If we want further clarity, we'll have to look elsewhere to see whether he ever resolves these ambiguities.

Of "Well-Endowed Men"

Talk of a "division of labour" within an insect community naturally invites further comparison between human and nonhuman societies. If appeal to selection at the level of communities might help to explain some otherwise puzzling features of insect societies, might not the same be true for understanding how human communities come to have the characteristics they do? Darwin took up this challenge in his major work on human evolution, *The Descent of Man* (1871), where he again invoked selection at the level of communities or groups in response to the problem of explaining how a characteristic apparently detrimental at the individual level could nonetheless evolve. The particular problem in question was "how within the limits of the same tribe did a large number of members first become endowed with [their] social and moral qualities, and how was the standard of excellence raised?" (Darwin 1871, vol. 1, p. 163). The problem is that this seems difficult to explain in terms of selective benefits for those individuals displaying exceptional levels of sociability and morality, as Darwin goes on to explain:

> It is extremely doubtful whether the offspring of the more sympathetic and benevolent parents, or of those which were the most faithful to their comrades, would be reared in greater numbers than the children of selfish and treacherous parents of the same tribe. He who was ready to sacrifice his life, as many a savage has been, rather than betray his comrades, would often leave no offspring to inherit his noble nature. The bravest men, who were always willing to come to the front in war, and who freely risked their lives for others, would on an average perish in larger numbers than other men. Therefore it seems scarcely possible . . . that the number of men gifted with such virtues, or that the standard of their excellence, could be increased through natural selection. (Darwin 1871, vol. 1, p. 163)

Just because we are all acquainted with acts of self-sacrifice, of parents for children, of comrades for their friends, and so on, the seriousness of the problem should not be underestimated. If natural selection favors characteristics exclusively of benefit to the individuals possessing them, then acts of self-sacrifice (i.e., of "altruism"), in which an individual's reproductive fitness is lowered, become genuinely puzzling from an evolutionary perspective.

As we might expect by this point, Darwin's solution to this problem lay in considering benefits accruing to *tribes* constituted by such virtuous men:

> It must not be forgotten that although a high standard of morality gives but a slight or no advantage to each individual man and his children over the other men of the same tribe, yet... an advancement in the standard of morality and an increase in the number of well-endowed men will certainly give an immense advantage to one tribe over another. There can be no doubt that a tribe including many members who, from possessing in a high degree the spirit of patriotism, fidelity, obedience, courage, and sympathy, were always ready to give aid to each other and to sacrifice themselves for the common good, would be victorious over most other tribes; and this would be natural selection. (Darwin 1871, vol. 1, p. 166)

Because the variation is between tribes, rather than between individuals within tribes, selection at the level of *groups* of individuals is apparently being proposed. This interpretation is strengthened by the recognition that the characteristic benefiting the group often involves the sacrifice of the individual – either literally, in cases in which an individual dies for the benefit of the group, or in the sense that individuals take greater risks and thus reduce their reproductive potential. It is this disadvantage for the individual that seemed to Darwin to require an explanation in terms of selection at the level of a more inclusive entity.

Some interpreters have flatly denied that Darwin entertained supra-individual selection to explain human morality. According to Ruse, for example, Darwin "saw the individual man or woman as being the crucial unit in the selective process. There was no question that, when faced with his own species, he was going to swing around suddenly and start to argue as a general policy that for *Homo sapiens* alone the group... is the key element in the evolutionary mechanism" (Ruse 1980, p. 626). Ruse insists that "apart from some slight equivocation over man, Darwin opted firmly for hypotheses supposing selection always to work at the level of the individual rather than the group" (Ruse 1980, p. 615). Yet as we have seen, there is nothing equivocal about Darwin's position. He was willing to entertain a group selectionist explanation, not only for the evolution of human morality, but for other puzzling biological phenomena as well. Indeed, Darwin was perfectly willing to generalize and extend this sort of explanation to account for other similarly puzzling social phenomena in a way that leaves little doubt that he believed that selection can and does operate at the level of communities:

> With strictly social animals, natural selection sometimes acts indirectly on the individual, through the preservation of variations which are beneficial only to the community. A community including a large number of well-endowed individuals increases in number and is victorious over other and less well-endowed communities; although each separate member may gain no advantage over the other members of the same community. (Darwin 1871, vol. 1, p. 155)

In this view, selection at the level of groups might still positively affect the individuals within the groups, because the individuals in one group may, on average, be more reproductively successful, precisely by being members of that group, than are the individuals belonging to other groups. But in such cases selection does not operate directly on the individuals *within* a given group, because the properties that selection operates upon are properties of the entire group, not of its individual members. Such a selection process is thus clearly distinct from the sort of selection that Darwin believed explains the properties of individual organisms. Consequently, in addition to selection operating on differences among individual organisms ("organism selection"), Darwin also recognized selection operating on differences among groups of organisms ("group selection").

The best guide to understanding what Darwin actually thought are his actual words, taken at face value if possible, and only reinterpreted if absolutely necessary. When Darwin writes that "certain mental faculties . . . have been chiefly, or even exclusively, gained for the benefit of the community" (Darwin 1871, vol. 1, p. 155), we should take this as a genuine expression of his thoughts on the matter. When we do this, it becomes evident that although Darwin preferred explanations in terms of selection operating on individual organisms, he was perfectly willing to entertain explanations in terms of selection at the level of groups when the situation warranted it.

Possibilities and Boundaries

As we have seen, an alternative to thinking of selection as operating exclusively among individual organisms is to think of it as operating as well on *groups* of organisms ("group selection"). I have been suggesting that there are good reasons to conclude that Darwin seriously entertained this idea, and attempted to apply it to solve several otherwise puzzling biological phenomena. It is noteworthy that the two instances in which Darwin most clearly appears to offer group selectionist explanations – for sterile castes among social insects and for the evolution of the human moral sense – both involve *social* phenomena. A brief consideration of the *development* of Darwin's thinking on such phenomena provides important clues to understanding why he believed that social phenomena merit a different sort of evolutionary explanation.

Social Evolution

One such clue appears for the first time in the fourth edition of the *Origin* (1866), as well as in a later work, *The Variation of Animals and Plants under Domestication* (1868). Once again, Darwin is considering the peculiar case of sterile neuters:

> With sterile neuter insects we have reason to believe that modifications in their structure have been slowly accumulated by natural selection, from an advantage having been thus indirectly given to the community to which they belonged over other communities of the same species; but an individual animal, if rendered slightly sterile when crossed with some other variety, would not thus indirectly give any advantage to its nearest relatives or to any other individuals of the same variety, thus leading to their preservation. (Darwin 1959, pp. 444–5; 1868, vol. ii, pp. 186–7)

The sixth and final edition of the *Origin* (1872) includes "fertility" along with "structure" and adds that the communities being discussed are *social* communities (Darwin 1959, p. 445). It is significant that in these additions Darwin explicitly contrasts the relevant explanations of sterile castes among social insects with that of interspecific and hybrid sterility. Whereas in the former case sterility can be explained by natural selection operating through advantages accruing to social communities, in the latter case no such socially mediated community-level advantage can be invoked. The addition of this passage, with its striking emphasis on an animal's membership in a "social community," suggests that Darwin considered sociality to be a distinct factor in evolution, one that in some cases perhaps licenses (or requires) the postulation of selection operating at a level more inclusive than that of the individual organism.

Selection and Individuality

This also suggests that in reading Darwin's frequent remarks to the effect that a given biological phenomenon "could not have been effected through natural selection [because] it could not have been of any direct advantage to an individual animal . . ." (Darwin 1959, p. 444) one has to take into account whether he is referring to a feature bearing only on the well-being of an individual organism or whether the feature in question is a structural property of a more inclusive social organization having a biological significance of its own. Darwin may well have been committed to explanations of the former solely in terms of selective advantages for individual organisms, while allowing for explanations of the latter in terms of selection operating on entities above the level of individual organisms, *precisely because he viewed these higher-level entities as "individuals" in their own right.* Significantly, in a letter to Wallace discussing his proposed explanation of hybrid sterility, Darwin wrote: "I believe, that Natural Selection cannot effect what is not good for the *individual, including in this term a social community*" (Darwin to Wallace, April 6, 1868; in F. Darwin and Seward 1903, vol. 1, p. 294; emphasis added; also in Wallace 1916, p. 170). What is especially striking about this remark is that Darwin explicitly includes in the denotation of the term "individual" a "social community." This means that it would be correct to attribute to Darwin the view that selection can only act upon "individuals," but a mistake to ascribe to him the view that only organisms can be individuals in the relevant sense. Social communities, too, can be individuals, and hence can be directly available for selection to act upon. In this interpretation, Darwin was indeed a strict "individual selectionist," but one whose conception of an "individual" included not just individual organisms but extended as well to certain other sorts of biological entities.[8]

What Natural Selection Cannot Do

Before leaving this topic it is important to note that, despite the ambiguity of some aspects of Darwin's treatment of natural selection, there is one related issue about which he could not have been clearer. Although he was at times willing to entertain the idea that selection might act upon and benefit some more inclusive entity than the individual organism, for example, the community, the variety, or even the species,[9] there is one issue concerning selection that was *never* an issue for Darwin, namely, whether selection might operate on one species for the good of another. He was absolutely clear that natural selection could never be understood to act in this way:

> Natural selection will modify the structure of the young in relation to the parent, and of the parent in relation to the young. In social animals it will adapt the structure of each individual for the benefit of the community; if each in consequence

profits by the selected change. What natural selection cannot do, is to modify the structure of one species, without giving it any advantage, for the good of another species; and though statements to this effect may be found in the works of natural history, I cannot find one case which will bear investigation. (Darwin 1859, pp. 86–87; Darwin 1959, p. 172)

Later in the same work Darwin put the point in the strongest possible terms:

Natural selection cannot possibly produce any modification in any one species exclusively for the good of another species; though throughout nature one species incessantly takes advantage of, and profits by, the structure of another. . . . If it could be proved that any part of the structure of any one species had been formed for the exclusive good of another species, it would annihilate my theory, for such could not have been produced by natural selection. (Darwin 1859, pp. 200–1)

For example, some authors had asserted that the rattlesnake's rattle is a mechanism for warning potential victims of danger, and thus of giving them a fair chance of escape. Darwin heaps scorn on this claim: "I would almost as soon believe that the cat curls the end of its tail when preparing to spring, in order to warn the doomed mouse" (Darwin 1859, p. 201). It is true that organisms of different species sometimes behave in ways that are mutually beneficial (e.g., symbiotic relationships between ants and acacia trees, termites and the cellulose-digesting bacteria that inhabit their guts, etc.), but all such cases can be explained as organisms acting for their own, rather than for their associate's, benefit. Darwin made his own view of the matter crystal clear: "Natural selection will never produce in a being anything injurious to itself, for natural selection acts solely by and for the good of each" (Darwin 1859, p. 201).

Summary: Darwin and Natural Selection

Natural selection is the central theoretical principle that distinguished Darwin's explanation of living things from all those that preceded him. In addition to offering a new explanation for the origin and nature of living things, Darwin proposed a new *kind* of explanation, based on a novel ideal of natural order according to which variation (rather than uniformity) is fundamental. Stated abstractly, how natural selection operates seems entirely unproblematic. Yet the very generality that gives the principle its broad explanatory power also raises difficult questions about its actual operation. On what sorts of "biological entities" can and does natural selection operate? How did Darwin approach this issue? As we have seen, for the vast majority of problems requiring a selectionist explanation, Darwin was content to appeal to selection at the level of individual organisms. He was, however, willing to countenance selection operating at some higher level of organization when the biological phenomenon under consideration did not lend itself to an analysis in terms of selection for individual benefit. So, for example, in the *Origin*, Darwin explained the otherwise puzzling case of sterility among certain members of insect communities by noting that "selection may be applied to the family, as well as to the individual, and may thus gain the desired end." In *The Descent of Man* (1871) he offered a similar explanation of the evolution of the human moral sense. Explaining the evolution of social behaviors, in particular, seemed to him to require extending the range of natural selection beyond the narrow compass of the individual organism but not necessarily beyond the scope of the individual understood as a biological entity having some significant degree of functional integration. In this way selection could, in principle, operate among communities as functionally integrated individuals. Interpretations of Darwin's thought that present him as strictly adhering to the view that selection only operates on individual organisms gain in simplicity but sacrifice appreciation of the subtlety of Darwin's attempts to solve some of the most difficult problems facing the theory of natural selection. His was an individualistic perspective at heart, but he refused to straightjacket himself into offering just one kind of evolutionary explanation. Darwin was too much of a pluralist for that.

2

The Group Selection Controversy

> Understanding what kind of variation is possible and at what level selection occurs over those variations is what has driven the conversation about evolutionary biology at least since Darwin.
>
> (Ahouse 1998, p. 370)

Introduction

In order to explain certain puzzling biological phenomena that seemed to make little sense on the assumption that natural selection operates exclusively at the level of individual organisms, Darwin toyed with the idea of selection operating at the level of entire communities. The implications of this idea were profound. If selection operated at this more inclusive level, then the "beings for whose good natural selection works" might include groups as well as individual organisms. Selection operating at the group level could forge adaptations that benefit the group rather than each organism considered separately. Consequently, not every property of an individual organism need benefit that organism. Indeed, some organismic properties might even be detrimental to their immediate possessors, so long as they were sufficiently advantageous at the group level. Thanks to Darwin's invocation of community-level selection, for any biological phenomenon or characteristic requiring an evolutionary explanation, one could now ask whether it was selected and had thereafter evolved for individual or for group benefit.

Darwin's bold move of introducing the idea of selection for group benefit significantly expanded his theory's ability to explain puzzling biological phenomena. It also created a troubling tension within his theory.

As we have seen, Darwin recognized that selection could never produce characteristics in one *species* for the good of another, and was even willing to deem his theory "annihilated" should anyone be able to produce a convincing counterexample. In Darwin's view, members of other species constitute either neutral components of the environment to be ignored, enemies to be avoided, or resources to be exploited. In no case, however, should one expect a member of one species to go out of its way to assist that of another in the absence of a fitness-enhancing recompense for the altruist. As he chillingly remarked in the chapter on "Instinct" in the *Origin*, "No instinct has been produced for the exclusive good of other animals, but... each animal takes advantage of the instincts of others" (Darwin 1859, p. 243). With regard to the characteristics of individual organisms, therefore, Darwin's position was clear. None of them exist for the benefit of members of other species. This stricture on cross-species altruism can be seen as a particular instantiation of a more general principle: "Natural selection will never produce in a being anything injurious to itself, for natural selection acts solely by and for the good of each" (Darwin 1859, p. 201). Yet, suggesting that selection might operate at the level of communities or groups introduced the possibility that selection could produce in organisms characteristics for the benefit of others, even though such characteristics would be detrimental to their possessors. But how is this possible? If the evolution of characteristics detrimental to their possessors is categorically excluded in the one case (between species), why not also in the other (within a given species)?

Either ignoring or unconcerned with this problem, some later biologists welcomed Darwin's expansion of evolutionary theory with open arms, and made group selectionist explanations the cornerstone of their theorizing. To other biologists, accepting the idea that selection could forge adaptations for the benefit of the group at the expense of the individual organism's interests seemed just as fatal to Darwin's theory as would be the discovery of characteristics exclusively of benefit to members of other species. Both seem to contradict the fundamental logic of Darwin's theory. These two fundamentally different assessments of the viability of group selectionist explanations coexisted in relatively quiet isolation from one another for much of the twentieth century, each sequestered in different subdisciplines of biology (e.g., ecology and population genetics, respectively), with minimal interdisciplinary contact to disturb their peaceful coexistence. They might never have reached a direct confrontation had it not been for certain key events in mid-twentieth-century evolutionary biology.

The tensions that simmered just beneath the surface of popular and professional presentations of evolutionary biology boiled over in the 1960s, a decade that was in many ways a watershed for evolutionary theory as a number of previously widely accepted ideas about how evolution operates were made explicit, challenged, and largely rejected in favor of ideas that continue to dominate evolutionary theory to the present. Prior to this decade explanations of biological phenomena appealing to the "good of the group" or the "preservation of the species" were common in the scientific literature. By the end of the decade such explanations had become rare. A number of factors intersected to bring about this development, but one especially salient event was the controversy surrounding the work of V. C. Wynne-Edwards, a British biologist stationed in a granite outpost of Scotland on the edge of the North Sea, who had spent a lot of time thinking about fishing, and was deeply perplexed by what he saw. Subsequent developments in evolutionary biology, including the surprising renaissance that group selectionism (in refurbished form) is currently enjoying, are inexplicable without an understanding of the controversy generated by his work.[1]

The story of the rise and fall (and miraculous resurrection) of group selectionism makes for a fascinating story. In this chapter I can only sketch some of its main features, focusing especially on those crucial turning points that illuminate the path from Darwin to the present. Wynne-Edwards's ideas did not develop in a historical vacuum. They arose out of his own assessment of earlier attempts to solve critical biological problems, and it is to these problems that we must first turn. The account begins apparently far afield, in the seemingly unrelated question of how (if at all) animals regulate their populations. As is so often the case, in order to understand the development of evolutionary ideas as well as their contemporary status, we have to begin with their original formulation in Darwin's work.

The Population Problem

Taking an insight from the Reverend Thomas Malthus as his point of departure, Darwin noted that organisms will tend to produce more offspring than can be supported by their environment, resulting in competition for limited resources and consequently a struggle for existence. This struggle, in turn, becomes the engine of natural selection, and hence of evolution. Recognition of the resource-limited nature of the living world also led to a minor puzzle. In the third chapter of the *Origin* Darwin noted that populations of organisms, if left unchecked, should grow at a geometric rate of increase; for example, two individuals giving rise to four, four giving rise to sixteen, and so on. Elephants, he noted, are reckoned to be the slowest breeders of all known animals. Yet a pair of elephants could produce fifteen million descendants in a mere five hundred years. That's a lot of elephants. Thought-experiments aside, we know from numerous actual cases since Darwin that when organisms are introduced into a favorable but previously unexploited niche (e.g., rabbits into Australia), they experience explosive population growth. "In such cases," Darwin noted, "the geometrical ratio of increase, the result of which never ceases to be surprising, simply explains the extraordinarily rapid increase and wide diffusion of naturalised productions in their new homes" (Darwin 1859, p. 65). This much is just a matter of simple arithmetic, but it leaves an important question unanswered. If organisms are physiologically capable of reproducing at such an explosive rate, and natural selection favors the more fecund, why then is it that under normal circumstances populations remain remarkably stable?

"Ten Thousand Sharp Wedges"

As Darwin recognized, the fact that we are not now buried under a sea of elephants (or emus, or echidnas) entails that "the geometrical tendency to increase must be checked by destruction at some period of life" (Darwin 1859, p. 65). He admitted that "What checks the natural tendency of each species to increase in number is most obscure," but ventured that predation, food shortage due to extremes of climate, and disease are among the main factors that limit population growth (Darwin 1859, p. 67). Were it not for such factors, natural selection, and hence evolution, could not occur.

In Darwin's view what we observe in nature as a more or less stable equilibrium (a "balance of nature") is in fact the consequence of two powerful dynamics operating in opposite directions, each threatening to overwhelm the other, but which coexist in an uneasy and unstable standoff. One is the natural tendency to increase. The other is the set of checks on this tendency. We must never forget, Darwin says, "that every single organic being around us may be said to be striving to the utmost to increase in numbers. . . . Lighten any check, mitigate the destruction ever so little, and the number of the species will almost instantaneously increase to any amount" (Darwin 1859, pp. 66–67). He deploys a "striking" metaphor to capture the reproductively driven competition that characterizes nature: "The face of Nature may be compared to a yielding surface, with ten thousand sharp wedges packed close together and driven inwards by incessant blows, sometimes one wedge being struck, and then another with greater force" (Darwin 1859, p. 67). Extinction is simply what results when one "wedge" (i.e., species) is forced out by another.

Darwin dealt with the question of population regulation by simply listing some of the more plausible limiting factors on population growth, without treating the issue in any great detail. His final conclusion could hardly be more noncommittal: "In the case of every species, many different checks, acting at different periods of life, and during different seasons or years, probably come into play; some one check or some few being generally the most potent, but all concurring in determining the average number or even the existence of the species" (Darwin 1859, p. 74). The contrast between the confident way in which Darwin poses the problem compared to the open-ended and uncertain way he attempts to resolve it is dramatic. Despite recognizing the problem, he apparently didn't feel a need to explore the issues further.

Planned Parenthood

David Lack (1910–73) was the first biologist after Darwin to give the problem of population regulation the serious attention it deserved.[2] In *The Natural Regulation of Animal Numbers* (1954) he noted, following Darwin, that natural populations are theoretically capable of a geometric rate of growth. A single breeding pair of robins could, assuming that their offspring also survived and reproduced, give rise to a population ten million times as large after ten years. Empirically, it was known that when fertile individuals were introduced into a favorable but previously unexploited environment, their numbers skyrocketed (Lack 1954, pp. 11–12). However, "where conditions are not disturbed, birds fluctuate in numbers between very restricted limits" (Lack 1954, p. 11). Thus, although birds (and other animals) can increase in numbers at a great rate, under normal circumstances they rarely do so. Why not? What factors limit population growth?

Lack argued that the comparative stability of animal populations must be because of density-dependent factors, that is, factors that influence reproductive and morality rates in relation to population density. Such density-dependent factors operate like a thermostat: as population density increases, these factors kick in, damping further population growth; as population density decreases, these factors are relaxed, allowing population growth to resume. Population stability would naturally result "if the reproductive rate is higher at low than at high densities, and if the death-rate is higher at high than at low densities" (Lack 1954, p. 19). Such a mechanism would account for population stability. The problem that remained was the identification of the precise factors governing differential birth and mortality rates.

Lack explored this question by utilizing the results of his own extensive research into the factors governing clutch size in birds. Individuals of each species lay a characteristic number of eggs in a given breeding season: petrals 1, pigeons 2, gulls 3, plovers 4, wagtails 5, leaf-warblers 6, and so on. On the assumption that the reproductive rate is closely related to the number of eggs in each clutch, one is led to ask why the individuals of particular species do not lay *more* eggs than they typically do. Given the reproductive imperative described by Darwin, it would seem that each individual ought to lay as many eggs as possible. Why, then, do birds stop when they have laid a certain species-typical number of eggs?

Lack considered four different hypotheses to account for clutch size in birds. One hypothesis was that birds *do* lay as many eggs as possible: Different species have different physiological capacities, and the individuals of each are always producing the greatest number of eggs of which they are capable. This hypothesis can be rejected, Lack showed, because if eggs are removed the bird will simply lay more. The restriction of clutch size is therefore not due to the inability to lay more, but rather "to a positive act, the cessation of laying" (Lack 1954, p. 21).

A second hypothesis suggested that clutch size is limited by the maximum number of eggs that the sitting bird can cover. While there must be some such limit, Lack pointed out that it does not correspond with the normal clutch size. Bird species with normal clutch sizes of 15, for example, can successfully hatch all the eggs in clutches of 20 just as well (Lack 1954, p. 21).

A third view suggested that clutch size has been adjusted by natural selection to balance the age-related mortality characteristic of the species. According to this idea, in species of long-lived individuals fewer eggs are needed to balance the loss through mortality, whereas in species of shorter-lived individuals more eggs are needed to replenish the population.[3] Clutch size is therefore a function of the needs of the species. This suggestion fails for both empirical and theoretical reasons. First, clutch size could be used to achieve population balance only if it were smaller when population density was high for a given species, and larger when population density was low for that species. But individuals of each species lay a species-typical number of eggs regardless of the population density characterizing the species as a whole. Second, this suggestion rests on a mistaken view of the operation of natural selection. "[N]atural selection operates on the survival-rate of the offspring of each individual genotype. If one type of individual lays more eggs than another and the difference is hereditary, then the more fecund type must come to predominate over the other (even if there is overpopulation)" (Lack 1954, p. 22). In other words, natural selection cannot act for the good of the species if this entails acting against the benefit of the individual organisms.

It was clear that none of the foregoing hypotheses proposed to explain clutch size was satisfactory, and that a different approach was needed. Lack's solution to the problem involved returning to the first hypothesis above but reconsidering what it might mean to claim that organisms produce "as many offspring as possible." The relevant issue is not how many offspring a mated pair can physically produce but, rather, *how many offspring likely to develop to reproductive maturity* a pair can produce. This is an important distinction. Natural selection favors individuals who lay more rather than fewer eggs, *unless* "for some reason the individuals laying more eggs leave fewer, not more, eventual descendants" (Lack 1954, p. 22). This proviso introduces Lack's own favored hypothesis: "Clutch size has been evolved through natural selection to correspond with the largest number of young for which the parents can on the average find enough food. In this view, the upper limit of clutch size is set by the fact that, with more young than this, some are undernourished, and so the parents tend to leave fewer, not more, descendants than those with broods of the normal size" (Lack 1954, pp. 22–23). Field studies showed that with a greater number of young to feed, parents made more frequent feeding visits, but that this increase in quantity of food was more than offset by the larger number of beaks to feed, with the result that each nestling in a larger brood received less food than each nestling in a smaller brood (Lack 1954, p. 23).

Common sense might dictate that larger broods (in which each nestling receives less food) would suffer greater mortality than smaller broods (in which each nestling receives more food), but this intuitive result needed to be demonstrated rather than just assumed. Data collected on swifts and starlings showed that this is indeed the case, and that "the most frequent clutch size is that which gives rise to the greatest number of eventual survivors among the young" (Lack 1954, p. 27). Lack's experiments provided powerful corroboration for the hypothesis that clutch size, and hence reproductive rate, is governed by availability of food. Although he hardly had to mention this, Lack goes on to remark that "It is reasonable to suppose that this correspondence is an adaptation due to natural selection" (Lack 1954, pp. 27–28).

Lack's research on clutch size concerned the mechanisms governing differential birth rates. Other chapters in Lack's book dealt with density-dependent mortality, food as a limiting factor, predation, disease, climatic factors, and population cycles. Summarizing the results from the large number of studies he surveyed, Lack concluded that "reproductive rates are a product of natural selection and are as efficient as possible. They may vary somewhat with population density, but the main density-dependent control of numbers probably comes through variations in the death rate. The critical mortality factors are food shortage, predation, and disease, one of which may be paramount, although they often act together" (Lack 1954, p. 276). Notice that this is precisely the same conclusion Darwin reached in the third chapter of the *Origin.* Having started with a problem posed by Darwin, Lack arrived at thoroughly Darwinian conclusions, albeit now backed by impressive experimental investigations.

Intrinsic Control of Population Density

By the time Lack's book was published, Vero Copner Wynne-Edwards (1906–97) had already been thinking about the problem of population regulation for some thirty years.[4] By his own admission, his experiences as a government consultant for the fisheries industry were fundamental in the formation of his views (Wynne-Edwards 1962, pp. 4–8; 1989, p. 503). The problem facing the fisheries industry was overexploitation of economically valuable fish populations because of unregulated commercial fishing. In the worst sort of case, overexploitation leads to entire populations becoming so depleted that they never recover. The obvious lesson to be learned from this problem, Wynne-Edwards realized, was that some system of regulation is necessary in order to sustain fish populations at optimum levels. In particular, the best way to avoid overfishing is to assign territories to commercial fishing operations such that the total catch per territory is strictly limited. By imposing artificially defined territories, fish stocks would be maintained at an optimum number for successful long-term harvesting.

These ideas found their first public expression in a brief review of Lack's book (Wynne-Edwards 1955), and then a few years later in a slightly longer review essay (Wynne-Edwards 1959). The problem Wynne-Edwards grappled with in these essays was disarmingly simple. Animals often exist in the midst of plenty, and starvation is rare.[5] How is this to be explained? Wynne-Edwards reasoned that animals must be managing the utilization of their food resources in some way. There is, in fact, a close correlation between the population-density of a species and the amount and quality of food available. "Such density differences," he wrote, "arise from the activities of the animals themselves, and this implies that population-density is subject to effective internal control, i.e., it is self-regulating" (Wynne-Edwards 1959, p. 440). He then hypothesized that each species maintains its population-density at a level which insures that food resources are not depleted. This conservation of food resources is achieved through the replacement of "conventional substitutes" for food, especially territory or social position. Because territories and high-ranking social positions are strictly limited, competition for such conventional rewards ensues. Establishing and defending territories or positions in a social hierarchy effectively excludes some individuals from partaking of an equal share of the available food resources. The function of such social behaviors is thus to regulate the species' population density, and thereby its utilization of its food resources, ensuring the continued survival of the species. The basic elements of his theory were now in place. Wynne-Edwards devoted the next few years to collecting and synthesizing additional data that would lend support to this theory, while also continuing to elaborate the theory itself.

The Theory of Animal Dispersion

The result was the appearance, in 1962, of *Animal Dispersion in Relation to Social Behaviour.*[6] In over six hundred smoothly flowing pages, Wynne-Edwards amassed evidence from a broad survey of the animal kingdom to show that animals actively regulate their population densities. Why such population regulation is necessary is explained early on: "[I]t must be highly advantageous to survival, and thus strongly favoured by selection, for animal species (1) to control their own population-densities, and (2) to keep them as near as possible to the optimum level for each habitat they occupy" (Wynne-Edwards 1962, p. 9).

In addition to presenting data supporting his theory, Wynne-Edwards also explored further the collateral requirements necessary for his theory to work. If animals are to regulate their population densities in relation to available resources, then they will need to have some kind of "homeostatic or self-balancing" control system analogous to the physiological systems that regulate the internal environment of the body (Wynne-Edwards 1962, p. 9). Such systems require two basic components. Just as a thermostat, if it is to regulate the temperature of a room, needs to be sensitive to temperature fluctuations, so, too, if animal populations are to homeostatically regulate their population densities in relation to food availability, they will need a mechanism for monitoring and assessing population size and density. According to Wynne-Edwards, "epideictic displays" serve this function. These are communal displays, often purely conventional and synchronized at dawn and dusk, which provide the necessary information about the current state of the population. Flocking behavior in birds, for instance, schooling in fish, swarming in insects, and the daily vertical migration of plankton in the water column were said to be means by which individuals assess the size and density of their populations.

The second component required by a homeostatic control system is some means of utilizing the information obtained to maintain (or restore the system to) equilibrium. A home or building thermostat must be connected to a furnace or air conditioning unit responsive (via the thermostat) to changes in temperature. Similarly, in a self-regulating homeostatic population system there must be some means of utilizing census information to adjust population density to available resources. Competition for conventional tokens (i.e., territories and social status) solves the same problem for animal populations that the assignment of territories to commercial fishing operations solves for the problem of overfishing. According to the evolved conventions of each animal society, reproductive and foraging rights go to those individuals who hold territories or assume positions of dominance in a social hierarchy. Because the number of such conventional goods is strictly limited, so too is the number of individuals permitted to breed. For those species that do not form territories or social hierarchies, increased population density leads to decreased reproductive output, providing a self-regulating means of population control.

At this point in the development of his theory Wynne-Edwards introduced what would prove to be its most controversial feature. He realized that natural selection operating at the level of individual organisms could not bring about the kinds of social adaptations central to his theory, adaptations that benefit the group and entail subordination of the interests of individuals to those of the community: "If intraspecific selection was all in favour of the individual, there would be an overwhelming premium on higher and ever higher individual fecundity, provided it resulted in a greater posterity than one's fellows. Manifestly this does not happen in practice" (Wynne-Edwards 1962, p. 19). In other words, selection operating on individuals would favor organisms that seek to maximize reproductive output without regard for group welfare, inevitably leading to overexploitation of the habitat and population crash.

Because such occurrences are rare, some other evolutionary force must be operative.

The mechanism for promoting the evolution of population regulation is identified as *group selection*.[7] Groups in which social conventions are not honored suffer from overcrowding, overexploitation of resources, and eventually population crash and extinction. Groups in which resource use is governed by homeostatic population-regulation systems will tend to persist longer, and may spread to occupy areas left vacant by groups lacking such systems. Wynne-Edwards thought this to be a widespread phenomenon in nature. The following quote conveys the essence of his view:

> Survival is the supreme prize in evolution; and there is consequently great scope for selection between local groups. . . . Some prove to be better adapted socially and individually than others, and tend to outlive them, and sooner or later to spread and multiply by colonising the ground vacated by less successful neighbouring communities. Evolution at this level can be ascribed, therefore, to what is here termed group selection – still an intraspecific process, and, for everything concerning population dynamics, much more important than selection at the individual level. . . . Where the two conflict, as they do when the short-term advantage of the individual undermines the future safety of the race, group-selection is bound to win, because the race will suffer and decline, and be supplanted by another in which antisocial advancement of the individual is more rigidly inhibited. (Wynne-Edwards 1962, p. 20)

In a later chapter on "The Social Group and the Status of the Individual," he reiterated that in his theory individual advantage is frequently subordinated to group welfare: "Under group-selection it is not a question of this individual or that being more successful in leaving progeny to posterity, but of whether the stock itself can survive at all" (Wynne-Edwards 1962, pp. 141–2).

Peppered throughout the rest of the book are descriptions of biological phenomena that can only, he says, be explained on the hypothesis of group selection. It is perhaps no coincidence that, like Darwin, the existence of sterile castes in eusocial insects and the evolution of human morality are singled out as two of the best examples of group selection at work. Social insects present an especially important case, because here "it has been possible to evolve castes of sterile individuals, something that is inconceivable in a world where the most successfully fecund were bound to be individually favoured by selection and the infertile condemned to extinction" (Wynne-Edwards 1962, p. 19). Such a biological feature "could only have evolved where selection had promoted the interests of the social group, as an evolutionary unit in its own right" (Wynne-Edwards

1962, p. 19). Of the essential role of group selection in this process, he concluded, there can be no doubt:

> [T]he evolution of sterility in a proportion of the individuals can only have been effected by selection at the group level, since it is self-evident that no agency can select in favour of sterility among organisms competing in status as individuals. Indeed in these closely-integrated societies it is, more than elsewhere, the group or colony that holds the spotlight as the vital evolutionary unit, undergoing intensive selection. (Wynne-Edwards 1962, p. 276)

The case of human morality is similar, in that "The manner in which the selfish advantage of the individual has thus been subordinated to the long-term welfare of the community can be noticed as a striking example of the over-riding power of group-selection" (Wynne-Edwards 1962, p. 190). The particular problem in this case is understanding how selection at the individual level could lead to the development of behaviors such as respect and care for the elderly, individuals who are no longer able to procreate. "In terms of group-selection, on the other hand, there is no difficulty in understanding the ascendancy of those human groups that are best able to benefit by the councils of their elder statesmen" (Wynne-Edwards 1962, p. 249). As with Darwin, selection at the level of the group finds one of its best examples in *Homo sapiens.*

Summary: Rival Theories of Population Control

Before going on to consider the critical reception of Wynne-Edwards's theory, it is important to see clearly how it differed from the theory advanced by Lack. Following Darwin's discussion in the *Origin*, Lack argued that population numbers, and hence densities, are ultimately limited by four factors, acting individually or in concert: viz., food shortage, predation, disease, and climate (Lack 1954, p. 276). Organisms engage in an unregulated ("scramble") competition for limited resources (food, mates, breeding sites), and population numbers are limited by the availability of such resources. Natural selection operates exclusively on individual organisms, especially on those organismic traits that bear on reproductive success. His studies on clutch size showed that organisms sometimes modulate their immediate reproductive output in relation to available resources in order to maximize their own individual long-term reproductive success.

Whereas Lack identified a number of different factors limiting population growth, Wynne-Edwards viewed the capacity of a habitat to provide a dependable supply of food as the ultimate factor limiting population growth. Population stability is achieved through a system of ritualized

TABLE 2.1: *Theories of Population Regulation*

	Lack's "Natural Regulation" Theory	Wynne-Edwards' "Animal Dispersion" Theory
Ultimate factors limiting population growth	Food shortage, predation, disease, climate	Capacity of habitat to yield reliable food supply
Proximate factors limiting population growth	Environmentally induced mortality; species-specific resource-modulated reproduction	Access to socially regulated distributions of conventional goods
Immediate objects of competition	Limited natural resources: food, mates, breeding sites	Conventional tokens: territories, social status
Type of competition	"Scramble": unregulated competition for limited resources	"Tournament": ritualized competition for access to conventional tokens
Explanation of resource-modulated reproduction	Optimization of individual reproductive benefits	Contribution to long-term population stability
Primary objects of natural selection	Individual Organisms ("organism selection")	Social Groups ("group selection")

("tournament") competition for conventional tokens (territories, social status) that serves to exclude some individuals from access to resources, preventing them from reproducing, and hence limiting population growth. Natural selection operates upon groups as evolutionary units, eliminating those groups that fail to develop self-balancing homeostatic mechanisms, and thus allowing stable, self-regulating populations to expand and colonize their habitats. The major differences between Lack's and Wynne-Edwards's theories are summarized in Table 2.1. What began as rival theories attempting to explain a minor puzzle left unresolved by Darwin eventually led to a provocative thesis about the evolutionary significance of social behavior and the level(s) at which natural selection can and does operate. As it turned out, this latter aspect would prove to be a potent catalyst for the subsequent development of evolutionary biology.

Group Selection Under Fire

In the Preface to *Animal Dispersion*, Wynne-Edwards remarks that "It has turned out to be an agreeable and characteristic feature of the theory [presented in the book] not to keep butting against widely held, pre-existing generalisations, but to lead instead into relatively undisturbed ground" (Wynne-Edwards 1962, p. v). Such a remark is revealing of the relationship between ecology and evolutionary biology in the early 1960s. Wynne-Edwards's emphasis on the subordination of individual advantage to group benefit reflected a long-standing and widely accepted explanatory tradition in ecology in which it was simply assumed that selection operates to insure the well-being of biological entities more inclusive than individual organisms.[8] Despite its taken-for-granted status in much ecology, this approach butted violently against the explanatory practices of population genetics and much of evolutionary biology. Rather than leading into "relatively undisturbed ground," as he had supposed, Wynne-Edwards had stumbled into a theoretical minefield.

Animal Dispersion stimulated controversy from the moment of its appearance. Reviews ranged from extolling its Darwin-like character to deriding its gullible author. The positive reviews tended to be effusive in their praise. Nicholson (1962), for instance, declared that in "this outstanding and richly illuminating book" a "very satisfactory balance has in fact been found between the factual and the theoretical elements," resulting in "convincing explanations of hitherto mystifying [animal] performances" (Nicholson 1962, p. 571). But such opinions were in the minority.[9] The numerous negative reviews of the book expressed either mild annoyance or open hostility. Either way, all were agreed that the book addressed important issues that needed to be further sorted out and clarified. Several biologists took Wynne-Edwards's book as a call to arms. The most important of these, in terms of their subsequent influence in evolutionary biology, were those of John Maynard Smith, David Lack, and George C. Williams.

Maynard Smith's Critique

In a review in *Nature* in 1964, John Maynard Smith attacked Wynne-Edwards's theory on a number of issues, but the point he returned to again and again was that the empirical data Wynne-Edwards's theory attempted to explain could be accounted for just as well by assuming that selection operates no higher than at the level of individual organisms. For example, if some individuals are willing to fight to secure and maintain territories in favorable areas, this will exclude others from access to those territories and whatever resources are associated with them (food, mates, etc.). To the victors go the spoils, and the losers are forced to do without – or to move on to more sparsely populated areas, thus accounting for population stability. "Thus there is no need to invoke group selection to explain the evolution of individual breeding territories, or the adjustment of territory size to food supply or to variations in the habitat" (Maynard Smith 1964, p. 1145).

A second argument tried to show that group selection of the sort that Wynne-Edwards supposed was pervasive in nature could occur only under the most unlikely circumstances. A basic problem with Wynne-Edwards's scheme, Maynard Smith argued, is that it postulates groups in which individuals altruistically participate in a system that limits their own reproduction for the sake of achieving population homeostasis. But such groups are always vulnerable to subversion by defectors with the antisocial (selfish) trait of seeking to maximize their own individual reproduction without regard for the good of the group. Such individuals will have an advantage over their altruistic rivals and their trait will quickly spread through the group. Thus, social arrangements of the kind required by Wynne-Edwards's theory are inherently unstable, and are thus unlikely to be realized very frequently in nature.

Lack's Critique

David Lack attacked Wynne-Edwards's theory in a number of contexts over several years, first in an article in *Nature* in 1964 (in which he essentially reminded readers of the results of his studies on clutch size), then in his 1965 Presidential Address to the British Ecological Society, and finally in an appendix to his book *Population Studies of Birds* (1966). A serious problem with Wynne-Edwards's theory, Lack noted, was that he had simply transferred his solution to the problem of human overfishing to the problem of the natural regulation of animal populations. Wynne-Edwards believed that the allocation of territories based on conventional substitutes (e.g., status in a social hierarchy) results in a more nearly optimal utilization of limited resources. Early in *Animal Dispersion* Wynne-Edwards wrote: "Ideally the habitat should be made to carry everywhere an optimum density, related to its productivity or capacity, without making any parts so crowded as to subject the inhabitants to privation, or leaving other parts needlessly empty" (Wynne-Edwards 1962, p. 4). References to "ideal" situations and "optimum" population densities appear throughout *Animal Dispersion.* It seems to be taken for granted that biological systems *are* optimally designed for long-term stability, and consequently that this fact should figure in biological explanations. This conviction was in stark contrast to Lack's view that it is enough to be able to explain how animal populations are limited by a variety of "external" factors operating on individual organisms, each of which is adapted for maximum reproductive success. Wynne-Edwards assumed that a *better* solution to the problem of population regulation is for populations to evolve self-regulating mechanisms, and that populations had achieved this solution through group selection. But no justification is given for why the better situation must obtain.

Fundamentally, Lack argued, the inference from a proposed solution to the problem of overfishing to the mechanism whereby natural animal populations regulate their own numbers is problematic. Whereas humans are sometimes able to anticipate the long-term consequences of their collective actions and to devise and enforce regulations to forestall these consequences, nonhuman animals lack the requisite cognitive and social capacities for doing so. In addition, there are important dissimilarities between the *problems* in the two cases. Whereas in virtue of technologically efficient harvesting methods little stands in the way of the commercial overexploitation of fish stocks, among naturally occurring predator-prey groups potential food items have had plenty of time to evolve defenses to being consumed by predators, with the result that overexploitation is less likely to occur. Finally, although there are good reasons for thinking that natural selection is responsible for the astounding feats of design we find among organisms, it is highly unlikely that if something *would* constitute an optimal solution to a biological problem, then natural selection must in fact have achieved this solution, especially if this involved shaping population-level entities. Lack spoke for the majority of biologists when he wrote, "Perhaps the most crucial difference between Wynne-Edwards's views and mine is his concept of the 'optimum population', which I regard as irrelevant to natural populations, though relevant to human fishing from which Wynne-Edwards derived it" (Lack 1966, p. 300).

Williams's Critique

As the history of this controversy has become canonized in recent years, G. C. Williams's book *Adaptation and Natural Selection* (1966) has often been seen as delivering the fatal *coup de grace* to Wynne-Edwards's theory (e.g., Wilson 1983).[10] As the title of the book suggests, it concerned the relationship between selection and adaptation, in particular the question of whether adaptations should be attributed to higher-level biological entities, such as groups. Whereas Wynne-Edwards and other ecologists were often willing to invoke group-level adaptations at the drop of a hat, Williams insisted that "Adaptation is a special and onerous concept that should be used only where it is really necessary. When it must be recognized, it should be attributed to no higher a level of organization than is demanded by the evidence" (Williams 1966, pp. 4–5). Wynne-Edwards argued that groups do display adaptations for group benefit. Williams countered that most (or perhaps all) such supposed group adaptations could be explained in terms of the adaptations of individual organisms, each behaving so as to maximize its own fitness.[11] For example, schooling in fish should be explained, not as a means for fish to assess the density of their population and to adjust their reproduction accordingly, as Wynne-Edwards supposed, but simply as the cumulative effect of the selfish behavior of individual organisms, each of which uses the bodies of its schoolmates to create a buffer between itself and any predators lurking nearby. Williams argued that fish do not swim in schools because doing so is good for the *school*, as Wynne-Edwards thought, but because each fish is doing what is good for *itself*. A good way to avoid being eaten by predators is to get lost in a crowd – the closer to the middle, the better. Schooling behavior, Williams argued, is explainable as a product of individual adaptations. Thus there is no need to postulate group adaptations when the same facts can be explained more parsimoniously in terms of individual adaptations (Williams 1966, pp. 212–17). Only if a population exhibits adaptations that promote group survival, which cannot plausibly be explained as an adaptation for individual reproductive success, can it be called an adapted population. If the group's continued survival is merely incidental to the operation of individual reproductive processes, however, then it is merely a population of adapted organisms (Williams 1966, p. 108).

The Fate of Animal Dispersion

Although not the last critique of *Animal Dispersion* to appear in the 1960s, for many biologists, Williams's book marked the end of one era in evolutionary biology and the beginning of another. Together with the other critical reviews, it convinced most biologists that group adaptations of the sort that Wynne-Edwards considered common were a chimera, and that group selection of the sort that he thought pervasive in nature was both unlikely and unnecessary. By the end of the decade a clear consensus had formed that Wynne-Edwards's theory was untenable. Yet Wynne-Edwards himself never accepted this verdict and continued to advance his theory in virtually unrevised form in a number of publications as late as 1993, including a substantial book, *Evolution Through Group Selection*, published in 1986. Although the titles of these later publications often suggested a defiant attitude (e.g., Wynne-Edwards 1991), an examination of his writings after Williams's critique makes for depressing reading (Wynne-Edwards 1968, 1970, 1971, 1977, 1978). Criticisms of his ideas go largely unacknowledged. One of the characteristic features of science is the interplay between creativity and criticism. It was in part the fact that Wynne-Edwards showed little interest in answering, or even seriously acknowledging his critics, that ultimately doomed his view.

At the beginning of the 1960s explanations of biological phenomena in terms of their benefits for supra-organismic entities were common. By the end of the decade, as one commentator has noted, "group selection rivaled Lamarckianism as the most thoroughly repudiated idea in evolutionary theory" (Wilson 1983, p. 159). Many scientific works hardly make a ripple in the ongoing development of science. Others are hailed for providing the critical insight needed for advance. Others achieve fame in ways that their authors could hardly have expected. *Animal Dispersion* is widely acknowledged as important in jolting biologists out of their vague, "group-adaptationist" slumbers, but is now cited chiefly as a shining example of how *not* to frame evolutionary explanations.

Group Selection Resurgent

Given the fate of Wynne-Edwards's theory in the 1960s, one might think that "group selectionism" was dead and buried, never to be taken seriously again. But like Lazarus, group selection did not stay dead for long. In the 1970s group selectionist thinking "mysteriously rose from the dead" (Wilson 1983, p. 159). Why? For one thing, there were some cases that really did seem best explained as cases of group, rather than organism, selection (always a remote theoretical possibility that even Wynne-Edwards's staunchest critics conceded). Second, reaching back to models originally developed by Sewall Wright in the 1930s (and 1945), mathematically adept biologists developed new models of group selection that were explicitly designed to avoid the sorts of problems that plagued Wynne-Edwards's approach. By the mid-1980s, group selection had once again become mainstream, such that one reviewer at the time was led to observe that "decent folk can once again discuss it as a viable mechanism" (quoted in Wilson 1983, p. 159). Much of the discussion centered on a fascinating case study of evolution in action that naturally lent itself to a group selectionist explanation.[12]

The Myxoma Case

The rabbit *Oryctolagus cuniculus,* originally introduced into Australia by European settlers in the nineteenth century, had by the mid-twentieth century reached pest proportions. In 1950, the myxoma virus was introduced to control the rabbit population. This strategy was at first immensely successful, as the virus killed 99.5 percent of the rabbits that were infected. But over the course of the next decade mortality rates fell drastically, so that by 1964 the virus killed only 8.3 percent of infected rabbits (Fenner and Marshall 1965). How was this to be explained?

One answer that appeals to the ordinary mechanism of selection at the level of individual organisms is simply that the rabbits evolved greater resistance to the virus. Rabbits which, for whatever genetic and physiological reasons, enjoyed any degree of resistance to the virus, fared better than those with less (or no) resistance. Consequently, rabbit genotypes conferring resistance spread rapidly in subsequent generations. This hypothesis was confirmed when wild rabbits and laboratory-maintained rabbits were innoculated with a pure strain of the virus. As was expected, the wild rabbits exhibited greater resistance to the virus than did the laboratory rabbits.

However, further tests showed that this was not the complete explanation for the lower mortality rate among wild rabbits. When wild and laboratory rabbits were innoculated with a *wild* strain of the virus, it was found that both kinds of rabbits showed fewer effects than when they were innoculated with the pure strain of the virus. This suggested that while the rabbits were evolving greater resistance to the virus, the virus was evolving lower virulence with respect to the rabbits. But whereas the evolution of resistance on the part of the rabbits is easily explained on the hypothesis of individual organism selection, the trend toward avirulence in the viruses is not. The extent to which the myxoma virus weakens or kills rabbits is a function of the number of viruses within any given rabbit. Viral strains with a higher rate of reproduction are more virulent than those with a lower rate of reproduction. Individual organism selection could account for greater virulence caused by higher rates of reproduction (since individual selection favors the more fecund), but not for lower virulence resulting from *lower* rates of reproduction. How, then, is the trend toward avirulence to be explained?

Lewontin (1970) proposed the following answer:

> The key is that the myxoma virus is spread by mosquitoes, which mechanically transfer a few virus particles to the rabbits they bite.... Each rabbit is a deme

> [i.e., a distinct population] from the point of view of the virus. When a rabbit dies, the deme becomes extinct since the virus cannot survive in a dead rabbit. Moreover, the virus cannot spread from that deme because mosquitoes do not bite dead rabbits. Thus there is a tremendously high rate of deme extinction, with the result that those demes are left extant that are least virulent. This causes a general trend toward avirulence of the pathogen despite the complete lack of selective advantage of avirulence within demes. (Lewontin 1970, p. 15)

In this account, each rabbit represents a resource exploited by a population of viruses. If a population overexploits its resource by failing to check reproduction, the resource fails (i.e., the rabbit dies), causing the extinction of that population. Viral populations that show reproductive restraint have a lower ecological impact on their resources than those who do not, and are thus in a better position to survive and spread to other rabbits. Although a maximum rate of individual reproduction is favored by individual selection, lower reproductive rates are favored by group selection operating via differential group extinction. Thus, an otherwise puzzling biological phenomenon is neatly explained in terms of selection operating at the level of populations.[13]

Wilson's Structured Deme Model

The myxoma case, as explained by Lewontin, would be an example of *interdemic* group selection because it involves the differential survival of distinct demes (populations). Recovering insights about the role of population structure developed by Sewall Wright in the 1930s and 1940s, biologists forty years later began to develop *intrademic* group selection models according to which the process in question need not involve differential survival and extinction of distinct demes (D. S. Wilson 1975, 1980; Wade 1976; Michod 1980, 1982). David Sloan Wilson has done more than anyone to revive group selection by defending his "structured deme" model. A version of it is worth considering in some detail.

According to standard population genetics models, individuals within a randomly mating population (a deme) are assumed to interact at random as well. That is, every individual is assumed to have an equal chance of encountering every other individual. This is a simplifying assumption, made for the sake of mathematical convenience, not because it realistically represents actual populations. As Wilson points out, on a daily basis individuals usually encounter only a small proportion of the population to which they belong, often just their immediate neighbors. Rather than being spatially homogeneous with respect to interactions, most demes display a good deal of internal structure. In order to capture this important aspect of population organization in biologically realistic models, Wilson argues that the homogeneity assumption implicit in most population genetics models must be replaced by a principle of spatial heterogeneity, by recognizing that demes are often subdivided in biologically significant ways.

For example, consider a set of organisms whose interactions with each other during some part of their life history take place within small local populations. Wilson calls such local populations "trait groups." A trait group is defined as the subpopulation within which the actual ecological interactions take place. Trait groups are typically much smaller than their containing demes. Mosquito larvae occupying a pool of water in a pitcher plant interact among themselves, but are effectively isolated from the larvae in other pitchers. Bark beetles excavating galleries in a tree likewise have frequent interactions with each other, but have little or no interaction with the bark beetles laboring in other trees. Trait group isolation is not, however, a permanent situation. Many species undergo an annual dispersal phase in which individuals leave their local trait groups and mix into the global population to mate in a way that is essentially random with respect to previous trait group membership. After mating, individuals (or their offspring) form new trait groups, and the cycle can then begin again. The aforementioned mosquitoes, for example, will lay their eggs in other pitcher plants, and new local populations of mosquito larvae will have been founded.

Wilson asks us to consider how gene frequencies in the global population might be affected by this cycle of within-trait group interaction followed by dispersal:

> Consider a genotype whose activities increase the productivity of its local population without, however, changing the gene frequency within the population. Populations with a high frequency of this genotype will be more productive than those with a low frequency, and will differentially contribute to the pool of dispersers. The genetic composition of the dispersers will be biased toward the genotype that increases the productivity of its group, and this bias is carried into all groups colonized by the dispersers. (D. S. Wilson 1980, p. 19)

How might this work? Begin by imagining a population of haploid (i.e., having one set of chromosomes) individuals genetically identical except at one locus, where there are two alleles, **A** and **B**. (An "allele" is an alternative form of a gene. Genes are located at particular regions of a chromosome, known as "loci." So in this case **A** and **B** differ by having different forms of the gene that occupies a certain locus on a given

TABLE 2.2: *Trait Groups*

trait group #1: 4**A**/1**B**	Trait group #3: 2**A**/3**B**
trait group #2: 3**A**/2**B**	Trait group #4: 1**A**/4**B**

chromosome.) These alleles are present in the deme in frequencies p and q (where p + q = 1). Assume further that there are equal numbers of **A**s and **B**s in the population as a whole (i.e., p = q = 0.5). This situation is instantiated in a hypothetical deme consisting of four trait groups, each of which contains five individuals, as illustrated in Table 2.2: Although the "objective" frequencies" of **A** and **B** in the entire deme are each 0.5, the "subjective frequencies" that **A**s or **B**s *actually experience* in their own trait groups are not. Each of the four As in trait group no. 1 experiences a frequency of **A**s equaling 0.8, whereas they each experience a frequency of **B**s of only 0.2. Each of the three As in trait group no. 2 experiences a frequency of **A**s equaling 0.6, and experiences a frequency of **B**s of 0.4. Parallel calculations can be carried out, of course, for the subjective frequencies of the **B**s in each group.

Now suppose further that **A** and **B** are alleles that make their possessors "altruistic" or "nonaltruistic," respectively, and that the deme in question is a population of birds divided into a number of flocks (one flock = one trait group). Individuals possessing allele **A** (hereafter **A**-individuals) remain vigilant and give a warning call upon sighting an approaching predator, thereby alerting the flock to danger. In doing so an **A**-individual incurs some small cost to itself (in terms of effort or increased conspicuousness to predators), but is more than compensated for these costs if he makes the flock take wing, because he is safer from predation in the flock than he would be if he flew off alone. By issuing a warning call, an **A**-individual increases its fitness, that is, it improves its own chances for survival and reproduction. But **A**-individuals are still less fit than **B**-individuals who neither watch for predators nor give warning calls. **B**-individuals readily benefit from the presence of **A**-individuals by taking flight at the first indication of danger. Because they enjoy all the benefits of being in flocks with callers but incur none of the costs of remaining vigilant and calling, **B**-individuals enjoy greater fitness than **A**-individuals.

Within any given flock, **A**-individuals and **B**-individuals can be expected to be taken by predators in equal numbers, since all benefit equally from the presence of callers. But flocks with a higher frequency of **A**-individuals (i.e., trait groups no. 1 and no. 2 above) are automatically more alert than those with fewer **A**-individuals (i.e., trait groups no. 3 and no. 4 above). As a result, flocks with a higher frequency of **A**-individuals suffer fewer total casualties (loss of both **A**- and **B**-individuals) from predation. Likewise, groups with higher frequencies of **B**-individuals suffer greater total casualties from predation. By the end of the year, predation can therefore be expected to have taken a higher toll of **B**-individuals than **A**-individuals in the global population. The result is that the frequency of **A**s will have risen in the deme as a whole.

In order to complete the story, one further detail must be added. Although interactions during most of the year take place between individuals within their flocks, once every year all the flocks in the population migrate to a common breeding ground to mate. Mating between individuals is random with respect to previous flock membership. After this mating phase, the individuals (or their offspring) then fly away to form new flocks, and the cycle begins again.

Notice that, unlike Wynne-Edwards's model of group selection, Wilson's model does not require that individuals forgo immediate gains in fitness for the sake of the long-term benefit of the group. By issuing an alarm call an individual is conferring an immediate fitness benefit on himself because he is safer from predation if the entire flock takes wing than he would be if he flew off alone. The fitness benefit in this case is only achieved by simultaneously conferring a fitness benefit on every other member of the flock. Wilson calls such behavior "weak altruism," to distinguish it from "strong altruism" according to which aiding others involves some sacrifice of fitness on the part of the actor. This explains why, in calculating above the subjective frequency of a particular type of individual experiencing its own type within its trait group, each individual's experience of itself was factored in along with its experience of others. On Wilson's model the benefits for other members of a trait group arising from an individual's actions are shared by the actor as well. He considers the emphasis on strong altruism in the form of spectacular displays of self-sacrifice on the part of individuals, and the difficulty of explaining how this can evolve, to have been the chief impediments to taking group selection seriously. By focusing instead on behaviors which are individually as well as group advantageous, group selectionism can shake off an unnecessary restriction. By introducing the concept of structured demes, Wilson argues, it can be seen how natural selection becomes sensitive, not only to the fitness of individuals relative to each other within their local populations, but also to the productivity of local populations relative to each other in the global population. "This latter component may be regarded as natural selection on the level of populations, or *group selection*" (D. S. Wilson 1980, pp. 19–20).

But Is It Group Selection?

Remarkably, critics have generally acknowledged that the sort of process Wilson describes could well be a significant factor in evolution, but have nevertheless denied that it constitutes a *bona fide* form of *group selection*. Wilson first put forward his model of intrademic trait group selection in 1975. Writing the following year, Maynard Smith (1976, p. 246) argued that the process described by Wilson is better understood as a special form of individual organism selection in which individuals interact non-randomly with respect to genetic relatedness. "In particular, we must be clear whether our theory asserts that the evolution of a trait requires the existence of groups, or merely that neighbors be relatives" (Maynard Smith 1976, p. 243). Wilson's model, he maintained, only requires that neighbors be relatives, not members of a distinct group. "The term group selection," by contrast "should be confined to cases in which the group (deme or species) is the unit of selection. This requires that groups be able to 'reproduce,' by splitting or by sending out propagules, and that groups should go extinct" (Maynard Smith 1976, p. 247), conditions which the process Wilson describes does not necessarily satisfy.[14]

Wilson's structured deme model does indeed fail to satisfy the criteria Maynard Smith identified. The critical issue, however, is whether this disqualifies the process Wilson describes as "group selection." Maynard Smith believed that it did. In response, Wilson wrote,

> Is the structured deme model a form of group selection? I agree with Maynard Smith (1976) in the desirability of sharpening distinctions between modes of selection, but I would sharpen them along conceptual, rather than historical lines. . . . It would be a pity to avoid calling it group selection simply because that term has been applied to a different conception of groups in the past . . . (D. S. Wilson 1979, p. 609)

He also pointed out that genetic relatedness between individuals is not a requirement of his model, which only requires that individuals interact with some limited segment of the population (Wilson 1979, pp. 606–7). This move, however, created an opening in which Maynard Smith could renew his claim that the process Wilson describes is best understood as a form of individual organism selection. Even if Wilson's model as one in which interactions occur between nonrelatives, it is still best regarded as a form of individual selection of the kind known as *frequency-dependent selection* (Maynard Smith 1982, p. 35). Frequency-dependent selection occurs when fitness of an allele (or genotype, or phenotype) is affected by its frequency in the population. In the illustration of Wilson's model sketched above, the fitness of the **A** allele (connected with alarm-calling behavior) depends on the frequency of **A**s in the group. If **A** is at high frequency, it is fitter than if it is at low frequency. That is, **A**-individuals are better off in a group with a large number of other **A**-individuals than they are in a group with a small number of **A**-individuals. In other words, frequency-dependent selection is simply individual-level selection in special environments composed partially of other individuals. Since selection in such cases can be understood to proceed entirely on the basis of fitness differences between individuals, and the process Wilson described can as well, Maynard Smith argued that Wilson's structure deme model is therefore not a model of group selection.[15]

There is no need to follow this debate into its later stages (Wilson 1983, p. 178; Maynard Smith 1987a, p. 123; Sober and Wilson 1998). Clearly the issue at stake is whether the process Wilson proposed is really distinct from individual-level selection processes, and merits being described as "group selection." Did Wilson succeed in resurrecting group selection after its demise in the 1960s, or did he merely describe another way in which selection can operate on individual organisms? If Wilson and Maynard Smith agreed about the essential features of Wilson's model, and that the process it describes could be a significant factor in evolution, why do they disagree so strongly about whether it should be considered a model of "group selection"? In the final analysis, why does it matter? What's in a name?

Summary: The Group Selection Controversy

Appreciating what was and is at stake in the controversies just reviewed requires reminding ourselves why the question of group selection mattered in the first place. The fundamental issue concerns the sorts of explanations Darwin's theory provides for rendering comprehensible the natural world. When Darwin maintained that "natural selection works solely by and for the good of each being," did he mean to propose a general and open-ended evolutionary mechanism according to which natural selection works for the good of biological "beings" in addition to individual organisms? It appears that he did. Although he preferred explanations in terms of selection operating among individual organisms, Darwin freely invoked selection at the level of communities in order to account for certain puzzling biological phenomena that made little sense on the supposition that selection operates exclusively at the level of individual organisms.

Many later biologists followed suit. Wynne-Edwards invoked group selection to explain what he took to be a significant anomaly for the individual selectionist perspective, viz., stable population densities in relation to food resources. In all such cases, appeals to group selection were made when it was believed that simple models of individual-level selection were inadequate to the explanatory task. But the acceptability of such forays was far from obvious to critics who believed that flirtations with "group selection" threatened to undermine the individualist foundation upon which Darwinism had (or should have) been built. Hence the dispute about whether the process Wilson described in his structured deme model should be described as a *bona fide* example of "group selection" eventually became reduced to a difficult metaphysical conundrum: When does an interacting set of individuals become a "group," that is, a "higher-level" entity in its own right, subject to natural selection? The question is critical. If there are supra-organismic entities subject to natural selection, then in principle selection can forge *adaptations* explicable as having evolved for group, rather than for individual, benefit, and the living world appears in a very different light from that presupposed by a strict individualistic perspective. If Darwin's theory succeeds in anything, then it *should* succeed in making the living world appear in a very different light. As we shall see in the next chapter, however, even the idea of selection working for the good of individual organisms, an idea that Darwin and almost all his followers have taken for granted, is far from the scientific *terra firma* it once seemed.

3

For Whose Good Does Natural Selection Work?

> Birds' wings are obviously "for" flying, spider webs are for catching insects, chlorophyll molecules are for photosynthesis, DNA molecules are for. . . . What *are* DNA molecules for. . . . [This] is the forbidden question. DNA is not "for" anything . . . all adaptations are for the preservation of DNA; DNA just *is.*
>
> (Dawkins 1982a, p. 45)

Introduction

Natural selection operates by favoring those individuals whose characteristics confer any slight advantage in what Darwin termed "the struggle for existence." As selection operates generation after generation, distinguishing the fit from the less fit, adaptations evolve and are passed on to offspring, which in turn engage in the struggle anew, resulting in "the improvement of each organic being in relation to its organic and inorganic conditions of life" (Darwin 1859, p. 84). Hence the astounding array of complex adaptations characterizing living things. On this all evolutionists agree.

Darwin maintained again and again that "natural selection works solely by and for the good of each being" (Darwin 1859, p. 489; 1959, p. 758). But there are many "beings" involved in the evolutionary process. By whose and for whose "good" does natural selection work? As we have seen, although Darwin generally thought of selection as operating on and for the good of organisms (i.e., as improving the adaptations characterizing the individual organisms constituting an evolving lineage), when the situation warranted it he was also willing to countenance selection operating and forging adaptations at the level of communities or groups. Many biologists in the century after Darwin followed suit, until the issue finally came to a head in the 1960s. By the end of that decade, "group selectionism" seemed all but dead, but in the 1970s and 1980s it enjoyed something of a rebirth. Since then, although critics reluctantly concede that selection might, under special circumstances, operate at the level of groups, they have continued to resist the idea that such selection is capable of forging group-level adaptations. Conceding that selection can be sensitive to group membership is one thing. Accepting the idea that there are group-level adaptations that benefit the group at the expense of the individuals comprising them, is quite another. According to these biologists, the "being for whose good natural selection works" is and can only be the individual organism.

Or is it? Telling the story of how Darwinian views of selection have themselves evolved since the *Origin of Species* is a bit like peeling the proverbial onion: As surface layers are stripped away, deeper levels, not at first apparent, gradually come into view. My aim in this chapter is to peel away the remaining layers by explaining current thinking about this issue, and the dialectic by which it came about. More ambitiously, my aim is to arrive at a conception of natural selection that captures the causal structure of this process.[1] Given our best current understanding of evolution, for whose "good" does natural selection work? Just as importantly, what might it *mean* to say that natural selection works for the good of some being? In order to answer these questions we will need to return to an issue touched on only briefly in previous chapters. The evolutionary problem of "altruism" is the key for understanding the recent development of Darwinism.

The Evolutionary Problem of Altruism

A bear rips apart a beehive with its powerful claws; hundreds of bees boil from the hive in defense and retaliation, driving their stingers into the intruder and then pulling away, leaving their barbed weapons in the bear, the attached poison sacs continuing to pump venom into the wounds. The bear is deterred but at a terrible cost to the defenders. The bees, their viscera ripped from their bodies, fly off a short distance to die. A leopard approaches a troop of baboons on the open savanna. A lone male, scarred from earlier encounters, rushes to the defense of the troop, repeatedly charging the leopard and baring his large canines, allowing the others to move to safer ground, but in the process putting himself in grave danger. Adult male musk oxen in the far north form a defensive circle in the presence of a predator, enclosing the more vulnerable females and juveniles inside, but exposing themselves to greater danger. Similar examples could be multiplied indefinitely. Many animals put themselves at risk, or even endure the ultimate sacrifice, in order to safeguard their cohorts. They subject themselves to increased danger in order to warn others of approaching predators; they forgo reproduction in order to help rear the offspring of others; they share food, territories, mates; they groom each other; occasionally they even endure death for the sake of their compatriots. The epitome of such other-regarding behavior may be the social insects with their castes of workers who toil their entire lives for the good of the colony, only to die without reproducing. A list of such behaviors would be very long indeed.

The Central Theoretical Problem

Despite their apparent diversity, what all of these behaviors have in common is that they can be considered *altruistic* from an evolutionary perspective. That is, each of these behaviors confers a (fitness) benefit upon a recipient while imposing a (fitness) cost upon the actor. E. O. Wilson judged altruism to be the central theoretical problem for a Darwinian understanding of social behavior. He put his finger on the key issue: "[H]ow can altruism, which by definition reduces personal fitness, possibly evolve by natural selection?" (1975, p. 3) The problem is acute. As Darwin noted, "Natural selection will never produce in a being anything injurious to itself, for natural selection acts solely by and for the good of each" (Darwin 1859, p. 201; see also pp. 84, 85–86, 95, 199, 233, 459, 485–6). The basic logic of Darwin's theory predicts that organisms that always act in their own self-interest will flourish at the expense of more community-minded individuals, and consequently that such selfish individuals will come to predominate in every population. Altruistic behavior should be ruthlessly eliminated in favor of selfish behavior. Yet a very different picture of nature is evident to anyone who has ever observed animals in the wild, or even just seen a nature documentary, in which cooperation seems to be the norm. The problem of altruism is the problem of explaining how such behaviors can possibly evolve by natural selection. Darwin's theory is spectacularly successful at explaining the obviously adaptive characteristics of living things (e.g., claws, fangs, fur, feathers, etc.). But what about such apparently maladaptive behaviors?

Challenging the Organismic Paradigm

As Darwin well realized, altruism is only a serious problem if one takes an exclusively organism-centered view of evolution. If one moves "up" a level to that of the family, group, or community, the problem can (in principle) be resolved. As we saw in the previous chapter, such a move was enthusiastically embraced by many biologists in the hundred years after the *Origin* but then fell into disfavor by the end of the 1960s. Despite subsequent ingenious attempts to rehabilitate this approach, since then a very different strategy has been ascendant in evolutionary biology, one that attempts to solve the problem, not by thinking of selection as operating on levels "above" (i.e., more inclusive than) that of individual organisms, but instead as operating on a level "below" that of individual organisms; viz., at the level of the gene. Understanding and evaluating this strategy, and considering viable alternatives to it, will occupy us in the sections that follow.

Genes versus Organisms

Just as it was the social insects that posed the most difficult version of the problem of altruism, so, too, it was theoretical work on the social insects that led to the resolution of the problem. The seeds of the solution can be found in the work of R. A. Fisher (1930, pp. 177–81) and J. B. S. Haldane (1932, pp. 10–131, 207–10), but it was William D. Hamilton (1963, 1964) who offered the first compelling explanation for the sort of altruism found in the eusocial insects. His idea was dubbed (by John Maynard Smith 1964) "kin selection" and despite some common misunderstandings (documented by Dawkins 1979), Hamilton's insight has generally been accepted as the correct explanation for a range of altruistic behaviors.

Kin Selection

Kin selection is the idea that under certain conditions natural selection can favor a behavior that imposes a cost on the actor if it confers a benefit upon another organism that shares many of the same genes (e.g., a close relative). For example, if an organism possesses a gene that causes it to give aid to relatives who also possess a copy of that gene, then by giving such assistance the organism is assisting in the transmission of that gene (copies of which reside in another organism's body) into the next generation. In a sense it doesn't matter whether the copies of one's genes are transmitted directly through producing offspring of one's own, or indirectly through the offspring of relatives. In either case, the genes in question are transmitted to the next generation. How the gene makes it into the next generation is secondary to the fact that it does so. It is important, however, that organisms confer benefits preferentially upon those possessing the same genes. If organisms perform altruistic behaviors indiscriminately for relatives and nonrelatives alike, then the gene for altruism has not gained any advantage. A bird that issues an alarm call indiscriminately amongst relatives and nonrelatives might be helping others to escape from a predator while drawing attention to itself, thus exposing itself to additional danger. Such a behavior imposes a cost without any commensurate gain. But a gene that causes a bird to issue an alarm call when surrounded by relatives may be increasing its representation in the next generation, even while putting its immediate "vehicle" at risk. There is substantial evidence that a range of organisms preferentially give alarm calls when their relatives are likely to benefit from such behavior (Dunford 1977; Sherman 1977).

Kin selection was developed as a way of solving the problem of altruism in the eusocial insects (which have an unusual genetic system that makes siblings more closely related than ordinary sibs). But its applicability is much wider. Its application to alarm calling has already been mentioned. Kin selection might also help to explain "helpers at the nest"; that is, cases in which an individual forgoes breeding and instead assists in the rearing of offspring that are not its own (Emlen 1984). In the majority of such cases the individuals helped are close relatives, and thus the offspring who benefit bear a strong genetic relationship to the helper. Under some conditions (e.g., where mates or breeding sites are rare), it may benefit an individual's genes more to help with the rearing of a close relative's offspring than to attempt (unsuccessfully) to mate or to sit around during the breeding season squandering one's energy. Kin selection thus provides a powerful explanation for how altruistic behaviors can be favored by natural selection. From the point of view of the behaving organism, such behavior is costly, but from the point of view of the gene, it can actually be advantageous.[2]

Selfish DNA

The gene's-eye point of view thus resolves a range of problems which persist if one thinks exclusively in terms of costs and benefits at the level of individual organisms. The value of the genic perspective is further increased when one considers cases that seem to defy explanation in terms of benefits for individual organisms. Consider "selfish DNA" (Doolittle and Sapienza 1980; Orgel and Crick 1980). Molecular geneticists have discovered numerous DNA sequences that carry information for their own replication, but which do not carry any information for the organism's phenotype (i.e., the organism's morphology, physiology, behavior). *Transposition* refers to the shift of a segment of DNA to a new locus in the genome. Transposable elements (transposons) are able to replicate themselves, insert themselves into other loci in the genome, and thereby increase their number of copies. Any variant that can transpose at a higher rate than others will increase in frequency in the genome, and may be said to have a selective advantage. Transposable elements do not, so far as we know, serve any organismic function, and in fact may persist in spite of any effects they might have on organisms (e.g., increasing an organism's load of mutations). In this sense they can be viewed as parasites of the genome in which they reside. Like all parasite-host associations, the host can be expected to evolve countermeasures to the parasite when the costs become significant. In this case a mutant sequence may repress transposition throughout the genome, thereby preventing large-scale mutations harmful to the organism. Consequently, selection at the organismal level favors DNA sequences that repress transposition, while selection at the gene level favors variant transposable elements that resist repression. It is possible to think of this as a coevolutionary "arms race" of the sort known to occur in predator-prey relationships (Dawkins and Krebs 1979), except here the struggle is between an organism and some of its genes. The point, however, is that such cases can only be understood as involving natural selection operating at the level of genes.

The Group Above and the Gene Below

It is worth briefly reviewing how significant the genic perspective has been for the development of Darwinism. According to some analysts, the triumph of modern evolutionary biology can be summarized as a triumph for the gene's-eye perspective over both the organismic paradigm and over group selectionism, with profound implications for how we think of the beings for whose good natural selection operates. According to Helena Cronin (1991, p. 275), for example, seemingly altruistic behaviors like alarm-calling look quite different from the perspectives of benefit for the sake of the group, for the individual organism, and for the gene. From a group selectionist perspective, behaviors like alarm calling are genuinely altruistic (i.e., entail costs to the actor), adaptive (at the level of the entire group), and are unproblematically explained (from

TABLE 3.1: *Apparently Costly (e.g., "self-sacrificing") Organismic Behavior*

From a Group-Centered View	Genuinely Altruistic	Adaptive (for the group)	Unproblematic
From an Organism-Centered View	Genuinely Altruistic	Nonadaptive (for the individual)	Problematic
From a Gene-Centered View	Merely Apparently Altruistic	Adaptive (for the relevant genes)	Unproblematic

their point of view) in terms of group selection. For organism-centered Darwinism, on the other hand, such behavior is altruistic, nonadaptive (for the actor), and problematic, as it is unclear how such behavior could arise or be maintained by selection operating exclusively at the level of individual organisms. Finally, for gene-centered Darwinism, the altruism is merely apparent (it is simply a case of genes helping themselves, albeit in different bodies), the behavior is adaptive (for the genes in question), and thus it poses no problem. Her claims are summarized in Table 3.1. As we saw earlier, there are serious problems with understanding how selection could favor groups in ways that are detrimental to the organisms comprising them. Likewise, altruism poses a serious problem for understanding selection at the level of individual organisms. As Cronin makes clear, however, these problems simply disappear if one adopts the genic perspective.[3]

Gene Selection versus Gene Selectionism

There is little doubt that adopting a gene-centered perspective helps to resolve some otherwise very thorny problems. The idea that at least *sometimes* natural selection operates directly at the level of individual genes is uncontroversial. But advocates of this view typically wish to claim much more than this. In addition to claiming that taking a gene-centered perspective solves a number of problems, they wish to assert that a gene-centered perspective is *always* the correct point of view to take. I will call this latter view, that genes are the only true "units of selection," *gene selectionism.* Gene selectionism has come in for a good deal of criticism from biologists and philosophers. In this section, I want to explain what it might mean to claim that genes are the true "units of selection," and why this claim is often made. In a later section, I will consider objections to this view, as well as a range of responses.

The A Priori *Argument for Gene Selectionism*

Gene selectionism has sometimes been defended using *a priori* arguments to the effect that genes, and only genes, have the requisite properties to function as units of selection, and thereby to be the ultimate beneficiaries of whatever adaptations exist. Such arguments form a cornerstone for the most impassioned defenses of gene selectionism (Cronin 1991; Dawkins 1989; Dennett 1995; Williams 1966). In his popular exposition of this approach, *The Selfish Gene* (1989), Richard Dawkins explores with great ingenuity the meaning and consequences of viewing single genes as the ultimate benefactors of natural selection. In this view, genes can be thought of as repositories of information for constructing bodies (or "vehicles," in Dawkins's terminology). Differences in genes give rise to differences at the phenotypic level, resulting in the differential survival and/or reproduction of the genes responsible for those phenotypes. Genes are perpetuated to the extent that they produce phenotypic effects that give their possessors advantages over genes producing less advantageous phenotypic effects. Natural selection operates directly on phenotypes, but the indirect effects on the fate of genes is what leads to evolutionary change.

Dawkins argues that the unique properties of genes qualify them as the genuine "units of selection." Genes replicate themselves faithfully, exist in large numbers in virtue of the many copies of the same gene in a population, and persist for long periods of time. Genotypes (larger or smaller combinations of genes), organisms, and groups, by contrast, are short-lived entities that quickly get broken down and reshuffled, exist in far fewer numbers, and "reproduce" themselves only in the most imperfect sense. Changes to an organism's body that are not encoded in the organism's genes exist only for the brief time that particular organism exists. Genetic changes, by contrast, can be passed on indefinitely. According to Dawkins, "[T]he individual is too large and too temporary a genetic unit to qualify as a unit of natural selection. The group of individuals is an even larger unit. Genetically speaking, individuals and groups are like clouds in the sky or duststorms in the desert. They are temporary aggregations or federations" (Dawkins 1989, p. 34). The point is that only genes get preserved from one generation to the next, and hence only genes have the properties necessary to be the units of selection and the "owners" of adaptations (Cronin 1991, p. 70).

The Explanatory Scope Argument

A second kind of argument in support of gene selectionism points to its immense explanatory scope. Whereas some biological phenomena requiring a selectionist explanation can be explained either in terms of selection operating at the level of organisms or at the level of genes, *every* such phenomenon (plus those that resist individualist explanations) can be explained in terms of selection operating at the level of genes. Therefore, gene selectionism provides a much more general, and hence considerably more powerful, explanatory perspective in evolutionary biology.

Consequently, the gene selectionist can claim that *all* cases of natural selection are really cases in which genes function as units of selection benefiting from adaptations. Consider a simple, and well-known, example. The *classic* example of natural selection effecting a change in the characteristics of organisms in a natural population concerns the evolution of melanism in moths in industrial England. The moth *Biston betularia* is a polymorphic species, existing in two distinct forms. Some individuals of this species are melanic (dark), while others are nonmelanic (light). Before the Industrial Revolution in England, the melanic form was virtually unknown. By 1895, however, it constituted roughly 98 percent of some populations in the vicinity of Manchester. The accepted explanation for this rapid increase in the frequency of melanic individuals is that the trees upon which the moth rests became covered with soot from nearby factories. Kettlewell (1955, 1956, 1961, 1973) was able to show that melanic moths, because of their cryptic coloration, enjoyed at least a 50 percent selective advantage over nonmelanic forms in avoiding predation by birds. A difference of this magnitude could be because of sheer chance, but the odds are strongly against it. In virtue of their greater fitness, melanic moths are selected for while nonmelanic moths are selected against. The units of selection in this case *seem* to be organisms.

From the perspective of gene selectionism, this initial judgment needs to be revised by the recognition that melanic and nonmelanic moths differ with respect to the genes they carry. Being melanic or nonmelanic is controlled by segments of the moths' DNA. Although melanism is probably, like most phenotypic traits, controlled by genes at several loci, suppose for the sake of illustration that melanism is the result of the possession of a specific allele at a single genetic locus. The presence or absence of a specific allele makes the difference between being melanic, and enjoying greater fitness, or being nonmelanic, and suffering greater risk of succumbing to predation. If we assume that causality is transitive, such that if **A** (the gene) causes **B** (a phenotypic character), and **B** (the phenotypic character) causes **C** (enhanced fitness), then **A** (the gene) ultimately causes **C** (enhanced fitness), then it looks like an individual gene is the ultimate cause of differential organismic survival. In other words, it appears that organismic fitness differences are directly caused by, and hence directly reducible to, fitness differences between individual alleles. What looked at first like an obvious example of organism selection begins, upon closer scrutiny, to look like an example of gene selection. In principle at least, the same sort of analysis could be used to reinterpret *any* purported case of organism (or higher-level) selection. Gene selectionism is thus advanced as an exceptionless thesis having universal applicability. Nature is essentially a contest in which genes vie with each other by constructing bodies with which they lever themselves into succeeding generations. Organisms are simply "vehicles" driven by their genes (Dawkins 1982a,b). In the evolutionary race, genes, not organisms, are in the driver's seat.

Causality and Representation

Gene selectionism has been a controversial thesis, and has thus met its share of objections. In considering such objections, it is worth identifying two different versions of gene selectionism, which in our discussion so far have not been distinguished. Each is associated with a particular thesis. According to *the causal thesis,* all selection is in fact selection at the level of individual genes. Expressed concisely, genes cause phenotypes, which then interact with the environment, resulting in the differential perpetuation of genes. Adaptations thus exist for the sake of genes. According to the *representation thesis,* by contrast, regardless of the identity of the entities upon which selection actually operates, all selection can be *represented* in terms of selection at the level of genes. Even if selection acts directly on phenotypes, it is still true that only genes get passed on to subsequent generations, and in doing so serve as repositories of information. This is why all evolutionary change can be represented as changes in gene frequencies. Both versions of genic selectionism will be considered below as the two chief arguments in support of genic selectionism are critically evaluated.

The A Priori *Argument Undermined*

Recall the chief *a priori* argument in support of gene selectionism. Gene selectionists point out that organisms and groups are too ephemeral to be the beneficiaries of selection, because each is broken up and destroyed, if not in each generation, then after only a few generations. Only genes are passed on intact and hence persist from one generation to the next; therefore only genes qualify as the beneficiaries of natural selection. This claim requires further scrutiny. If pressed, gene selectionists admit that it is not literally specific bits of genetic material that are passed on, but rather the *information* encoded in such bits of genetic material (Williams 1992). But if so, then it can be claimed with equal justice that *phenotypic properties* persist and get passed on from one generation to the next. That is, an objection that could be leveled at gene selectionism at this point is that certain (but not all) *properties* of organisms have every bit as much right to be considered that which gets passed on to offspring as do their genes (Sober 1984). In fact, they might have more right to be considered that which gets preserved and passed on, because in many organisms (e.g., birds, primates, humans) there is intergenerational transmission of learned behaviors. Thinking of properties, rather than genes, as that which gets passed on would include all those properties coded for by genes, and then some. It would therefore be the foundation for a more comprehensive account of evolutionary change. Quadrupeds pass along their quadrupedalism; animals with binocular vision pass on their binocularity; and so on. Darwin's theory is supposed to be a very general theory of evolution, applicable to life forms anywhere in the universe. Conceivably, there might be life forms elsewhere that do not use DNA or any genetic material but instead pass along their characteristics via some other mechanism. Were we to encounter such creatures, we would recognize them as having evolved by a process of natural selection, despite the fact that they lack "genes" in the normal sense. All that is needed for the transmission of information or properties from one generation to the next is suitable physical embodiment and a reliable copying mechanism. Genes represent one such possibility. There is no reason there couldn't be others. But if so, then the chief *a priori* argument in support of gene selectionism collapses.

Thrust and Parry re the Causal Thesis

Whereas the representation thesis is often grudgingly conceded by critics of gene selectionism, the causal thesis has more frequently been vigorously challenged. They argue that although all evolutionary change can be *represented* in the currency of selection for or against individual genes, it is nonetheless false that all selection is causally *explainable* in terms of selection for or against individual genes. Gould puts the matter bluntly: "Selection simply cannot see genes and pick among them directly. It must use bodies as an intermediary. A gene is a bit of DNA hidden within a cell. Selection views bodies" (Gould 1980, p. 90). According to this view, the whole organism, rather than the individual gene, is the unit of natural selection. A similar argument points out that in some cases it is not individual genes that are selected, but rather *pairs* of genes that form the basic functional unit (Sober and Lewontin 1982). Heterozygote superiority occurs when a heterozygous condition (e.g., *Bb*) is more fitness-enhancing than is either alternative homozygous condition (e.g., *BB* and *bb*). To take the stock example, consider the sickle-cell allele found in some human populations. Individuals homozygous with two copies of the sickle-cell allele (*bb*) cannot produce normal hemoglobin and thus suffer from severe anemia and usually die in childhood. Individuals homozygous for the normal allele (*BB*) produce normal hemoglobin and hence do not suffer from anemia, but are susceptible to malarial infection. Heterozygous individuals (*Bb*), however, with one sickle-cell allele and one normal allele, produce normal hemoglobin and are resistant to malarial infection. Heterozygous individuals are fitter in malarial regions than are homozygotes. But being heterozygous is not a property of individual genes – it is a property of pairs of genes. So, it is argued, in such cases gene selectionism fails. Gene selectionism should therefore be thought of as merely a "bookkeeping" technique (Wimsatt 1980), with the fitness values of genes as "artifacts" that fail to identify the *causes* of evolutionary phenomena.[4]

In response, however, the gene selectionist can point out that just as phenotypic characteristics only have the fitness-enhancing effects they do in certain environments, so, too, genes only have their fitness-enhancing effects in the presence of other genes, which form part of that gene's immediate environment. The gene's environment includes everything that affects the fate of the gene in question. The sickle-cell allele does well in the presence of both malarial conditions and a normal allele at that genetic locus, which explains why it is favored by selection under these conditions. Cases of heterozygote superiority thus pose no problem at all.

Another common objection to the causal thesis of gene selectionism points out that the idea that there is a "gene for" some particular phenotypic characteristic is pure fiction (and not very good fiction, at that). There is no "gene for blue eyes" because (i) blue eyes are the effect of a number of genes, and (ii) each gene can have multiple phenotypic effects. Between genes and phenotypes there are "one-many" (pleiotropic) and "many-one" (epistatic) effects. Consequently, selection cannot be discriminating among individual genes.

Gene selectionists respond that such objections miss the mark. What an expression like "gene for" means is that the presence or absence of a particular gene makes a difference at the phenotypic level. Dawkins develops this point by distinguishing between blueprints and recipes. The individual words in a cake recipe do not map onto particular bits of the finished cake, but replacing one word with another – salt for sugar, for example – will result in a very different final product. Any given gene may have multiple phenotypic effects. The selective advantage of having any particular gene is the net benefit conferred when all the effects (positive, negative, or neutral) of the gene are taken into account. Consequently, even if a simplistic understanding of the expression "gene for" is false, there is still an important sense in which all selection is selection for or against individual genes. Gene selectionism cannot be defeated so easily.

A different tack is taken by Brandon, who argues that Dawkins's claim that all adaptations are for the good of genes is mistaken: "If we agree with Dawkins and say that adaptations are for the good of [genes], then we will be unable to distinguish group and individual adaptations, thus depriving such talk of all explanatory significance" (Brandon 1985, p. 91). Consider an example. Typically it is thought that cryptic coloration exists for the sake of the cryptically colored organism, because it provides protection from being detected by predators, or allows the cryptically colored predator to employ stealth while hunting. In Dawkins's view, however, the cryptic coloration is an adaptation for the sake of the organisms' genes, as are all other properties of the organism, as are whatever beneficial group-level properties that might exist. Doesn't this way of understanding evolution erase the distinction between organismic and group adaptations?

In response the gene selectionist can argue that this worry is unfounded. As Dawkins notes, emphasizing the causal primacy of genes "does not mean, of course, that genes . . . literally face the cutting edge of natural selection. It is their phenotypic effects that are the proximal subjects of selection" (Dawkins 1982b, p. 47). In other words, the objection can be diffused by distinguishing between proximate and ultimate beneficiaries of adaptations. The proximate (i.e., immediate) beneficiary of cryptic coloration would be the organism that escapes predation by being cryptically colored, and we could say (speaking loosely) that cryptic coloration is an organismic adaptation that confers a fitness benefit on such organisms. But the ultimate beneficiary would be the genes that produced the coloration, and thus (speaking strictly) the cryptic coloration is an adaptation that evolved because it conferred a fitness benefit on such genes. Thus, even if all adaptations are ultimately for the good of genes, it might still be possible to distinguish organisms and groups as the proximate beneficiaries of adaptations.

Assigning Functional Roles

At this point, the debate over gene selectionism seems to have reached a stalemate, with some biologists arguing that genes are the true "units of selection," whereas others deny this assertion. Everyone agrees that genes are involved in natural selection in some way, and that genes are closely related to the organisms they "build." Part of the problem involves disentangling the respective causal roles of the various biological entities involved. This is complicated by the fact that often the fate of genes and that of organisms are closely linked, such that in many cases organism-centered and gene-centered approaches will coincide. Genes are likely to do well in the company of other genes also working to promote the survival and reproduction of the organism that houses them. By sharing a common fate in virtue of residing in the same organism, and passing through the "bottleneck" of the organism's reproduction, phenotypes which benefit individual genes will also benefit the organism of which they are a part. So in most instances, organism-centered and gene-centered approaches both predict that genes will cause phenotypic effects that are conducive to the survival and reproduction of the organisms that possess them. How can such ideas be used to resolve the controversy about the status of gene selectionism?

As David Hume wisely observed long ago, "From this circumstance alone, that a controversy has been long kept on foot, and remains still undecided, we may presume, that there is some ambiguity in the expression, and that the disputants affix different ideas to the terms employed in the controversy" (Hume 1777, p. 53) The idea that some terminological housekeeping can clarify and perhaps even resolve the "units of selection controversy" informs several proposals that are worth examining in this section.

Replicators and Interactors

Recall Dawkins's clarification that asserting the causal primacy of genes "does not mean ... that genes ... literally face the cutting edge of natural selection. It is their phenotypic effects that are the proximal subjects of selection" (Dawkins 1982b, p. 47). Although it is typical to think of organismic properties (e.g., cryptic coloration) as a "phenotypic effect" of genes, in principle phenotypes can exist at any level of the biological hierarchy. For example, a bee hive might possess the "phenotypic" characteristic of having a certain percentage of workers, in virtue of which the hive is more efficient than hives with a different percentage of workers.

When the hive fissions and sends out a swarm in search of a new home, the new hive formed might have the very same percentage of workers. In other words, hives could display variation, differential fitness, and heritability, just the general requirements for selection to operate (Lewontin 1970). Selection could operate on variations between hives as distinct "superorganisms" (Seely 1989).

According to David Hull (1980, 1981, 1988a), distinguishing between the entities that replicate their structures and produce bodies, on the one hand, and those that directly face selection, on the other, is essential for resolving the units of selection problem. That is, two kinds of entity are important for describing the operation of natural selection. First, for evolution by natural selection to take place, spatiotemporal sequences of replicates are necessary (Hull 1981, p. 149). In Hull's terminology, a *replicator* is defined as "an entity that passes on its structure largely intact in successive replications" (Hull 1988a, p. 408). Replication by itself, however, is insufficient. Some entities must interact causally with their environments in such a way as to bias their distribution in later generations. An *interactor* is defined as "an entity that interacts as a cohesive whole with its environment in such a way that this interaction *causes* replication to be differential" (Hull 1988a, p. 408; emphasis in original). The relationship between an interactor and its environment is mediated by phenotypic properties (or "traits") that affect the interactor's biological success (as measured by survival and reproduction). "Selection" is then defined as "a process in which the differential extinction and proliferation of interactors cause the differential perpetuation of the relevant replicators" (Hull 1988a, p. 409).[5] Dawkins (1978, 1982a,b) offers a similar distinction, using somewhat different terminology.[6]

What are the advantages of drawing such distinctions? Hull holds that controversies have flourished because of ambiguity in the term "unit of selection." Sometimes this term is used to refer to the entities responsible for replication, at others to the entities responsible for interaction. Such ambiguities breed confusion. "When gene selectionists like Dawkins argue that genes are *the* units of selection, they mean to claim at the very least that genes are the only entities capable of replication. When organism selectionists like Mayr argue that organisms are *the* unit of selection, they mean to claim at the very least that organisms are an important focus of interaction with the environment" (Hull 1981, pp. 150–1). When the distinction between replication and interaction is kept in mind, Hull believes, conceptual disagreements concerning *the* unit of selection are seen to be only apparent.

The replicator/interactor distinction is not intended to resolve all questions about the units of selection, however. Empirical questions will remain concerning which biological entities play each of these evolutionary roles. The distinction between replicators and interactors suggests that there are really two central empirical questions at issue in discussing the levels of selection: "[A] t what levels does replication occur, and at what levels does interaction occur?" (Hull 1980, p. 318). There is widespread agreement that genes are paradigmatic replicators. Can other entities function as replicators? Hull notes that "replication is concentrated at the lowest levels, primarily at the level of the genetic material," although perhaps also at the level of organisms and possibly colonies, but rarely higher (Hull 1988a, p. 419). Organisms and even single cells can sometimes function as replicators. "Cells exhibit structure of their own and can pass on this structure quite directly and largely intact. For example, when a paramecium splits longitudinally to produce two new paramecia, both the genetic material and the organism itself are replicated" (Hull 1988a, p. 414). One reason that organisms are not typically replicators is because in those organisms with sexual reproduction, in order for the structure of the organism to be copied, it must first pass through a gametic stage which is radically different from the adult organism. There is replication of a sort, but it is indirect rather than direct. This is quite unlike the way in which genes replicate, according to which their structure is passed on directly in the form of copies.

Interactors, too, must meet certain ontological criteria. Foremost among these is that the entity must interact with its environment as a "cohesive whole." That is, interactors must be *individuals*, understood in a very specific sense. "By 'individual' I mean any spatiotemporally localized entity that develops continuously through time, exhibits internal cohesiveness at any one time, and is reasonably discrete in both space and time" (Hull 1981, p. 145). Organisms are paradigmatic individuals, but other biological entities displaying cohesiveness and internal organization may qualify as well. Consequently, "Interaction occurs at all levels of the organizational hierarchy, from genes and cells, through organs and organisms, up to and possibly including populations and species" (Hull 1988a, p. 409).[7]

How does this analysis bear on the issue of gene selectionism? In most selection processes, the replicators are genes.[8] But, if so, then selection processes must be distinguished on the basis of the interactors involved. In point of fact, this is the way that selection processes are usually distinguished. Organism selection is a process in which organisms interact with their environments (including each other), resulting in differential reproductive success among organisms. Group selection, at least as Wynne-Edwards conceived it, involves differential interaction between groups and their environmental resources. The question of "group selection" then becomes whether groups have the requisite properties to function as interactors. This is an important start, but further questions must be addressed as well if we are to identify the beings "for whose good" natural selection works. Knowing that certain biological entities can function as interactors does not by itself tell us for whose good adaptations exist. In order to answer this fundamental question, further distinctions must be introduced.

Getting Serious About Functional Roles

This is just what Elisabeth Lloyd sets out to do. Incorporating the Hull's distinctions, Lloyd (2000) distinguishes four different kinds of entities that play a role in the evolutionary process: interactors, replicators, beneficiaries, and manifestors of adaptations. "Interactors" as those individuals that respond directly to selection pressures via their phenotypic properties. "Replicators" are those entities that pass on their structures directly through replication. "Beneficiaries" are the entities that benefit in evolution; for example, get more copies of themselves into the next generation; or those entities that benefit from adaptations. Finally, "manifestors of adaptations" are, as the name suggests, the entities that manifest adaptations.

Lloyd argues that deploying these distinctions helps to further clarify questions about "the units of selection." For example, in the debates over group selection there are really at least two issues at stake: (a) Does selection ever operate on groups as cohesive wholes? (That is, do groups ever function as *interactors*?) (b) Do groups ever manifest group-level adaptations that are not better understood as simply a summation of the adaptations of lower-level entities, for example, organisms? (That is, are groups ever *manifestors of adaptations*?) Whereas Wynne-Edwards proposed a model of group selection according to which groups function as both interactors and as manifestors of adaptations, Wilson's structured deme model of group selection suggests that groups can sometimes function as interactors (i.e., that group-level properties can be important in evolution), without also claiming that groups are either the beneficiaries of selection or the manifestors of adaptations. This much seems to be generally accepted. The contemporary units of selection debate tends to focus on the question of whether organisms or genes should be thought of as the beneficiaries of selection or the manifestor of adaptations. Lloyd's distinctions help to clarify these issues as well. In particular, her distinctions show that some gene selectionists conflate the various functional roles entities play in evolution. For example, Cronin (1991) tends to collapse the identity of interactors, replicators, manifestors of adaptations, and beneficiaries of adaptation, identifying genes as the all-purpose "units of selection" that serve all of these distinct functional roles. Lloyd's analysis makes clear why this is problematic. Genes replicate, but then so do *some* organisms and *some* groups. Genes typically interact with the environment to influence their own survival only via adaptations associated with organismic phenotypes. It therefore makes at least as much sense to say that organisms are the manifestors of adaptations as it does to say that genes are. Finally, the question of the ultimate beneficiaries of adaptations remains open. The genes of a biologically successful organism benefit from that organism's success, but then so, too, does that organism. Identifying genes as "the units of selection" *simpliciter* provides a misleadingly simple account, and fails to acknowledge that there are alternative, equally plausible, ways of describing the dynamics of evolution.[9]

It is time to take stock. As useful as Lloyd's distinctions are, they still leave open questions about *which* entities do, as a matter of act, play each of the functional roles she identifies. The problem is acute. As noted above, genes seem to be paradigmatic replicators. But organisms, and in some cases colonies (e.g., of bees) seem to replicate after a fashion as well. Likewise, organisms and groups interact with their environments, but then so, too, do genes with their environments. Adaptations can be ascribed to organisms just as easily as they can be ascribed to genes. Finally, gene selectionists take genes to be the ultimate beneficiaries of natural selection, because they persist from one generation to the next, whereas phenotypes do not. But even if entire phenotypes, that is, the total collection of properties characterizing a given individual – does not persist across the generations, particular phenotypic properties, especially those that are fitness-enhancing, do. For example, if a particular pattern of camouflage is fitness-enhancing, then this phenotypic pattern will get passed on and hence persist from one generation to the next. So one could as well focus on particular phenotypic properties as on individual genes.

The attractive but overly simplistic sparse desert landscape view of evolution provided by gene selectionism has become a lush but impenetrable rainforest in light of the discussion earlier. Is there any way to transcend both of these approaches in order to resolve the issues with which we began? At this point there seem to be just two options: (1) adopt a "pluralistic" approach according to which selection can be modeled equally well as operating and forging adaptations at any of a number of distinct biological levels; or (2) abandon the idea that selection is properly conceived as acting on distinct levels, and embrace instead a "holistic" vision in which processes at various levels are integrated. We'll explore both options in the following section.

Pluralism and Holism

Standard attempts to resolve the units of selection problem take for granted that the entities upon which selection operates can be identified with one or another of the entities constituting the biological hierarchy (e.g., genes, organisms, groups, etc.), and that a satisfactory analysis of a given selection process requires the identification of the unit(s) and level(s) of selection causally responsible for that process. Sterelny and Kitcher (1988) christen this view *Hierarchical Monism*, defined as follows: "Hierarchical monism claims that, for any selection process, there is a unique level of the hierarchy such that only representations that depict selection as acting at that level are maximally adequate. (Intuitively, representations that see selection as acting at other levels get the causal structure wrong.)" (Sterelny and Kitcher, 1988, p. 359). Thus defined, Hierarchical Monism embodies two distinct claims: (1) for any selection process there is just *one representation* that correctly captures the causal structure of that process, and (2) this representation depicts selection as operating on just *one level* of the biological hierarchy.

Sterelny and Kitcher reject this view. Not only is there no uniquely correct way of describing any selection process, but talk of "units of selection" is itself philosophically suspect. "Monists err," they write, "in claiming that selection processes must be described in a particular way, and their error involves them in positing entities, 'targets of selection', that do not exist" (Sterelny and Kitcher, 1988, p. 359). Asking about the "real unit of selection" in a given case is based on a confusion, because selection events can be modeled in any of a number of different, equally correct ways (e.g., as acting on, or as benefiting, genes or organismic phenotypes, etc.). The way that one chooses simply depends upon one's methodological interests.

The Pluralist Option

In place of Hierarchical Monism, they propose a view they call *Pluralism*. Unlike the "Monist" who claims that for each process there is just *one* adequate representation, the Pluralist maintains that for any process there are *many* adequate representations but that processes are diverse in the kind of representations they demand. Whereas Hierarchical Monism recommends a "plurality of processes," each with its own model, Pluralism recommends "a plurality of models of the same process" (Sterelny and Kitcher, 1988, p. 359).

Consider spider webs. They are clearly biological adaptations ("extended phenotypes" in Dawkins's terms). An adequate story can be told about the evolution of spider webs in terms of competition and selection amongst organisms that display web-building behaviors. But an equally adequate story can be told in terms of competition and selection amongst *genes* that cause the behaviors in question. Neither of these stories has any monopoly on correctness. They are equally correct descriptions. Hence neither genes nor organisms can be claimed as *the* units of selection. From such considerations Sterelny and Kitcher conclude that, "There is no privileged way to segment the causal chain and isolate the (really) real causal story. . . . We are left with the general thesis of pluralism: there are alternative, maximally adequate representations of the causal structure of the selection process" (Sterelny and Kitcher, 1988, p. 358).

Sterelny and Kitcher explicitly link their Pluralism with various "antirealist" doctrines in the philosophy of science. Specifically, their view is "instrumentalist" in the sense that scientific theories are to be understood as tools for introducing order into our representations of the often chaotic flux of phenomena, rather than as laying bare the actual causal structure of the events in question. Just as hammers can be useful (or not) for a particular carpentry task, so, too, a particular scientific perspective may (or may not) be useful for the scientific task at hand. (Scientific theories, like hammers, are to be judged exclusively on their usefulness, not on their "truth," whatever that might mean.) They also align their view with "conventionalism": "Another way to understand our pluralism is to connect it with conventionalist approaches to space-time theories. Just as conventionalists have insisted that there are alternative accounts of the phenomena which meet all our methodological desiderata, so too we maintain that selection processes can usually be treated, equally adequately, from more than one point of view" (Sterelny and Kitcher, 1988, p. 359). Although some previous analyses have hinted at a conventionalist solution to the units of selection problem (Cassidy 1981; Buss 1987; Maynard Smith 1987; see also Waters 1991), Sterelny and Kitcher provide the most explicit argument in support of such an interpretation.[10]

Sterelny and Kitcher's Pluralism states an important fact: Many selection processes can be equally well represented by models depicting selection as operating on any of a number of distinct biological levels. A description of the evolution of spider webs in terms of selection for the properties of spiders need not be inferior to a description in terms of the properties of genes. It is a mistake to think that if a description of a selection process in terms of organismic properties is correct, then a description in terms of genic properties cannot be. As they note, for any given selection process, there might be any number of *equally good* descriptions. This is one of the lessons from the discussion of functional roles, earlier.

Nonetheless, Sterelny and Kitcher overstate their case when they claim that for any biological phenomenon in need of a selectionist explanation, there are *alternative, maximally adequate representations* of the causal structure of the selection process. To see this, consider again the spider web example. A well-constructed spider web benefits both the spider that made the web as well as the genes responsible for the relevant behaviors. On Sterelny and Kitcher's view, the evolution of spider webs could be represented with equal adequacy as a result of selection operating on spider-genes, *or* as selection operating on spiders. These are held to be alternative, maximally adequate representations of the causal structure of the selection process, each framed in terms of the entities on just one level of the biological hierarchy. But are they? Undoubtedly one can provide a partial account of the evolution of spider webs solely in terms of selection operating on the entities on just one level of the biological hierarchy, but a *maximally adequate representation* of the causal structure of this process would necessarily have to take into account (at a minimum) the causal connections *between* entities on different levels of the biological hierarchy, without which the phenomenon under consideration could not have occurred at all. This is because *the causal connections between entities on different biological levels constitute part of the causal structure of the selection process.* Spider-genes partially cause spiders; spiders build webs; these webs assist the spiders in passing along their genes to subsequent generations; and so on. Spider-genes, spiders, and webs exist in an interconnected causal nexus that taken as a whole results in the differential survival and/or reproduction that is natural selection. This suggests that for any selection event there is just one maximally adequate representation that correctly depicts the causal structure of each selection process, one which does so by taking into account the causal contributions of (and causal connections between) whatever entities are involved in the process. Once all such contributions are taken into account, then at most one such representation will be possible. Consequently, there will be nothing left with which to generate another, alternative representation, much less another one that is itself maximally adequate. But if so, then Pluralism of the sort defended by Sterelny and Kitcher is mistaken.[11]

To sum up, while the claim there is a uniquely correct representation of any selection process that depicts selection as operating on just *one* level of the biological hierarchy is indeed incorrect, the claim that for any selection process there is not, or cannot be, a single maximally adequate representation that correctly captures the causal structure of that process is mistaken. A maximally adequate representation of the causal structure of a selection process would take into account the causal contributions of biological entities on multiple levels, as well as their interactions.

The Holist Option

Such considerations suggest a very different perspective on selection, adaptation, and the evolutionary process as a whole. These insights are developed most fully in the *developmental systems approach* to evolution (Gray 1992, 2000; Griffiths and Gray 1994, 1997; Oyama 2000; Oyama, Gray, and Griffiths 2001; Sterelny and Griffiths 1999). The essential challenge to both gene selectionism *and* the attempt to distinguish distinct functional roles in evolution is straightforward: "Developmental systems theorists claim that there is no privileged class of replicators among the many material causes that contribute to the development of an organism – that the entire replicator/interactor representation of evolution is refuted by the facts of developmental biology" (Sterelny and Griffiths 1999, p. 94).[12] According to this alternative view, "Rather than replicators passing from one generation to the next and then building interactors, the entire developmental process reconstructs itself from one generation to the next via numerous interdependent causal pathways" (Sterelny and Griffiths 1999, p. 95). Genes are replicated, but the replication of genes is just one aspect of the replication of a life cycle. "Every element of the developmental matrix which is replicated in each generation and which plays a role in the production of the evolved life-cycle of the organism is inherited. . . . The process of evolution is the differential reproduction of variant life-cycles" (Griffith and Gray 1997, p. 474). Entire developmental systems, rather than genes (for example), are thus the "units of natural selection."

Consider again the spider-web example. Spider genes may be thought of as embodying the "design instructions" for both spiders and webs. Webs are constructed by organisms executing these instructions. Web designs are in competition with one another for effectiveness in catching prey. Spider-genes are causally connected with spiders, web-building behavior, webs, and to the capture of prey, resulting in the differential propagation of genes, the spiders, and webs. Selection in this case is operating on the entire gene-spider-web causal process. A maximally adequate representation of the causal structure of this selection process will include not only a description of how each of the entities involved in the process fares relative to other entities of the same kind (i.e., at the same level), but also how the entire causal chain contributes to the differential representation of genes, spiders, and webs in successive generations.[13]

Illustrating a view is one thing. Showing why it is superior to other views is quite another. The basic arguments for this view are straightforward. Consider first an argument against gene selectionism. Genes can be the uniquely correct units of selection *only if* genes play some distinctive, privileged role in the development of the organism. But it is *false* that genes play such a role. Hence, genes are not the uniquely correct units of selection. A variety of considerations can be brought to bear to show that genes do not play some privileged role in development. First, organisms inherit more than genes from their parents. Besides the non-nuclear DNA that appears in mitochondria (Jablonka and Lamb 1995), some organisms inherit behaviors learned from parents or from other conspecifics. The distinctive song dialects of some birds, for example, are passed on from generation to generation, and are acquired only by juvenile birds being exposed to adult renditions of these dialects. Preferences for nest sites and nesting materials, as well, are acquired through early exposure as nestlings and juveniles. Such cases of "cultural transmission" are by no means restricted to vertebrates like birds (Keller and Ross 1993). Genes obviously play a role in the development of adult song repertoire and nesting behavior, but not the only, or even necessarily the most important, role.

Second, the causal thesis associated with genic selectionism is simply false. Strictly speaking, genes, by themselves, produce nothing. Only genes working within a context of a complete developmental system can produce phenotypes. But if so, then genes, by themselves, ought not to be given privileged status.

Finally, the very idea of genes as the "replicators" in evolution breaks down when one considers that other biological entities (e.g., organisms) replicate in much the same way. In any case, it is entire developmental processes, not individuated biological entities, that replicate themselves. But if so, then the "replicator/interactor" distinction is also suspect.

In summary, in contrast to gene selectionists, developmental systems theorists see the entire "life cycle" as the fundamental unit of evolution. "A life cycle is a developmental process that is able to put together a whole range of resources in such a way that the cycle is reconstructed.... Organisms have life cycles, and so do groups like ant colonies. Variants on these life cycles compete with one another" (Sterelny and Griffiths 1999, p. 108). This approach provides a very different perspective than that insisted upon by gene selectionists. Recall (from the quotation that opens this chapter) that according to Dawkins, asking what birds' wings, spider webs, and chlorophyll molecules are *for* are perfectly legitimate questions. Asking what DNA molecules are for, on the other hand, is "the forbidden question. DNA is not 'for' anything... all adaptations are for the preservation of DNA; DNA just *is*" (Dawkins 1982b, p. 45). In the Developmental Systems perspective, however, "the forbidden question" can intelligibly be asked. Genes are just one component of the complex developmental system. It can as well be said that genes exist in order to assist in the transmission of developmental systems (or of phenotypic properties) as it is to say that developmental systems exist for the sake of transmitting genes. All are bound together in integrated systems that fail to replicate should any essential component fail to function as designed by natural selection. Genes play an important, but not a privileged, role in this process.

Replicators Strike Back

Critics have generally applauded the insights of the Developmental Systems approach, while drawing attention to its weaknesses. Among the latter, the "boundary problem" has attracted special attention. Developmental Systems Theory emphasizes the "connectedness" of all parts of the developmental system. But what principled way is there to distinguish what is and is not relevant to a given developmental system? As some critics put it, "Everything causally connects with everything else.... So if developmental systems include everything causally relevant to development, they are too ill-defined to be a coherent active unit; they are too diffuse to be units of selection" (Sterelny, Smith, and Dickinson 1996, p. 382). Likewise, this approach requires the generation by generation reproduction of developmental systems, which presupposes that "generations" have distinct boundaries. But critics ask: "So when do generations begin and end: do we count from bird to bird, egg to egg, or nesting hole to nesting hole? Cycles of developmental resources are not necessarily in sync. . . . Is an ant-plant mutualism a single developmental system or several?" (Sterelny, Smith, and Dickinson 1996, p. 383).

Developmental Systems theorists are not without resources to respond at this point. In response to the boundary problems described above, they can choose to "bite the bullet" and take the holist perspective to its logical conclusion by agreeing that, in principle, anything might be relevant to a given developmental cycle. However, "interconnectedness of everything" no more counts against the Developmental Systems approach in biology than it does against the Newtonian research program in physics. One is always forced to be selective in deciding which factors to include in one's model. A similar response is available for deciding where one generation ends and another begins. Noting that there is no given line demarcating one generation from another simply reinforces the point that biological systems are integrated, continuously developing wholes. Finally, asking whether an ant-plant mutualism is a single developmental system or several, rather than constituting a *reductio ad absurdum* of the Developmental Systems approach, actually highlights its potential for considering familiar biological phenomena in an entirely new light. For example, perhaps termite mounds with their elaborate air conditioning system, their termite builders, and the cellulose-digesting protozoa that inhabit their guts should be viewed as components of the same developmental system that gets replicated generation after generation. Rather than asking which component of this system is the "real" unit of selection, the entire system can be fruitfully viewed as subject to selection.

So there are some advantages to the Developmental Systems approach. Nonetheless, it could be argued, whereas developmental processes are part of a causally complete account of selection, the specific developmental details are really of very little interest for understanding evolution. Selection operates upon phenotypes; the developmental processes that produce such phenotypes can be treated like "black boxes." As Amundson observes, "Some black boxes need never be opened because their insides really are uninteresting. It would be futile to argue, for example, that statistical thermodynamics suffers from insufficient attention to the actual paths of the individual molecules of a gas as they strike the walls of a container. Such a detailed causal account would be uninteresting even if it were attainable" (Amundson 2001, p. 316). Likewise, perhaps the facts of development are not important to our understanding of evolution. The causal completeness argument only shows that developmental processes intervene between genotypes and phenotypes.

It doesn't show that these processes are important for understanding evolution.

Clearly replication of *something* is essential for evolution to proceed. Advocates of the rival "Extended Replicator" view agree with Developmental Systems theorists that entire developmental systems are necessary for replication to take place, and that genes cannot replicate themselves in any meaningful sense. Yet they still see a privileged role for genes in this sense: While it takes an entire developmental system to replicate, only genes are specifically adapted to cause similarities between one developmental cycle and its successor: "The genome is one of the designed mechanisms in virtue of which phenotypes and genotypes duplicate themselves. . . . This idea of a designed copying mechanism is the key to understanding the privileged role of the replicators in the total developmental matrix" (Sterelny, Smith, and Dickinson 1996, p. 387). Genes thus play the role in relation to the developmental cycle that a blueprint plays in relation to a building. A blueprint is not the only or even necessarily the most important cause of the building that results. But it does "represent" the building in a way that the materials and construction workers do not. In the Extended Replicator view, genes are not to be uniquely identified as replicators (the whole developmental system is a replicator), yet within this developmental system genes play a special role: they represent developmental outcomes.

Summary: For Whose Good Does Natural Selection Work?

Darwin maintained that "natural selection works solely by and for the good of each being" (Darwin 1859, p. 489; 1959, p. 758). But there are many "beings" involved in the evolutionary process. *By* whose and *for* whose "good" does natural selection work? As we have seen, the problem is much more difficult than it might at first appear. In order to resolve it, a number of closely related questions need to be distinguished. Upon what sorts of biological entities can (and does) selection operate? What sorts of biological entities manifest adaptations? Which biological entities benefit from such adaptations? It has commonly been assumed that typically selection operates on individual organisms (and perhaps some groups), that adaptations "belong" to organisms (and perhaps some groups), and that genes (as measured by their representations in subsequent generations) are the ultimate beneficiaries of adaptations. Gene selectionists take precisely this position. They maintain that although organisms (and perhaps some groups) directly face the cutting edge of natural selection by interacting with the environment, nonetheless the adaptations that evolve as a result of this process ultimately serve genes, which uniquely benefit from the entire process by being passed on intact to subsequent generations. As critics have pointed out, however, this perspective faces a number of serious difficulties, not the least of which is that upon closer analysis, genes do not seem to play quite the privileged role that is claimed for them. Entire life cycles, not to mention specific phenotypic properties, are also "replicated," reappear generation after generation, and thus have equal right to be considered the beneficiaries of natural selection.

It is tempting to declare at this point that selection can be "represented" in any of a number of equally adequate ways. Which sort of representation is chosen can be decided simply in light of methodological considerations (e.g., simplicity, scope, predictive power, etc.). Because gene selectionism offers a perspective that is simple, has broad applicability, and provides an excellent basis for predicting the sorts of biological adaptations one will find among living things, it ought to be preferred on instrumentalist grounds. An alternative would be to adopt a realist philosophy of science, according to which the aim of scientific theories is to accurately capture the causal structure of natural processes (McMullin 1984). Because biological entities are causally connected in complex ways, perhaps the only truly accurate account of natural selection includes biological entities and their causal interrelations at a number of different functional levels, and treats entire biological systems as subject to selective forces. What this approach gains in completeness, however, is lost in simplicity and usefulness (Cartwright 1981). Gene selectionism provides a powerful general perspective from which to view (virtually) all evolutionary change. Yet it does so at the expense of recognizing the complex interrelations among various biological entities that characterize every actual selection event. In evolutionary biology, as in other areas of science, the so-called Symmetry Thesis is false: Good explanations are not necessarily logically identical to equally good predictions. (Two models of the solar system may be equally adequate for predicting the next solar eclipse, but may differ in their adequacy for explaining such an event.)

Evolutionary theory, like all scientific theories, faces a dilemma of competing values. Scientific theories are judged (among other things) on the basis of theoretical generality and their empirical specificity. Both are required of any scientific theory, but they exist in tension with one another. To generalize is to look for recognizable patterns in the often chaotic flux of phenomena. To do so requires abstracting from the particulars of specific cases. Empirical specificity, by contrast, is concerned with the precision of fit of theory to actual, highly specific, situations. The most empirically accurate theory would be one that simply describes, in minute detail, each event occurring in its intended domain. But to do so would hardly resemble science as we know it at all. As a rule, scientists are interested in predictive generalizations (e.g., natural laws), rather than in the explanation of particular, often unrepeatable, events. The distinction between predictive power and detailed explanation of the particular is one way in which science differs from history. Between the competing values of generality and specificity, the former often takes priority. But the latter value can be just as important, if our aim is to achieve a deeper understanding of the world that science seeks to explain.

PART II

ADAPTATION

4

Darwin (and Others) on Biological Perfection

Slow though the process of selection may be . . . I can see no limit to the amount of change, to the beauty and infinite complexity of the coadaptations between all organic beings, one with another and with their physical conditions of life, which may be effected in the long course of time by nature's power of selection.

(Darwin 1859, p. 109)

Introduction

It would be difficult to find a more optimistic expression of the power of natural selection to shape living things to any imaginable degree of biological perfection. Such claims abound in Darwin's writings. Consider the famous closing words of the *Origin*:

It is interesting to contemplate an entangled bank, clothed with many plants of many kinds, with birds singing in the bushes, with various insects flitting about, and with worms crawling through the damp earth, and to reflect that these elaborately constructed forms, so different from each other, and dependent on each other in so complex a manner, have all been produced by laws acting around us. . . . There is grandeur in this view of life, with its several powers, having been originally breathed into a few forms or into one; and that, whilst this planet has gone cycling on according to the fixed law of gravity, from so simple a beginning endless forms most beautiful and most wonderful have been, and are being, evolved. (Darwin 1859, pp. 489–90)

The image Darwin presents to the reader in such passages is that each kind of living thing is, or is becoming, exquisitely adapted both to other living things and to its physical environment. Beauty, harmony, and perfection of design are the hallmarks of life. Indeed, one of his chief aims in the *Origin*, he tells his readers, is to show "how the innumerable species inhabiting this world have been modified, so as to acquire that perfection of structure and coadaptation which most justly excites our admiration" (Darwin 1859, p. 3).

Yet, however attractive this vision of nature might be, it is deeply problematic. First, there seem to be good reasons (given Darwinian principles) for predicting that living things will *not* attain biological perfection, and that in many instances they will fall far short of this ideal – reasons of which Darwin was fully aware. Second, even a cursory examination of nature suggests that many (perhaps all) living things do in fact fall far short of perfection. Darwin certainly had more than merely a cursory acquaintance with nature, and thus in many passages explicitly *denies* that living things are as perfect as they could be. Why, then, does he so often describe living things as displaying "perfection of structure and co-adaptation"? How could he claim, again and again, that thanks to the unceasing work of natural selection, "all corporeal and mental endowments will tend to progress towards perfection" (Darwin 1859, p. 489; 1959, p. 758)? Just how "perfect" did Darwin consider living things to be?

Sorting out these issues is the aim of this chapter. Subsequent chapters will examine how the idea of "perfect adaptation" fared in the century after Darwin, as well as how this issue is, and should be, understood at present. But first, in order to chart the evolution of Darwinian thinking about adaptation, and to properly understand these later developments, it is essential to understand how Darwin himself approached this issue, how his own view developed over time, and how he differed from the co-discoverer of natural selection, Alfred Russel Wallace. Understanding Darwin's view, in turn, requires acquaintance with some key elements of pre-Darwinian natural history. It is to these elements that we turn next.

Biological Perfection and Imperfection in Pre-Darwinian Natural History

That living things display remarkable complexity and integration of parts had been evident from the earliest times, and found its first serious appreciation in the biological works of Aristotle (384–322 B.C.E.) who, unlike his teacher Plato, found beauty and order in the natural world: "Every realm of nature is marvellous . . . so we should venture on the study of every kind of animal without distaste; for each and all will reveal to us something natural and something beautiful. Absence of haphazard and conduciveness of everything to an end are to be found in Nature's works in the highest degree, and the resultant end of her generations and combinations is a form of the beautiful" (*Parts of Animals* 645:17–26). Not only are nature's productions functional and beautiful, but they cannot be surpassed: "Nature creates nothing without a purpose, but always the best possible in each kind of living creature by reference to its essential constitution. Accordingly if one way is better than another that is the way of nature" (*Progression of Animals* 704b15–17). Likewise, the second century Roman physician Galen (c. 130–c. 200 C.E.) showed a remarkable depth of understanding of the intracacies of biological organization, emphasizing purpose and harmony, the aptness of structure to function, and of function to usefulness in meeting the requirements of life. In the design of living things, he noted that nature does "everything for some purpose, so that there is nothing ineffective or superfluous, or capable of being better disposed" (*On the Natural Faculties*, Book I, Section 6). The idea, formulated by Aristotle and reinforced by Galen, that "nature does nothing in vain," guided centuries of natural history from ancient times right through the eighteenth century. Its effect was still being felt in the late eighteenth and early nineteenth centuries, when the immediate precursors of Darwin's views were being formulated.

Utilitarian-Creationism

How did naturalists immediately prior to Darwin view the "perfection" of living things? As Cronin (1991, p. 23ff) notes, there were two main schools of thought on this issue. According to the "utilitarian-creationists," every feature of every organism, no matter how minute and seemingly insignificant, was believed to be of use to that organism. Emphasis was placed on the adaptive integration of the entire organism, as well as the utility of each particular part. In his *Lectures in Comparative Anatomy* (1805), Baron Georges Cuvier (1769–1832) christened the former idea the principle of the "Correlation of Parts," according to which an animal must have all of its body parts coordinated (i.e., "correlated") with one another so as to make a given way of life possible. Cuvier saw organisms as integrated wholes, in which each part's form and function were integrated into the entire body. No part could be modified without impairing this functional integration. Carnivores, for example, tend to have forward facing eyes, sharp teeth, and a skeletal structure suited for running down prey. Herbivores, by contrast, tend to have eyes on the sides of their heads, flat teeth for grinding, and a skeletal structure for evading predators. Each kind of organism appears to be designed as a suite of functionally coordinated parts. In addition, each part of each organism is perfectly constructed to carry out the function for which it was designed.[1]

This idea was epitomized in William Paley's *Natural Theology* (1802). A favorite example of writers in the natural theological tradition, and one on which Paley dwells extensively, is the vertebrate eye, which appears to be perfectly designed for carrying out the function of seeing. Use of the word "design" in such cases is, of course, intended literally. Paley's point was that each organism *has* been designed, by God, with its present structure, unchanged since its creation, and therefore unconstrained by the vagaries of history. Perfect adaptation implied a close, causal connection between environmental conditions and the structures of organisms. According to Charles Lyell, for instance, God calls into existence those organic forms that are best fitted for every given set of environmental conditions. Since only one form can be the best of all possible forms for a given environment, God, acting through external conditions, determines that only those forms are created. "What is here called necessity," Lyell said, "may merely mean that it pleases the Author of Nature not simply to ordain fitness, but the greatest fitness" (quoted in Wilson 1970, p. 6).

Despite the widespread popularity of the "utilitarian-creationist" school of thought, not every observer of nature considered every creature to be perfectly designed. Indeed, in some cases it was argued that particular organisms are not even remotely close to exemplifying such a lofty state, leading to serious differences of opinion even with regard to the same organism. For example, in describing the "dodo" (*Rhaphus cucullatus*), the great French naturalist Georges Louis Leclerc, Comte de Buffon (1707–1788) invited his readers to:

> Imagine a large, almost cubical body, barely held up by two very thick, very short pillars. The head . . . mounted on a thick and goitrous neck, consists almost entirely of an enormous beak. . . . All this results in a stupid and voracious appearance. . . . The first Dutchmen who saw this creature . . . called it a "walck-vogel" or disgusting bird. . . . Bulk which, in animals, usually suggests strength, only produces weight in this case. The ostrich and the cassowary are no more able to fly than the dodo; but at least they run quickly, while the dodo is overcome by its own weight, and is barely able to drag himself. This is in birds what laziness is in four-footed animals; one might say it is composed of brute material, inactive, where the living molecules are too few. It has wings, but they are too short and too weak to propel it into the air; it has a tail but this tail is disproportionate and in the wrong place. You would think it was a tortoise which was dressed up with the cast-off skin of a bird. Nature, giving him such useless ornaments, seemingly wanted to add insult to weight, awkwardness of movement to the inertia of the mass, and make the heavy thickness even more shocking, by making it look like a bird. (Buffon 1770)[2]

A harsh assessment, but it should not be thought that Buffon had singled out the dodo alone for critique. In other places he is no kinder to sloths and other animals. Buffon evidently had no qualms at all about pointing out the gross deficiencies and imperfections of animals. His was neither a world of divinely created, perfectly designed creatures, nor a world of exquisitely constructed organisms with adaptations honed to perfection over eons of relentless natural selection. Instead, it was a world in which nature, like humans themselves, had "fallen" from an earlier, more perfect state. In Buffon's view, life does indeed have a history, but this should be understood as a history of *degeneration* and *decline* from better designs to less well fitted organisms (Lyon and Sloan 1981; Roget 1997).

A diametrically opposite view of the dodo, and indeed of all of life on earth, was expressed by Hugh Edwin Strickland in his 1848 treatise on this bird. Conceding to its critics that the dodo appears to us as "a massive clumsy bird, ungraceful in its form, and with a slow waddling motion," he nevertheless insisted that God created each creature, including the dodo, with features optimally designed for its particular way of life:

> [L]et us beware of attributing anything like *imperfection* to these anomalous organisms, however deficient they may be in those complicated structures which we so much admire in other creatures. Each animal and plant has received its peculiar organization for the purpose, not of exciting the admiration of other beings, but of sustaining its own existence. Its perfection, therefore, consists, not in the number or complication of its organs, but in the adaptation of its whole structure to the external circumstances in which it is destined to live. And in this point of view we shall find that every department of the organic creation is equally perfect. (Strickland and Melville 1848, p. 34)[3]

In the particular case of the dodo, he admits that the purpose of the bird's useless wings present a bit of a puzzle, but nonetheless maintains that they "are really the indications of laws which the Creator has been pleased to follow in the construction of living beings," even if we cannot at present decipher their exact meaning. Unfortunately, Strickland's spirited defense of the dodo appears to be more an expression of his fundamental theological conviction that "every department of the organic creation is equally perfect" than a convincing empirical demonstration that the dodo itself is optimally adapted to its circumstances.

Biological Idealism

A very different view from that of Strickland was articulated by Richard Owen (1804–92). In his 1866 monograph on the dodo, he followed Buffon in cataloging the creatures's inherent inferiorities, focusing especially on its pathetically small brain which (he speculated) earned it the mistaken but commonly used scientific name *Didus ineptus.* Owen evaluated the dodo on purely functional grounds, and attempted to explain how, despite its obvious imperfections, the dodo could survive for so long in this condition:

> [T]here would be nothing in the contemporaneous condition of the Mauritian fauna to alarm or in any way to put the Dodo to its wits; being, like other Pigeons, monogamous, the excitement, even, of a seasonal or prenuptial combat, might, as in them, be wanting: we may well suppose the bird to go on feeding and breeding in a lazy, stupid fashion, without call or stimulus to any growth of cerebrum proportionate to the gradually accruing increment of the bulk of its body. (Owen 1866, p. 39)

Owen, of course, rejected Darwin's theory of natural selection, so he was not implying that the dodo's features resulted from an absence of selection pressures. Instead, he presumes that because the dodo was not put to the test by the harsh conditions of life that most nonisolated species face, it degenerated into a form that allowed it to just get by in making a living on a remote island habitat, and nothing more. According to Owen,

> The Dodo exemplifies Buffon's idea of the origin of species through departure from a more perfect original type by degeneration; and the known consequences of the disuse of one locomotive organ and extra use of another indicate the nature of the secondary causes that may have operated in the creation of this species of bird, agreeably with Lamarck's philosophical conception of the influence of such physiological conditions of atrophy and hypertrophy. (Owen 1866, p. 49)

He then explicitly attacked Strickland's interpretation, quoting the passage from Strictland given above in which he warns against "attributing anything like *imperfection* to these anomalous organisms." The fact of the matter, Owen points out, is that the dodo, "through its degenerate or imperfect structure, howsoever acquired, has perished" (p. 50), thus confirming its inferior, imperfect structure.

Owen represented the "idealist" or "transcendentalist" approach to natural history according to which God creates organisms, not directly, but through the operation of natural laws. Organisms are thus only as perfect as possible within the limits set by these laws.[4] Like the utilitarian-creationists, idealists saw nature as permeated with intentional design. But unlike the utilitarian-creationists, this design is most evident, not in the intricate adaptive details of each living thing, but instead in the basic structural plans that underlie the diversity of organic forms. Organisms as diverse as whales, bats, horses, and primates share a basic structural plan, yet occupy distinctly different environments. This commonality of structure ("homology") seemed inexplicable on the utilitarian-creationist view. As Darwin (no idealist, yet impressed by the homologies to which idealists drew attention) later observed,

> What can be more curious than that the hand of a man, formed for grasping, that of a mole for digging, the leg of the horse, the paddle of the porpoise, and the wing of the bat, should all be constructed on the same pattern, and should include the same bones, in the same relative positions? . . . Nothing can be more hopeless than to attempt to explain this similarity of pattern in the members of the same class, by utility or by the doctrine of final causes. (Darwin 1859, pp. 434–5)

These same facts, however, fall neatly into place if one assumes that God created all living things according to a small number of basic blueprints, which then adapt (imperfectly) to each different environment. This outlook was epitomised by Owen's theory of "archetypes" – the divinely created groundplans for the major groups of organisms. In this view, there is no reason to expect all structures to be perfectly adapted to their functions, and indeed one would expect, given the different environments occupied by organisms with essentially the same structural plans, significant imperfections in adaptive fit. For utilitarian-creationists imperfections could not exist, thus none were expected, thus none were acknowledged. For idealists, however, biological imperfections were to be expected. Unsurprisingly, such imperfections, once one was attuned to finding them, were not difficult to find.

Biological Perfection in the *Origin of Species*

There were, then, two main approaches to understanding biological perfection in natural history immediately preceding Darwin. On the one hand, utilitarian-creationists saw organisms as perfectly fitted for their ways of life. Idealists, on the other hand, emphasized the underlying structural plans of organisms, and viewed organisms as adapted within the constraints imposed by these divinely created archetypes. It was in the context of such ideas that Darwin's view of biological perfection was initially formulated. In his private "transmutation" notebooks, dating from the late 1830s (i.e., just as he was formulating the theory of natural selection), Darwin operated within the same framework of beliefs about the harmony and overall perfection of nature common amongst utilitarian-creationist naturalists at that time. Like them, he considered living things to be perfectly adapted to their environments, and the whole of nature to be a well-adjusted mechanism exhibiting harmony and order.

Towards the end of September 1838, he read Malthus's *Essay on the Principle of Population* (1798). In his autobiography Darwin suggests that this was a critical event in his formulation of the theory of natural selection (Darwin 1958, pp. 119–20). A great deal of speculation has been devoted to determining precisely what effect this had on his theorizing. One effect it apparently did *not* have, however, was to change his belief in the perfection and harmony of nature. Shortly thereafter, Darwin wrote: "Now my theory makes all organic beings perfectly adapted to all situations, where in accordance to certain laws they can live" (in Barrett et al. 1987, p. 633).[5] However, there is evidence that Darwin's view was beginning to change, as he gradually moved away from a conception of nature as a harmonious whole with organisms perfectly fitted for their environments. For example, in the "Essay of 1844," natural selection is presented, not as an ongoing process constantly improving organisms, but as a force that acts only when the situation so permits it. Variation occurs only as a result of changed conditions of existence. Because perfectly adapted organisms have no need to change, they do not vary. Thus, if organisms are typically already perfectly adapted to their circumstances, then unless there is some change in these circumstances, further change in the organisms themselves will not occur. Athough Darwin had by now embraced an evolutionary view of living things, there is still a "static" element in his view. Organisms change only if forced to do so, and the conditions for such change are not always present.

Fifteen years later in the *Origin*, however, we find a different, more dynamic view of evolution, emerging. A reader of the *Origin* might be forgiven for concluding that "perfection" was one of Darwin's favorite topics. The word "perfect" (or its cognates, "perfection," "perfected," etc.) occurs at least 155 times in (the sixth edition of) this work, far more than in any of his other major published works.[6] It was not, of course, one of Darwin's tasks to convince his readers that innumerable marvelous adaptations and contrivances characterize the living world. His readers would hardly have needed to be persuaded of this. Indeed, it was the ubiquity of adaptation and astounding contrivances that constituted the *problematic* for his theory: How to convincingly explain such features without appeal to divine agency? It was in the context of this problem that his pronouncements about "perfection" must be understood. When Darwin, in the "Introduction" to the *Origin*, refers to "that perfection of structure and coadaptation which most justly excites our admiration" (Darwin 1859, p. 3), he can take it for granted that his readers will likewise be as astonished at this feature of the world as he was, but will also demand that he provide an explanation at least as plausible as the one offered by natural theologians. Providing such an explanation is one of his primary goals in the *Origin.*

Perfect Adaptation

Many biological structures seem so exquisitely adapted for their specific functions that describing them as "perfect" seems apt. Sometimes Darwin writes in such a way as to suggest that he thought that at least some organisms or their structures were indeed perfect, or nearly so. For example, in Chapter III of the *Origin* ("Struggle for Existence"), he asks: "How have all those exquisite adaptations of one part of the organisation to another part, and to the conditions of life, and of one distinct organic being to another being, been perfected?" (Darwin 1859, p. 60). Likewise, in the chapter on "Instinct" Darwin devotes twelve pages to providing a reconstruction of the evolution of the cell-making instinct of hive-bees. Bees have apparently succeeded in solving a difficult mathematical problem: How to construct a hive that will hold the greatest amount of honey while using the least amount of wax. The hexagonal cells of the honeycomb fit together with one another perfectly to solve the problem. No better solution is even possible in this case. As Darwin remarks, "Beyond this stage of perfection in architecture, natural selection could not lead; for the comb of the hive-bee, as far as we can see, is absolutely perfect in economising wax" (Darwin 1859, p. 235).[7]

Perfectly Useless Structures

The case of the bee-hive is significant precisely because it is exceptional. One would not usually expect to find "absolute perfection" in living things, even in truly astounding biological structures. Indeed, some structures serve no adaptive purpose, and could therefore hardly be considered "perfect." Vestigial and rudimentary organs (e.g., the human appendix and male nipples, respectively) are prime examples. A few pages after noting that "Organs or parts in this strange condition, bearing the stamp of inutility, are extremely common throughout nature" (Darwin 1859, p. 450), Darwin remarks that there can be no doubt that such organs pose a "strange difficulty" for the strict utilitarian-creationist style of explanation favored by the natural theologians, because "the same reasoning power which tells us plainly that most parts and organs are exquisitely adapted for certain purposes, tells us with equal plainness that these rudimentary or atrophied organs, are imperfect and useless" (Darwin 1859, p. 453). Far from posing a difficulty for his own theory, however, one should actually expect such imperfections as a consequence of the laws of inheritance. Rudimentary organs "may be compared with the letters in a word, still retained in the spelling, but become useless in the pronounciation, but which serve as a clue in seeking for its derivation" (Darwin 1859, p. 455). Descent with modification, rather than divinely designed utility, easily explains such imperfections.

Vestigial and rudimentary characteristics drive home the point that it is not necessary to attribute an adaptive function, much less perfection, to every biological chararacteristic. In Chapter IV of the *Origin* ("Natural Selection") Darwin notes that some characteristics are neither "useful" (a product of natural selection) nor "attractive" (a product of sexual selection); for example, "the tuft of hair on the breast of the turkey-cock, which can hardly be either useful or ornamental to this bird" (Darwin 1859, p. 90). Likewise, in Chapter VI ("Difficulties on Theory") he advises caution in simply assuming that because one can always concoct an adaptive explanation for any given characteristic, that therefore one has correctly explained that characteristic. For example, the naked skin on the vulture's head could be explained as a direct adaptation for feeding on decomposing carcasses. And so it may be. But the fact that the head of the clean-feeding male turkey is likewise naked should give us pause. Likewise, some naturalists explained the sutures in the skulls of young mammals as a beautiful adaptation for aiding parturition, and it is quite conceivable that they facilitate, or are even indispensable, for this act. But the fact that similar sutures also appear in in the skulls of young birds and reptiles, which have only to escape from a broken shell, should make us circumspect in assuming that such characteristics require an adaptive explanation. Indeed, the fact that such characteristics appear in distantly related species suggests that they have arisen from "the laws of growth," and that mammals have simply taken advantage of a characteristic that already existed in their nonmammalian ancestors (Darwin 1859, p. 197). The fact of the matter, however, is that "many structures are of no direct use to their possessors" (Darwin 1859, p. 199). Such structures might be due to the effect of physical conditions, to correlation of growth, to sexual selection, or due to inheritance. Clearly Darwin was no knee-jerk adaptationist who insisted that all features of all organisms *must* serve some adaptive function.

Exquisite but Imperfect Organs

Perfect adaptations (e.g., the hexagonal structure of the cells in a bee hive) and perfectly useless structures (e.g., vestigial and rudimentary organs) occupy endpoints on a continuum of biological characteristics. Some characteristics are as perfect as they can be. Others serve no adaptive purpose whatsoever. Most characteristics, however, will occupy the middle ground of serving some adaptive purpose, but doing so in a way that is less than perfect. For example, "Can we consider the sting of the wasp or the bee as perfect, which, when used against many attacking animals, cannot be withdrawn, owing to the backward serratures, and so inevitably causes the death of the insect by tearing out its viscera?" (Darwin 1859, p. 202). Even the vertebrate eye, that paradigmatic example of astounding design, could be better than it is. Against those who would deny that "an organ as perfect as the eye could have been formed by natural selection," Darwin argued on the contrary that "if we know of a long series of gradations in complexity, each good for its possessor, then, under changing conditions of life, there is no logical impossibility in the acquirement of any conceivable degree of perfection through natural selection" (Darwin 1859, p. 204). There is no logical impossibility, true enough. But, as a matter of fact, the eye as it actually exists is far from perfect: "The correction for the aberration of light is said, on high authority, not to be perfect even in that most perfect organ, the eye. If our reason lead us to admire with enthusiasm a multitude of inimitable contrivances in nature, this same reason tells us, though we may easily err on both sides, that some other contrivances are less perfect" (Darwin 1859, p. 202). Darwin returned to this theme at the end of the *Origin*:

> Nor ought we to marvel if all the contrivances in nature be not, as far as we can judge, absolutely perfect; and if some of them be abhorrent to our ideas of fitness. We need not marvel at the sting of the bee causing the bee's own death; at drones being produced in such vast numbers for one single act, and being then slaughtered by their sterile sisters; at the astonishing waste of pollen by our fir-trees; at the instinctive hatred of the queen-bee for her own fertile daughters; at ichneumonidæ feeding within the live bodies of caterpillars; and at other such cases. The wonder indeed is, on the theory of natural selection, that more cases of the want of absolute perfection have not been observed. (Darwin 1859, p. 472)

Such remarks provide an important clue about Darwin's mature view of the "perfection" of living things. Throughout the *Origin* he does not hesitate to describe living things as "perfect," in the sense that they rightly inspire admiration and wonder; but they are generally not so perfect that no improvements are conceivable. "Perfection" in this sense does not preclude improvement: "No country can be named in which all the native inhabitants are now so perfectly adapted to each other and to the physical conditions under which they live, that none could anyhow be improved" (Darwin 1859, pp. 82–83). The explanation has to do with the way in which natural selection operates, namely, by pitting individuals against other actually existing organisms with whom they have to contend, in the context of whatever conditions obtain locally, rather than against some abstract standard of perfection. The standard of superiority is always a *local* one: "As natural selection acts by competition, it adapts and improves the inhabitants of each country only in relation to the degree of perfection of their associates" (Darwin 1859, p. 472). Consequently, "Natural selection tends only to make each organic being as perfect as, or slightly more perfect than, the other inhabitants of the same country with which it has to struggle for existence. And we see this in the degree of perfection attained under nature" (Darwin 1859, p. 201). In summary, "Natural selection will not produce absolute perfection, nor do we always meet, as far as we can judge, with this high standard under nature" (Darwin 1859, p. 202).

From Absolute to Relative Adaptation

The foregoing account is taken from the first edition of the *Origin* (1859). In subsequent editions Darwin's shift from absolute to relative adaptation becomes increasingly clear. In the fourth edition (1866), responding to a criticism of the chief "weakness" of his theory, namely, that it treats all organic beings as imperfect, he responds that by calling organic beings "imperfect" he was simply drawing attention to the fact that not all are as perfect as they could be in relation to the conditions under which they live, as demonstrated by the fact that native forms are so often displaced by intruding and naturalized foreigners. Even if organic beings did achieve perfect adaptation to their conditions of life, they could not long remain so, because the conditions themselves are always liable to change, thus rendering their adaptations outdated (Darwin 1959, pp. 226–7).

That Darwin's view had undergone a change is clear not only from successive editions of the *Origin* but also from what he had to say elsewhere *about* how his view had changed. For example, in the first edition of the *Descent of Man* (1871), published between the fifth (1869) and final (1872) editions of the *Origin*, he admitted that in the earlier editions of the *Origin* he had perhaps attributed too much power to the action of natural selection, as he "had not formerly sufficiently considered the existence of many structures, which appear to be, as far as we can judge, neither beneficial nor injurious." He considered this to be one of the greatest oversights as yet detected in his work. As an "excuse," he confessed that he was not able to "annul the influence of his former belief, then widely prevalent, that each species had been purposely created; and this led to my tacitly assuming that every detail of structure, excepting rudiments, was of some special, though unrecognized, service." The result of maintaining this perspective, he reflected, was almost inevitable: "Any one with this assumption in his mind would naturally extend the action of natural selection, either during past or present times, too far" (Darwin 1871, vol. 1, pp. 152–53).[8]

In response to further objections, Darwin continued to clarify his view right through the sixth edition of the *Origin* (1872). For example, given the potency of natural selection for forging the exquisite adaptations of some species, why hasn't it likewise provided other species with characteristics that would certainly seem to be of benefit to them? Why is natural selection so powerful in some instances, and apparently so weak in others? Darwin noted that in some few cases it may be possible to answer with confidence but, considering our ignorance of the past history of each species, and of the precise nature of its present conditions, in most cases only very general sorts of reasons, based on our knowledge of how evolution works, can be provided. For instance, "To adapt a species to new habits of life, many co-ordinated modifications are almost indispensable, and it may often have happened that the requisite parts did not vary in the right manner or to the right degree" (Darwin 1959, p. 263). Likewise, "In many cases complex and long-enduring conditions, often of a peculiar nature, are necessary for the development of a structure; and the requisite conditions may seldom have concurred" (Darwin 1959, p. 264). Finally, it is entirely possible that what we believe would be an advantageous characteristic for a species to possess would be nothing of the sort, given the range of problems individuals of that species face, and the unavoidable tradeoffs between different characteristics. In sum, "The belief that any given structure, which we think, often erroneously, would have been beneficial to a species, would have been gained under all circumstances through natural selection, is opposed to what we can understand of its manner of action" (Darwin 1959, p. 264).

Wallace on Adaptation

Taken to an extreme, "adaptationism" is the claim that every characteristic of every organism, no matter how seemingly inconsequential, serves some significant role in the functioning of the whole, which may therefore be considered perfectly adapted. Ironically, just as Darwin moved away from a youthful "adaptationist" view to a more qualified "nonadaptationist" perspective as he pondered the operation of natural selection more deeply, Alfred Russel Wallace's overall intellectual movement was in the opposite direction. Given his importance in the development of Darwinism, an examination of Wallace's change of heart on this issue is essential. Not only is it interesting in its own right, but an examination of the ensuing disagreement between the co-discoverers of natural selection prefigures controversies that have characterized evolutionary biology from their day to the present.

Wallace's Early Nonadaptationism

Although in later life Wallace became a strict adaptationist, he certainly didn't begin as one, as evidenced by some of his writings in the 1850s. In these he was adamant that it is a serious mistake to assume that every characteristic of an organism serves some adaptive purpose. For example, his essay "On the Habits of the Orang-utan of Borneo" (1856) was not only an occasion to reflect upon the physical characteristics and behavior of these fascinating creatures but also an opportunity to issue a sweeping condemnation of uncritical adaptationism. He noted that despite the fact that male orang-utans live exclusively on fruits and other soft foods, they nonetheless have huge canine teeth which are never used either in attacking other animals or in defending themselves from predators. Females, on the other hand, who when carrying offspring might be thought to be a much more tempting object of attack by predators, have small canines. This seemed to him inexplicable on the assumption that all characteristics serve some adaptive function. Wallace concluded that animals are sometimes provided with organs of no use to them, and consequently that it is a serious mistake (albeit a common one) to assume that every feature of every organism has adaptive significance.

In Wallace's view, large canines in male Bornean orang-utans were merely one example of "animals provided with organs and appendages which serve no material or physical purpose." There are many others. The brilliant colors of some insects, the extravagant feathers of some birds, the excessively developed horns of antelopes, and the beautiful myriad forms of flower petals provide further confirmation of his claim. None can be considered as necessary for their possessors. In a sentence calculated to raise the hackles of the die-hard utilitarian-creationists, Wallace declared it to be "a most erroneous, a most contracted view of the organic world, to believe that every part of an animal or of a plant exists solely for some material and physical use to the individual, – to believe that all the beauty, all the infinite combinations and changes of form and structure should have the sole purpose and end of enabling each animal to support its existence" (Wallace 1856, p. 30).

Part of the problem with adaptationism, it seems, is that it cheapens or degrades organic nature by insisting that every part of it be "useful," as if the beauty of a flower couldn't be reason enough for its existence. Even when the parts of organic beings do have functions, it is mere hubris, Wallace argued, to believe that we are capable of discerning the one sole end and purpose of every characteristic. Naturalists are guilty of simply imagining a use for everything when they cannot discover why a given characteristic exists.

Not that "beauty" cannot function in a proper explanation of *some* sort, however. At this point Wallace was vague as to what sort of alternative, nonutilitarian explanation he had in mind, but ventured that: "The separate species of which the organic world consists being parts of a whole, we must suppose some dependence of each upon all; some general design which has determined the details, quite independently of individual necessities" (Wallace 1856, p. 31). This sounds a bit like the idealist view discussed earlier, but from the little he says in this context, it is impossible to tell precisely what he had in mind. One thing, however, is abundantly clear. Wallace was convinced that an emphasis on the usefulness of every characteristic of organisms was a dire mistake: "[T]he constant practice of imputing, right or wrong, some use to the individual, of every part of its structure, and even of inculcating the doctrine that every modification exists solely for some such use, is an error fatal to our complete appreciation of all the variety, the beauty, and the harmony of the organic world" (Wallace 1856, p. 31).

A similar, albeit less developed, position is evident in his 1858 paper (the one presented along with Darwin's to the Linnean Society, announcing the principle of natural selection) in which he referred twice to the fixation of "unimportant" parts: "Variations in unimportant parts might also occur, having no perceptible effect on the life-preserving powers; and the varieties so furnished might run a course parallel with the parent species" (Wallace 1858, p. 59). Examples of such "unimportant parts" are "colour, texture of plummage and hair, [and the] form of horns or crests" (Wallace 1858, p. 62). In the 1850s, at least, Wallace was an ardent and articulate nonadaptationist.[9]

Wallace's Adaptationism

From the 1860s until his death in 1913, however, a very different Wallace is in evidence, one who espoused a virtually uncompromising adaptationism. This is the Wallace well-known to historians of evolutionary biology, the tireless advocate for "Mr. Darwin's theory" who prided himself on being "more Darwinian [i.e., more adaptationist] than Darwin." For example, in his 1867 paper, "Mimicry and other Protective Resemblances among Animals," he articulated a principle which he claimed to be a "necessary deduction from the theory of Natural Selection, namely – that none of the definite facts of organic nature, no special organ, no characteristic form or marking, no peculiarities of instinct or of habit, no relations between species or between groups of species – can exist, but which must now be or once have been *useful* to the individuals or the races which possess them" (Wallace 1867, p. 3). It would be hard to find a clearer, or less compromising, statement of adaptationism. As Malcolm Kottler (1985, p. 412) points out, "such a principle is a 'necessary deduction' from natural selection *only if* one holds the view that all evolutionary change has resulted from the action of natural selection." This is precisely the view that Wallace adopted in the 1860s and maintained to the end of his life. Natural selection is the primary force responsible for all evolutionary change. If natural selection could have produced a given characteristic, then it did produce that characteristic. Consequently, all traits must be considered useful (i.e., adaptive). Likewise, in a stunning reversal of his earlier condemnation of the common practice of simply assuming that all characteristics serve some definite purpose, Wallace maintained that "the assertion of 'inutility' in the case of any organ or peculiarity which is not a rudiment or a correlation, is not, and can never be, the statement of a fact, but merely an expression of our ignorance of its purpose or origin" (Wallace 1889, p. 137). So much for the "fatal error" he earlier ascribed to adaptationism!

Darwin and Wallace on the Power of Selection

By this point, Wallace's transformation from impassioned critic to ardent defender of adaptationism was complete. But why had his view changed so dramatically? Perhaps a better question is why Darwin resisted strict adaptationism. After all, natural selection was the key explanatory principle in Darwin's theory, and succeeded marvelously in explaining a wide and diverse range of biological phenomena. It had proven its value time and again. Why *not* simply apply it in the explanation of whatever biological phenomena needed explaining?

It is useful at this point to compare the mature views of both Darwin and Wallace, to make clear how and why they disagreed, and to try to understand the significance of their differences. As we have seen, from the 1860s on Wallace was a committed adaptationist. Darwin himself, on the other hand, recognized the ubiquity of characteristics which either serve no function, or else serve some function less well than one might expect if natural selection were truly a perfecting agent. It is true that at times Darwin's view sounds very close to that of Wallace, for example, when he writes in the first edition of the *Origin* that "every detail of structure in every living creature . . . may be viewed, either as having been of special use to some ancestral form, or as being now of special use to the descendents of this form – either directly, or indirectly through the complex laws of growth" (Darwin 1859, p. 200). But here Darwin was careful to qualify his adaptationism by including the idea that some details of structure may have been of use to some *ancestral* form, thus leaving open the possibility that such details serve no special use *now*. In addition, the reference to "the complex laws of growth" in this context signals Darwin's conviction that organisms must be treated as integrated systems in which a change in one characteristic might necessitate specific changes in others. But if so, then there arises the possibility that some especially useful characteristics might entail other characteristics which do not, by themselves, directly serve any adaptive function. Wallace, by contrast, tended to treat organisms more as clusters of characteristics, each of which could be directly modified by natural selection independently of the rest. Whereas Wallace was an uncompromising adaptationist, Darwin saw an important but limited role for adaptation.

The Question of Interspecific Differences

A key locus of disagreements between Darwin and Wallace about the pervasiveness of adaptations concerned *interspecific differences*. Closely related species often differ from one another in just a few characters, and are indeed indistinguishable to the nonexpert. Are such differences adaptive, or are they mere differences, without adaptive significance? Darwin believed that some of the most important characteristics recognized by systematists were in fact nonadaptive. In the sixth edition of the *Origin*, he noted that through convergent evolution distantly related organisms could nonetheless come to resemble one another in remarkable ways, but such similarities would be useless for classification:

> On the view of characters being of real importance for classification, only in so far as they reveal descent, we can clearly understand why analogical or adaptive characters, although of the utmost importance to the welfare of the being, are almost valueless to the systematist. For animals, belonging to two most distinct lines of descent, may have become adapted to similar conditions, and thus have assumed a close external resemblance; but such resemblances will not reveal – will rather tend to conceal their blood-relationship. (Darwin 1959, p. 664)

Darwin was therefore concerned to show that not all specific differences are adaptive. If they were, a natural system of classification by descent would be difficult or impossible to establish. Wallace, by contrast, believed that Darwin had underestimated the effectiveness of natural selection. According to Wallace,

> [I]t has not... been proved that any truly "specific" characters – those which either singly or in combination distinguish each species from its nearest allies – are entirely nonadaptive, useless, and meaningless; while a great body of facts on the one hand, and some weighty arguments on the other, alike prove that specific characters have been and could only have been, developed and fixed by natural selection because of their utility. (Wallace 1889, p. 142)

In an 1896 paper focusing on precisely this issue, Wallace argued that, despite ingenious arguments to the contrary by some of Darwin's most faithful disciples (e.g., George John Romanes [1848–94]), *all* the characters distinguishing one species from another have adaptive significance. After arguing for this claim by way of a detailed analysis of how species form, he concludes that "whether we can discover their use or no, there is an overwhelming probability in favour of the statement that every truly *specific* character is or has been useful, or, if not itself useful, is strictly correlated with such a character" (Wallace 1896, p. 496).

The Power of Selection

Recognizing a disagreement is one thing. Explaining it is another. Why did Darwin and Wallace differ on this issue? The disagreement between Darwin and Wallace can be traced largely to their differing views about the power and sufficiency of selection to account for biological phenomena. Wallace was more thoroughly convinced of the power of natural selection, and thus adopted the more consistently adaptationist position. For example, in the context of his disagreement with Darwin concerning the explanation of sterility, Wallace remarks, "I am deeply interested in all that concerns the powers of Natural Selection, but though I admit there are a few things it cannot do I do not yet believe sterility to be one of them" (Wallace 1916, p. 167). Darwin, by contrast, while maintaining that natural selection is the "main" mechanism of evolutionary change, also recognized several other evolutionary forces. Indeed, he ends the Introduction to the *Origin* by explicitly stating: "I am convinced that Natural Selection has been the most important, but not the exclusive, means of modification" (Darwin 1959, p. 75). Sexual selection, for example, might result in characteristics (e.g., the long tailfeathers of the peacock), which were a hindrance to survival. Darwin also was much impressed by the effects of "the laws of growth" and, in particular, Cuvier's principle of the "Correlation of Parts," which treated organisms as integrated systems. An adaptive change in one part might entail nonadaptive changes in other parts. Finally, as we saw in Part I, Darwin postulated selection at the level of "families" to account for otherwise inexplicable features of social insects. While adaptive at the level of families or communities, such features appear to be nonadaptive at the level of individual organisms. Darwin was genuinely pluralistic in his approach to explaining biological phenomena. As a result, he was more willing to grant the existence of nonadaptive characteristics.[10]

"A Matter of Chance"?

Another part of the explanation for the difference of opinion between Darwin and Wallace concerns the availability of the variation needed to "fuel" selection-driven adaptive change. Darwin was far less convinced than Wallace that the favorable variations would always be present when needed. The *Origin* is sprinkled with phrases such as "if variations useful to any organic being do occur." In Wallace's view, by contrast, variation is abundant and pervasive. As he maintained years after Darwin's death, "We now know that variations of every conceivable kind occur, in all the more abundant species, in every generation, and that the material for natural selection to work upon is never wanting" (Wallace 1896, p. 482). For Wallace this was the "grand fact" that rendered modification and adaptation to conditions (almost) always possible.[11]

The difference between Darwin and Wallace on this issue points to an even deeper difference of perspective. A key component of Darwin's theory, which has proven to be perhaps *the* conceptual obstacle to its widespread acceptance, is that the variations upon which selection operates occur "by chance." What did he mean by this? Although he sometimes speaks (as Lamarck did) of "chance" as ignorance of causes, this meaning is secondary to the notion of chance as "accident." In *The Variation of Animals and Plants Under Domestication* (1868), he put this idea to work in explaining the nature of the materials upon which natural selection works:

Let an architect be compelled to build an edifice with uncut stones, fallen from a precipice. The shape of each fragment may be called accidental; yet the shape of each has been determined by the force of gravity, the nature of the rock, and the slope of the precipice, – events and circumstances, all of which depend on natural laws; but there is no relation between these laws and the purpose for

> which each fragment is used by the builder. In the same manner the variations of each creature are determined by fixed and immutable laws; but these bear no relation to the living structure which is slowly built up through the power of selection. (Darwin 1868, vol. 2, pp. 248–49)

The variations upon which selection operates arise by "accident" in the sense that they do not arise *because* their presence is likely to help their possessors; that is, they are not the direct result of an organismic need. This is in sharp contrast with Lamarck's theory, according to which adaptive variations *do* arise in direct response to the needs of organisms as caused by environmental conditions, and hence are in principle predictable. For Darwin, by contrast, variations arise by "accident," and hence the course of evolution is radically unpredictable.

Darwin's view was also in sharp contrast with the natural theological tradition. A problem facing theistic accounts of organic nature is explaining the apparent imperfections to be found in living things. Many organisms seem, at least from the perspective of structural efficiency, to be very badly designed. By suggesting that nature produces organic beings with the materials (i.e., variations) it has at hand, rather than with foresight, Darwin was easily able to explain such imperfections. It is a matter of "chance" that particular variations arise at a given time; yet, selection works on whatever variations are available, not on whatever variations would be *ideal* in a particular circumstance. To his contemporaries, to say that variations arise "by chance" was to deny that variations arise through the providential ordering of a Deity who foresees which variations will be needed to fill the various preordained "stations" in nature. Darwin substitutes "blind" efficient causality for far-sighted final causality in his account of the generation of organic variability. It is in this sense that Darwin is correctly thought to have substituted "chance" for purpose in the living world.

"Chance" enters Darwin's theory in another important way, one that captures the *historical* nature of biological facts. Darwin's desire to explain the biogeographical observations made while on the *Beagle* voyage was a major catalyst for his acceptance of the idea of evolution. As he came to realize from his biogeographical studies, the characteristics of present-day organisms show the idiosyncratic influence of their ancestral migratory and dispersal histories. It was a purely contingent event that one or a few birds got blown off course during a storm long ago and ended up on an island (no law of nature dictates that this must happen). But such contingencies are the stuff of which evolutionary novelties are made.

They provide opportunities for organisms to develop specializations for the colonization and exploitation of new habitats. Given sufficient time and conditions, such colonizers may evolve into distinct species.[12] The existence of any particular species may be crucially determined by just such environmental contingencies.

The unpredictability of such particular "once-only" events means that the particular paths evolution has taken are essentially unrepeatable, in much the same way that human history is unrepeatable. Each stage in the process depends on the materials it inherits from the previous stage. A contingent event at any given stage alters the trajectory of the remainder of the process. Species, like the individuals that compose them, are unique historical entities. As such, they will be governed by natural laws but will not be predictable simply from knowledge of such laws. Because of the sheer complexity such historical (and ecological) contingency introduces into all evolutionary processes, Darwin realized that prediction in such cases can at best be conjectural. "Throw up a handful of feathers, and all must fall to the ground according to definite laws; but how simple is the problem where each shall fall compared to the action and reaction of the innumerable plants and animals which have determined, in the course of centuries, the proportional numbers and kinds of trees now growing on the old Indian ruins!" (Darwin 1959, p. 75). What is true for trees growing on Indian ruins is true in spades of the evolution of species over millions of years of undirected ecological change.

Summary: Darwin (and Others) on Biological Perfection

In place of the absolute or perfect adaptation both he and many of his contemporaries embraced in the 1830s and 1840s, by the time he composed the *Origin of Species* Darwin had come to accept a notion of relative adaptation. Forms that are successful in the struggle for existence need not be perfectly adapted to their specific conditions. They need only be slightly better adapted than their direct competitors. In addition, because organisms are locked in an unceasing struggle both with others of their kind and with an ever-changing environment, perfect adaptation will be a fleeting condition, if indeed it ever arises at all. More commonly, organisms will be well adapted to conditions that have long since ceased to exist. Hence Darwin's claim that "all corporeal and mental endowments will tend to progress towards perfection." Evolution is an ongoing process with no end in sight.

In developing his view, Darwin explicitly sought to combine key elements from both utilitarian-creationism and from biological idealism but reworked in light of the theory of natural selection and descent with modification. Like the utilitarian-creationists, he was impressed by the degree to which organisms are exquisitely adapted to their "conditions of existence," but proposed that natural selection, not divine design, provides the correct explanation for this fit. Like the idealists, he was keenly aware that imperfections abound, but proposed that historically constrained descent with modification, not conformity to transcendental "archetypes," explains both "unity of type" and lack of perfect fit. By borrowing elements from each of the previously dominant views of living things, while rejecting their questionable metaphysical assumptions, Darwin succeeded in formulating a novel account that made sense of the full range of biological phenomena. In Darwin's theory, one should expect neither perfect adaptation, nor the absence of adaptive fit but, rather, "good solutions within constraints" (Cronin 1991, p. 24).[13] This is, of course, precisely what one does find.

5

Adaptation After Darwin

> [Discovering] the use of each trifling detail of structure is far from a barren search to those who believe in Natural Selection.
>
> (Darwin 1862, pp. 351–52)

Introduction

The two most striking facts about the living world that Darwin attempted to explain were that organisms are so often superbly fitted for survival and reproduction, and that living things display a staggering diversity of different forms. In principle, the two facts could be explained independently of one another (as Lamarck believed they should), but if a significant part of the explanation of the latter fact is that species have diversified in the course of *adapting* to different environmental challenges, then *adaptation* becomes the central Darwinian concept for explaining both good design and diversity. Familiarity with the career of adaptationist explanations, therefore, becomes critical to understanding the evolution of Darwinism.

As we saw in the previous chapter, Darwin's view of adaptation underwent a significant shift from his earlier view (influenced by theological considerations) that organisms were perfectly designed, to his later view (developed in light of his understanding of the operation of natural selection) that organisms are at best only relatively well adapted to their circumstances. As Darwin was moving from a notion of absolute to a notion of relative adaptation, his ally and intellectual sparring partner Alfred Russel Wallace was becoming a strict selectionist-adaptationist for whom every feature of every organism (with one notable exception, to be discussed in Chapter 10) has resulted from present or past utility. Darwin, by contrast, maintained that there are a range of factors involved in the evolution of any given characteristic, and that we need not assume that every characteristic bestows some adaptive advantage on its possessor. Given the success of "Darwinism" for explaining so many otherwise puzzling aspects of the living world, one might think that Darwin's views on adaptation would have simply steadily gained acceptance to the present, with Wallace's more extreme position being relegated to the status of a mere historical curiosity. But the history of evolutionary thinking tells a very different (and much more interesting) story.

My aim in this chapter is tell part of this story by charting the course of adaptationist thinking in the twentieth century, focusing especially on the careers of Darwin's and Wallace's different views of adaptation as the century progressed. To anticipate the discussion that follows, at the dawn of the twentieth century there were a number of rivals to Darwinism, that is, evolutionary theories that did not embrace natural selection as the most (or even an) important cause of evolutionary change. Instead, they postulated various nonselective evolutionary forces. Each of these theories eventually failed to account for the main facts of evolution, and were therefore discarded by professional biologists. Among Darwinians, Wallace's selectionist-adaptationist view became orthodoxy. By the 1930s and early 1940s, however, the tide turned slightly in Darwin's favor as biologists began to attribute considerable importance to nonselective (e.g., "chance") factors in evolution. From the late 1940s to the 1970s, the pendulum then gradually swung back toward Wallace's view as such factors were invoked less and less frequently, and as natural selection came once again to be seen as the most important agent of evolutionary change. Reactions against this hyperselectionism, and responses to this reaction, characterize contemporary discussions of adaptation (as discussed more fully in the next chapter). To understand how the present situation developed, however, it is essential to see how it began. We pick up the story right after Darwin.

Evolutionary Alternatives After Darwin

Darwin succeeded in convincing most of his contemporaries that evolution (i.e., "descent with modification") is a fact. He was less successful in convincing them that natural selection plays the central role in bringing about evolutionary change (Mayr 1982, pp. 506–25). The result was what has been called "the eclipse of Darwinism" (Bowler 1983; Huxley 1942), a period lasting roughly from the time of Darwin's death in the early 1880s to the 1930s but peaking from about 1890 to the turn of the century. While biologists generally accepted Darwin's claim that life had evolved through a process of descent with modification, many could not accept the idea that natural selection was the chief agent of evolutionary change. Instead, they embraced various non-Darwinian theories of evolution that were thought to remedy the (perceived) defects of Darwin's theory.

Standing Paley on His Head

Two problems, in particular, motivated biologists to develop non-Darwinian alternatives. One problem was the sense that Darwinians had gone too far in their unrestrained adaptationism. Julian Huxley characterized the state of Darwinian thinking in the period immediately following Darwin's death (i.e., the 1890s; Darwin died in 1882) as follows:

> Darwinism grew more and more theoretical. The paper demonstration that such and such a character was or might be adaptive was regarded by many writers as sufficient proof that it must owe its existence to Natural Selection. Evolutionary studies became more and more merely case-books of real or supposed adaptations. Late nineteenth-century Darwinism came to resemble the early nineteenth-century school of Natural Theology. Paley *redivivus*, one might say, but philosophically upside down, with Natural Selection instead of a Divine Artificer as the *Deus ex machina*. (Huxley 1942, p. 23)

Although not all Darwinians took such a cavalier attitude toward explanation, in basic orientation, at least, many followed Wallace's strict selectionist-adaptationist approach rather than Darwin's more pluralistic perspective.

The other feature of Darwin's theory that led many to reject it was its dependence on "chance." As we saw in the previous chapter, Wallace objected to what he perceived to be Darwin's appeal to chance factors (i.e., the availability of useful variations, environmental contingencies leading to historically constrained structures, etc.), and responded by simply reasserting the power of natural selection. Others not committed to selectionist explanations saw the chance element in Darwin's theory as sufficient reason to jettison it entirely in favor of theories which construed evolution as more "lawlike" in its behavior. In one way or another, each of the major alternatives to Darwinism sought to avoid either or both of these perceived flaws.

As Bowler (1983) discusses at length, at the turn of the century there were essentially four serious alternatives to a Darwinian account of life: theistic evolutionism, neo-Lamarckism, orthogenesis, and the mutation theory. With hindsight it might be possible to determine that each of these alternatives was seriously defective in some way, and that Darwinism would eventually emerge victorious. But at the time it was far from obvious to many biologists that Darwinism would be remembered as anything more than an interesting episode in the prehistory of biological science. Some decreed that its heyday was essentially over, and that if natural selection operated at all, it was at best a marginal force of little real significance. Clearly they were wrong. But to understand the eventual resurgence of Darwinism, it is worthwhile to review each of these non-Darwinian evolutionary theories in turn, focusing especially on the role of adaptation in each.[1]

Theistic Evolutionism

As a student at Cambridge in the late 1820s, Darwin had read, and been much impressed by, William Paley's *Natural Theology* (1802). When he was working through the ideas that would become the theory of natural selection, the problem Paley sought to explain – the obvious fit of living things to their ways of life – was never far from his thoughts. Darwin's mechanism of natural selection was intended to play much the same role that the Creator had played for Paley, namely, as the agent primarily responsible for organic design. Although religious opposition to the idea of *evolution* had diminished by the turn of the century, the idea that the major cause of evolution was natural selection driven by undirected environmental change coupled with random variations was still morally and theologically repugnant to many people, including many scientists. Once again, it was the notion of *chance* that proved unacceptable to critics of Darwin's theory. The idea of natural selection operating on *chance variations* seemed to undermine the belief that nature is a purposeful system designed and directed by a wise and benevolent Creator. Consequently, some biologists sought to retain a primary place for divine intervention in the evolutionary process. (Theistic evolutionists included the Duke of Argyll [1868], St. George Mivart [1871], Asa Gray [1876], and William B. Carpenter [1889].)

The key claim of theistic evolutionism was that the variations upon which natural selection operates are not random, as Darwin had supposed, but instead are brought about by the Creator in such a way as to direct the evolutionary process toward predetermined ends. Divinely guided progress, rather than adaptation to local conditions, was the primary cause of evolutionary change. Adopting a theistic rather than naturalistic account of evolution had interesting consequences for explaining organic diversity. According to the Darwinian account, living things diversify in the course of adapting to different environments. According to the Duke of Argyll (1868), however, the diversity of life is a reflection of the Creator's desire to maximize the variety of living things. Nonadaptive characteristics like the brilliant plumage of tropical birds or the colorful carapaces of beetles were indicative of a *higher* purpose than mere utility, namely, divine concern for order and beauty.

Such a view was not without its problems. For one thing, it seemed incredible to many biologists that the seemingly random paths that evolution had taken, with countless twists, turns, and dead-ends, could be the working out of a rationally ordered divine plan. Evolution appears much too haphazard for that to be the case. The most basic problem facing theistic evolutionism, however, was that it seemed to place a central feature of evolution – the origin of the variations upon which selection operates – outside the scope of scientific investigation. If all variations are introduced directly by God, how could they be amenable to scientific study? Theistic evolutionism could be (and often was) simply dismissed as "nonscientific." However, other non-Darwinian approaches that restricted themselves to purely natural causes could not be dismissed so easily.

Neo-Lamarckism

Recall that Lamarck had postulated two distinct (but interacting) forces determining the course of evolution. First, and most important, there is an inherent tendency for living things to become more complex. This accounts for the "chain of being" from simple organisms (like invertebrates) at the bottom, all the way up to highly complex organisms (e.g., vertebrates, especially humans) at the top. Second, environmental contingencies deflect organisms from a simple progression as they adapt to their specific conditions through a process of use and disuse. Useful characteristics acquired during one individual's lifetime could be passed on to offspring, thus accelerating evolutionary development. By the turn of the century, "Lamarck's Theory" came to be virtually synonymous with the doctrine of the inheritance of acquired characteristics.

"Neo-Lamarckism" (a term coined in 1885 by E. Ray Lankester) represented the revival of Lamarck's ideas. Herbert Spencer (1887) was a particularly enthusiastic proponent. Part of the attraction of this strategy was that embracing such ideas seemed to bypass what was thought to be an outstanding problem for Darwin's theory, namely, that natural selection was a purely "negative" force that could perhaps eliminate the grossly unfit, but was powerless to bring about new characteristics without which species change would be impossible. Lamarck's theory, by contrast, was all about how new, beneficial characteristics arise and are passed on. In addition, neo-Lamarckians could point to many phenomena that seemed to clearly validate the doctrine of use and disuse, and the inheritance of acquired characteristics. The reduction in size of the wings of domestic fowl, and the loss of eyes in many cave animals, were favorite examples. According to Lamarckians, the environment itself caused such beneficial variations to appear, thus bypassing the appeal to chance altogether. Another, perhaps more powerful, attraction of neo-Lamarckism was that if the environment could directly influence adaptation in the way postulated, then evolution could proceed much more rapidly than seemed possible by the slow, blind, trial-and-error method of random variation and natural selection. Applied to human beings, neo-Lamarckism suggested that rather than being helpless pawns of natural selection, we can play an active and significant role in our own evolution.

Given its (apparent) considerable attractions, why did neo-Lamarckism eventually fade away as an acceptable theory of evolution? A commonly heard explanation is simply that it succumbed to experimental refutation. August Weismann's famous experiments in which he showed that cutting off the tails of mice had no effect on their offspring sounded the death knell for neo-Lamarckism (Weismann 1891–92). However, crucial experiments that succeed in convincing one and all that a theory is mistaken are rare in science. Weismann's experiments could be dismissed by neo-Lamarckians as irrelevant, because they were not committed to the idea that all bodily changes are necessarily passed on to offspring. Because the bodily change in question did not arise from the organism's striving to satisfy some need, there is no reason to suppose that it would be passed on to offspring. The fact that neo-Lamarckians could not articulate a coherent theory of heredity to undergird the supposed examples of inheritance of acquired characteristics proved to be a more serious problem. Neo-Lamarckism eventually was abandoned due to a number of additional factors, among them the lack of consistent experimental evidence showing that acquired characters were, in fact, inherited (those experiments that seemed to demonstrate Lamarckian mechanisms were later shown to have been fabricated), and the availability of Darwinian accounts of the data to be explained. Later attempts to revive this approach (e.g., Steele 1981) have not met with much success.

Orthogenesis

This is not to suggest, of course, that Darwinism did not face its own difficulties. In 1898, Theodor Eimer published a book with the revealing title *On Orthogenesis and the Impotence of Natural Selection in Species-Formation.* Eimer's study of butterfly coloration had convinced him that there were trends in evolution with no adaptive significance, and thus difficult or impossible to explain in terms of natural selection. Other orthogeneticists posed similar arguments (e.g., Edward Drinker Cope [1871, 1873]). A particularly serious problem for Darwinism was to account for long-term nonadaptive evolutionary trends that involve the accumulation of disadvantageous characteristics that eventually drive a species to extinction. The enormous antlers of the "Irish Elk," and the elongated teeth of the saber-toothed cat, were routinely identified as characteristics that developed so far beyond anything that could be considered adaptive that they became positively harmful to their possessors, eventually leading to their demise. The consistency of such trends seemed to demand postulation of an internal predisposition of organisms to vary in a single direction, independently of environmental factors. Thus, it was argued that evolutionary trends unfold without reference to the demands of the environment, which explains why they might sometimes lead to extinction for entire species. Darwinians could respond to such examples, however, by pointing out that they are the exceptions rather than the rule, and that in the vast majority of cases natural selection provides a better explanation of the exquisite fit between organisms and environment than does the orthogenesists' postulation of mysterious "internal predispositions," which seemed by comparison a scientific dead-end.

Mutation Theory

In the contemporary Darwinian understanding of evolution, mutations are understood to be small genetic changes that, when expressed at the phenotypic level, may occasionally result in characteristics that aid the organism in survival and/or reproduction. But the term "mutation" was originally introduced in the first decade of the twentieth century to denote *large-scale* genetic changes, occurring at a number of genetic loci simultaneously, and capable of creating entirely new species in a single step. According to "the mutation theory," evolution proceeds by the sudden appearance of significantly new forms without reference to the demands of the environment. The changes that occur appear randomly, and are nonadaptive. At some level, of course, the environment must exert minimal control on the course of evolution. If a large-scale mutation produced a rhinoceros without a heart, for instance, obviously such a creature would not survive to pass on its genes. But this simply meant that those organisms with grossly defective characteristics would be eliminated. Those with neutral or conceivably beneficial characteristics would not be eliminated. The degree of control exerted by the environment was understood to be such that even mutations that did not confer any advantage on organisms could still direct the course of evolution. Mutationism was advanced as yet another alternative to what some biologists considered the sterile utilitarianism of Darwinism, according to which every character of every organism was assumed to have some adaptive significance.

The English-speaking world became acquainted with the mutation theory primarily through the work of Hugo DeVries (1906, 1910). DeVries believed that the mutation theory solved one of the outstanding problems facing Darwin's theory, namely, the problem of explaining how minute changes in individual organisms could resist being "swamped" in a population by those individuals interbreeding with other individuals lacking the new characteristic. This was the nagging problem of "blending inheritance" that decades earlier Darwin had struggled with and attempted (unsuccessfully) to resolve with his "provisional theory of pangenesis" (Darwin 1868). How could any favorable characteristic that happened to arise by chance be preserved in the population? What would prevent it from being diluted as individuals with that characteristic mated with individuals lacking it? DeVries insisted that a mutation might appear in many individuals simultaneously, thus forming a new variety or subspecies which through interbreeding could then be sustained. Because varieties were not formed by natural selection, there was no reason to believe that the characteristics distinguishing one variety from another were of any adaptive significance. So long as a characteristic was not harmful, it could be established and maintained in a population.

An influential proponent of the mutation theory, and critic of Darwinism, was Thomas Hunt Morgan. In his *Evolution and Adaptation* (1903) he rejected natural selection as the main force guiding the course of evolution in favor of mutations of the sort DeVries postulated. Organisms had to be more or less well-suited to their environments in order to survive, he argued, but this was consistent with the vast majority of characteristics distinguishing one species from another having no adaptive significance. According to Morgan,

> Animals and plants are not changed in this or that part in order to become better adjusted to a given environment, as the Darwinian theory postulates. Species

exist that are in some respects very poorly adapted to the environment in which they must live. If competition were as severe as the selection theory assumes, this imperfection would not exist. In other cases a structure may be more perfect than the requirements of selection demand. We must admit, therefore, that we cannot measure the organic world by the measure of utility alone. If it be granted that selection is not a moulding force in the organic world, we can more easily understand how both less perfection and greater perfection may be present than the demands of survival require. (Morgan 1903, p. 464)

Hence at the level of individual organisms, adaptation should be considered only a minor factor in evolution.[2] This left the way open for mutations to determine the course of evolution. Or so it seemed. Unfortunately, mutationism was more plausible for the sort of organisms DeVries studied (i.e., certain plant species) where new species could arise in a single generation (e.g., by "polyploidy," a doubling of the number of chromosomes) than for animals, in which major genetic changes are almost inevitably fatal. Nonetheless, despite a general lack of scientific support, in one form or another the mutation theory survived well into the twentieth century.

Summary: Neo-Darwinism and Its Rivals

By the mid-1940s, the various alternatives to Darwinism discussed above were almost nowhere to be found. Why? The simple answer is that Darwinism had been set on a new foundation, based on the rediscovery of the genetics Gregor Mendel had worked out in the previous century. The resulting "synthetic" or "neo-Darwinian" theory proved far more powerful than any of its rivals. Starting from the view that populations may be thought of as "gene pools" subject to various evolutionary forces (mutation, migration, inbreeding, selection, drift), mathematically sophisticated models were created in order to calculate the changes in gene frequencies resulting from different combinations of evolutionary forces. The work of R. A. Fisher (1890–1962), J. B. S. Haldane (1892–1964), and Sewall Wright (1889–1988) typifies this approach. Their work showed how the principles of Mendelian genetics could be integrated with a Darwinian understanding of natural selection to produce rigorous models of changes of gene frequencies in populations. By the 1940s and 1950s, this approach had attained a high degree of sophistication, and it began to appear that the earlier objections to the theory of natural selection were groundless. Just as mathematics had transformed physics centuries earlier, it was finally doing the same for biology. The final steps in the synthesis occurred in the 1940s as biologists like Theodosius Dobzhansky (1937), Ernst Mayr (1942), George Gaylord Simpson

(1944), and G. Ledyard Stebbins (1950) attempted to bridge the gap between the highly theoretical work of Fisher, Haldane, and Wright with detailed empirical research into the structure and genetic composition of natural populations. The result was the "modern synthesis" in evolutionary biology, christened as such in Julian Huxley's magisterial overview, published as it was still unfolding (Huxley 1942).

Table 5.1 summarizes some of the main differences between the victorious neo-Darwinian understanding of evolution with that of its main rivals earlier in the century.

Wright's Shifting Balance Theory

A substantial number of biologists in the early decades of the twentieth century considered many of the most significant characteristics of organisms to be nonadaptive. Robson and Richards (1936, pp. 314–15) expressed the belief of many biologists when they claimed that "A survey of the characters which differentiate species . . . reveals that in the vast majority of cases the specific characters have no known adaptive significance." As we have seen, a major selling point for some of the non-Darwinian evolutionary theories then current (whatever their faults) was precisely that they had ways of explaining such characteristics. For Darwinians, however, this posed a more serious problem. Shouldn't selection eliminate nonadaptive characteristics, leaving only optimally adapted suites of characters? How could the widespread existence of nonadaptive characters be squared with natural selection a perfection-producing mechanism? Many in the 1930s and thereafter believed that they had found a satisfactory Darwinian explanation of nonadaptive characters in Sewall Wright's "shifting balance theory."

The Fisherian Background

To understand Wright's theory, we have to consider the quite different approach taken earlier by R. A. Fisher, to which it was a reaction. For the sake of mathematical ease, in his calculations Fisher had treated populations as structurally homogeneous and effectively infinite in size, thus simplifying derivations of the effect of selection on fitness differences. He also applied with a vengeance Wallace's conviction that selection is by far the most powerful evolutionary force. Consequently, as Provine notes, "Fisher believed . . . that selection so dominated evolution that non-adaptive characters were virtually non-existent, except for some secondary sexual characteristics produced by sexual selection" (Provine 1985, p. 856). By 1936,

TABLE 5.1: *Neo-Darwinism and its Rivals*

	Theistic Evolutionism	Neo-Lamarckism	Orthogenesis	Mutation Theory	Neo-Darwinism
Contains a mechanism adaptation	yes	yes	no	no	yes
Views the "shape" of evolution as:	linear adaptive trends	linear adaptive trends	linear patterns followed in parallel by groups of related forms	an irregular, constantly branching process of diversification	an irregular, constantly branching process of diversification
Evolution is guided by a force outside of nature	yes	no	no	no	no
Variations are random with respect to needs	no	no	yes	yes	yes
Primary cause of evolutionary change is:	external: divine guidance of variations	external: effect of environment on individual development	internal: laws of growth and development	internal: random mutations	internal & external: variation & natural selection
Explanation of nonadaptive characters and trends is central	no	no	yes	yes	no

the sort of view Fisher had been developing for many years had crystallized into a distinctive Fisherian doctrine. According to Fisher, there are essentially two kinds of evolutionary theories, viz., those that account for adaptations and those that don't. The basis for this judgment is simple: "[E]volution is progressive adaptation and consists of nothing else. The production of [supposedly nonadaptive] differences recognizable by systematists is a secondary by-product, produced incidentally in the process of becoming better adapted" (Fisher 1936, p. 58). True, others after Darwin had promoted a range of alternative non-Darwinian evolutionary theories, but "for *rational* systems of evolution, that is for theories which make at least the most familiar facts intelligible to the reason, we must turn to those that make progressive adaptation the driving force of the process" (Fisher 1936, p. 59; emphasis in original). When we do this, he believed, Darwin's theory (correctly interpreted) wins hands down.

The Fisherian approach, in which evolution is driven by the gradual selection of slight variants within an undifferentiated global population, leading to ever-better adaptation, seemed to Wright unable to account for the rapidity with which much evolutionary change has occurred. In particular, Wright argued that on Fisher's continuous population theory, major evolutionary novelties, of the sort that characterize the history of life, are difficult to achieve, because selection acting alone tends to favor characteristics that depart little, if at all, from the current average. Major evolutionary novelties associated with speciation events, by contrast, would require that entire genetic systems be capable of reorganization.

The question that guided Wright's theorizing embodied these central concerns: What combination of factors is most conducive to the adaptation, survival, and multiplication by the splitting of species, and hence to significant evolutionary change? Fisher's approach emphasized progressive adaptation of already existing species, but fell short in explaining both nonadaptive characteristics and the evolution of new species. Wright therefore concluded that a radically different approach to evolution was necessary, one that would explain both the nonadaptive diversification of groups and the rate of adaptive evolution and speciation. The key ideas Wright was developing in constructing an alternative approach are reflected in a 1931 letter to Alfred Kinsey, in which Wright confided, "I am especially interested in the question as to how far there is subdivision of species into small local strains differentiated in the random fashion expected of inbreeding (instead of in adaptive ways by natural selection). My results seem to indicate that such a condition is the most favorable for progressive evolution of the species as a single group" (quoted in Provine

1986, p. 291). These ideas found fuller expression in his “shifting balance theory,” the central idea of which was “random genetic drift.”

Random Genetic Drift

Clearly, if natural selection is operating within a less than optimally adapted biological population, one might expect gene frequencies in that population to vary from one generation to the next. But even apart from the effects of natural selection, gene frequencies may vary from generation to generation simply because of “sampling error.” To grasp the basic idea, consider the classic way of modeling this concept. Let a jar consist of one thousand beans, with five hundred white and five hundred brown beans. Suppose that the beans are distributed evenly in the jar. Then (with eyes closed) draw out two hundred beans at random. It is possible that the proportion of white and brown beans in the randomly chosen sample perfectly matches the proportion of white and brown beans in the jar (i.e., one hundred white and one hundred brown beans). But it is more likely that the proportion of beans drawn at random will not perfectly match the proportion of each color bean in the jar. Suppose that upon inspection one sees that one has drawn 150 white and 50 brown beans. If these beans are mixed together and a second blind sampling is performed on this collection of beans, it is possible (and now even more likely) that the disproportion between white and brown beans could increase even further. If the process is iterated a number of times, it is even possible that one will end up with only white beans, the brown beans having been eliminated, not because there was deliberate selection against brown beans, but rather simply by the iterated effects of “chance” events.

The iterated effects of the blind sampling of beans from a jar are analogous to what Wright suggested sometimes occurs in nature. Starting from a population consisting of certain frequencies of alleles, simply by chance some demes (i.e., subpopulations) within a biological population may come to have gene frequencies that are atypical of the population as a whole. As individuals within these demes reproduce, allelic frequencies among the demes may drift even further apart. Over a number of generations the results of these nonrepresentative sampling events can accumulate, yielding a significant change in gene frequencies in a population.

Adaptive Peaks and Intergroup Selection

Wright used the idea of random genetic drift to solve two problems. One was explaining the apparently nonadaptive characteristics of some organisms (e.g., banding patterns on snail shells). Natural selection can account for adaptive differences between local strains by assuming that each strain has undergone adaptation with respect to its slightly different environment. But there seem to be cases in which subspecific differences cannot be attributed to adaptive divergence, and hence cannot be explained simply on the basis of selection. Wright suggested that such characteristics might be the result of a random drifting of gene frequencies, rather than the products of natural selection: "The actual differences among natural geographic races and subspecies are to a large extent of the nonadaptive sort expected from random drifting apart" (Wright 1931, p. 127). Nonadaptive variation among populations could be explained simply as the result of random genetic drift, without the need to appeal to any other mechanism.

The other major biological phenomenon Wright was interested in explaining is the rate of evolutionary change, especially the rate at which one could expect evolutionary novelties to arise. This includes both the rate of adaptive evolution within species as well as rates of speciation (i.e., the creation of new species). One of the common results of random genetic drift is that a species comes to consist of many demes (interbreeding local populations) differing in genetic makeup. Some of these demes may have especially favorable interaction systems of genes (genes that work well together), causing them to expand and send out migrants that interbreed with the members of other demes, thus transforming the latter groups into demes genetically similar to the original colonizing subpopulation. Wright described this process in a 1932 paper using the fertile metaphor of "adaptive peaks":

> With many local races, each spreading over a considerable field and moving relatively rapidly in the more general field about the controlling peak, the chances are good that one at least will come under the influence of another peak. If a higher peak, this race will expand in numbers and by crossbreeding with the others will pull the whole species toward the new position. (Wright 1932, p. 363)

Thus, just as evolution can proceed more rapidly when there is greater phenotypic variation on which to operate, so too, Wright suggested, evolution can proceed more rapidly when there is higher-level variation for selection to operate upon, that is, variation between demes within a population. If a species is divided into a number of small, semi-isolated demes, then gene frequencies within each population can undergo random genetic drift, thus allowing a population to cross an "adaptive valley" and then be pulled up a different, and perhaps higher, "adaptive peak." In this way the evolution of a species can proceed much faster than if a species is affected only by selection tending to hold it tightly to whatever adaptive peak it is at presently.

Clearly, the role of *groups* in this theory is critical. Through drift, groups within a population come to display different combinations of gene frequencies which do not reflect differences in adaptedness between individuals within each group. As a result, differences between groups come to be more important than differences within groups, and there arises the possibility for selection to act on groups as such. Selection among groups was then held to be responsible for increasing the average adaptiveness of the species to which these groups belong. As Wright put it in his 1932 paper, "The average adaptiveness of the species thus advances under intergroup selection, an enormously more effective process than intragroup selection" (Wright 1932, p. 363).

In the same paper, however, Wright went on to argue that intergroup selection could form the basis for speciation (i.e., the formation of new species), a process that is not necessarily adaptive from the viewpoint of individual organisms:

> It need scarcely be pointed out that with such a mechanism complete isolation of a portion of a species should result relatively rapidly in specific differentiation, and one that is not necessarily adaptive. The effective intergroup competition leading to adaptive advance may be between species. . . . That evolution involves nonadaptive differentiation to a large extent at the subspecies and even the species level is indicated by the kinds of differences by which such groups are actually distinguished by systematists. It is only at the subfamily and family levels that clear-cut adaptive differences become the rule. The principle evolutionary mechanism in the origin of species must then be an essentially nonadaptive one. (Wright 1932, pp. 363–64)

According to Wright, it is the nonadaptive differentiation of local populations that ultimately results in the greater adaptedness we so often encounter: "The nonadaptive differentiation of small subgroups and the great effectiveness of subsequent selection between such groups as compared with that between individuals seem important factors in the origin of peculiar adaptations and the attainment of extreme perfection" (Wright 1931, pp. 153–54). Ironically, because of this wedding of nonadaptive chance elements in evolution with the adaptation-producing power of natural selection, Wright's shifting balance theory could later be embraced even by those squarely in the Fisherian (adaptationist) tradition. In fine Hegelian fashion, simple adaptationist and nonadaptationist perspectives had been synthesized into something far more interesting.[3]

Adaptation in the Modern Synthesis

Views on the extent and nature of adaptation continued to develop during the "Modern Synthesis," with a pronounced swing toward the "Wallace" end of the Darwin-Wallace spectrum of views. This trend has been dubbed "the hardening of the synthesis" by Stephen Jay Gould (1983).[4] According to Gould, at the beginning of the Synthesis in the late 1930s and early 1940s, natural selection was recognized as just one of a range of evolutionary processes shaping evolution, with other agents (e.g., random genetic drift) also recognized as important. By the late 1940s and thereafter, however, the dominant view in evolutionary biology "hardened" in favor of the view that selection-driven adaptation was the preeminently important factor: "As the synthesis developed, the adaptationist program grew in influence and prestige, and other modes of evolutionary change were neglected, or redefined as locally operative but unimportant in the overall picture" (Gould 1983, p. 78). Gould collects interesting evidence to support his claim, especially by comparing successive editions of important texts by Theodosius Dobzhansky, George Gaylord Simpson, and David Lack. Dobzhansky and Simpson, in particular, were initially pluralists, accepting the importance of a range of evolutionary forces, but gradually came to espouse the view that selection is the most important evolutionary force.

Dobzhansky, Simpson, and Lack

Arguably the most important text in this period was Dobzhansky's *Genetics and the Origin of Species,* which appeared in three editions in 1937, 1941, and 1951. The central claim of the book in all editions (a claim that is perhaps the defining element of the Modern Synthesis) is that large-scale macroevolutionary changes (changes above the species level) can be understood and explained as arising from microevolutionary processes (changes taking place within species), namely, in terms of known (or knowable) genetic mechanisms. As Gould demonstrates, while Dobzhansky continued to accept the importance of other evolutionary forces alongside selection (e.g., genetic drift), the importance of selection, and of adaptive differences between species, became increasingly pronounced in each subsequent edition, representing his "increasing faith in the scope and power of natural selection and in the adaptive nature of most evolutionary change" (Gould 1983, p. 78).

Whereas Dobzhansky was primarily interested in integrating the results of theoretical population genetics with empirical data on gene frequencies in natural populations, Simpson's concerns were with the overall rate and shape of evolutionary change as reflected in the fossil record, concerns reflected in the title of his seminal 1944 work, *Tempo and Mode in Evolution.* Simpson embraced the emerging evolutionary synthesis as a crucial antidote to the "mysticism" (as he saw it) that previously infected evolutionary theorizing. Appeals to obscure notions like "entelechy," "aristogenesis," and so on, simply referred to unknown causes of known phenomena, and always result in "stultification" (Simpson 1944, p. 76). The only credible scientific approach to evolution attempts to fit together what is known about the history of life with principles derived from mathematical population genetics. Reflecting the paleontologist's awareness that it is often impossible to know everything about events of long ago that are only imperfectly preserved in fossil evidence, Simpson took a moderate view of the question of whether all characters have been, or are, adaptive: "In the nature of things it is quite impossible to establish that every single genetic difference between two populations has selective value, and probably some distinctions differ in this respect; but neither is it possible to prove that any are really indifferent, and this is certainly untrue of many and probably untrue of most" (Simpson 1944, p. 78).

For Simpson (no less than for Dobzhansky), Wright's metaphor of the adaptive landscape proved a fertile source of thinking about adaptation. He explicitly adopted but modified Wright's image, pressing it into service to represent, not gene frequencies, but structural variations: "[T]he field of possible structural variation is pictured as a landscape with hills and valleys, and the extent and directions of variation in a population can be represented by outlining an area and a shape on the field" (Simpson 1944, p. 89). Simpson pictured the adaptive landscape in static terms, rather than as a rolling and changing sea of crests and troughs. As Ruse notes, "One consequence of this way of thinking was that Simpson was inclined to suppose that there are ecological niches waiting to be occupied as soon as a group has climbed a particular adaptive peak. In other words, adaptive success is not just a relativistic phenomenon; it really represents something 'out there,' in nature" (Ruse 1996, p. 422). The adaptive landscape, so conceived, represents the possibilities and limitations of evolutionary change. As Simpson noted in a later, "popular" treatment of evolution, on the one hand the possibilities are enormous: "Over and over again in the study of the history of life it appears that what can happen does happen." But, by contrast, evolution does not allow any and all conceivable developments: "What can happen is always limited and often quite strictly limited. Boundless opportunity for evolution has never existed" (Simpson 1949, pp. 160–61). Driven by natural selection, evolution ranges over an enormous, but ultimately bounded, landscape of structural possibilities.

A somewhat more adaptationist perspective is evident in Simpson's book *The Major Features of Evolution* (1953), which was essentially a revision of *Tempo and Mode in Evolution.* Chapter VI of this book is devoted to adaptation. While once again expressing agnosticism on the issue of whether all characteristics are adaptive, he nonetheless sounds distinctly more adaptationist than in his earlier work. Summing up, he writes: "The preceding discussions have led to the conclusion that most evolution involves adaptation. Absolutely or relatively inadaptive phases occur and organisms develop nonadaptive and inadaptive characteristics, but overall patterns of evolution are predominantly adaptive and adaptation has been seen to be the usual orienting relationship even in minor details of pattern" (Simpson 1953, p. 199).

The works of Dobzhansky and Simpson are suggestive of a shift toward a more adaptationist perspective as the Synthesis developed. Perhaps the most striking evidence in support of Gould's "hardening" thesis, however, appears in the works of David Lack. In 1939 he began a research monograph on the Galapagos finches. Because of the outbreak of war, it would not be published for another six years (Lack 1945). In it he supported the then-common view that small-scale differences (e.g., in beak morphology) between birds on different islands were largely nonadaptive. In his 1947 book *Darwin's Finches,* however, he continued to defend nonadaptationism for many small-scale differences, but also introduced an adaptive interpretation for many others, including the claim that beak differences are adaptive responses to the different food items available on each island. When the book was reissued in 1960 as part of the Darwin centennial, Lack added a one-page preface to renounce his earlier nonadaptationist view:

> The reader may . . . be reminded that this text was completed in 1944 and that, in the interval, views on species-formation have advanced. In particular, it was generally believed when I wrote the book that, in animals, nearly all of the differences between subspecies of the same species, and between closely related species in the same genus, were without adaptive significance. I therefore specified the only exceptions then known and reviewed the various ideas as to how nonadaptive differences might have been evolved. Sixteen years later, it is generally believed that all, or almost all, subspecific and specific differences are adaptive, a change of view which the present book may have helped to bring about. Hence it now seems probable that at least most of the seemingly nonadaptive differences in Darwin's finches would, if more were known, prove to be adaptive. (Lack 1960, p. v)

The shift from toward a more adaptationist perspective could hardly be more striking.

Other Synthesists

Gould builds his case primarily on the works of Dobzhansky, Simpson, and Lack. Provine (1983) has shown a similar trend with regard to the views of Sewall Wright and Ernst Mayr. In later life Wright became convinced that many features he had previously thought of as nonadaptive actually had adaptive significance, and even went so far as to claim that he had never emphasized the nonadaptive aspects of the shifting balance theory, despite the fact that, as we have seen, statements from his papers of the early 1930s indicate clearly that this was one of his main claims. As for Mayr, in *Systematics and the Origin of Species* (1942), he wrote:

> It should not be assumed that all the differences between populations and species are purely adaptational and that they owe their existence to their superior selective qualities. . . . Many combinations of color patterns, sports, and bands, as well as extra bristles and wing veins, are probably largely accidental. . . . We must stress the point that not all geographic variation is adaptive. (Mayr 1942, p. 86)

By 1963 his view had become slightly more adaptationist:

> Each local population is the product of a continuing selection process. By definition, then, the genotype of each local population has been selected for the production of a well-adapted phenotype. It does not follow from this conclusion, however, that every detail of the phenotype is maximally adaptive. . . . Yet close analysis often reveals unsuspected adaptive qualities even in minute details of the phenotype . . . (Mayr 1963, p. 311)

Interestingly, Mayr himself later took exception to Gould's hardening thesis (Mayr 1988, p. 528). Nonetheless, his earlier and later remarks support Gould's general thesis.

Even general theses, however, may admit of exceptions. An apparent exception to Gould's "hardening thesis" is Julian Huxley. The only difference between the first (1942) and second (1963) editions of Huxley's *Evolution: The Modern Synthesis* is a very brief new introduction, containing no hint of a change in his view of evolution. Another apparent exception to Gould's general thesis is G. Ledyard Stebbins, who maintained that on the basis of Dobzhansky's work biologists are "no longer justified in assuming either that all characters are adaptive and can be demonstrated as such or that character differences must be considered nonadaptive and not influenced by natural selection until the basis of selection has been discovered and proved" (Stebbins 1950, p. 119). Instead, "neither the adaptive nor the nonadaptive quality of a particular character should be assumed unless definite evidence is available concerning that character" (Stebbins 1950, p. 119).

"The Cutting Edge of Adaptationism"

Identifying a trend is one thing. Explaining it is another. Gould himself admits to being puzzled by the trend he identifies, but part of the explanation is undoubtedly the fact that a number of celebrated cases of "nonadaptive evolution by random drift" were later successfully reinterpreted as adaptations forged by natural selection. For example, in 1940 Charles Diver had concluded from his study of differences among closely related species of the land snail *Cepaea* living in the same area that "selective forces and adaptive values have played little direct part in these specific differentiations," and conjectured that the most probable explanation was random differentiation of the sort that Wright described (Diver 1940, p. 327). In other words, interspecific differences in banding and color patterns were governed mainly by drift, and hence were nonadaptive. Divers's interpretation was criticized by A. J. Cain and Phillip Sheppard in the early 1950s, who argued that the seemingly insignificant differences found among snails were strongly correlated with different local (micro-)environments, such that the particular colors and patterns provided protective camouflage providing a selective benefit (Cain and Sheppard 1950, 1952). Based on a more extensive study of the European land snail *Cepaea nemoralis*, they concluded that "The proportions of different varieties vary considerably from one colony to the next. It has been claimed that this situation is due to genetic drift. We have shown that this snail is subject to strong visual selection by birds, which results in a correlation between the varietal composition of each colony and the exact background on which it lives" (Cain and Sheppard 1954a, p. 114). What these studies appeared to show was that even seemingly trivial differences between species (or varieties) could be given plausible adaptationist explanations.

In the 1950s, the work of Cain and Sheppard represented "the cutting edge of adaptationism" (Ruse 1996, p. 457). It can, of course, be a very real problem to determine whether a given characteristic is an effect of selection or of drift. Cain admitted that drift occurs and that given certain ecological conditions, it will produce an apparently random pattern of variation. But he lamented the fact that some authors had employed drift to explain every example of variation for which they could not envision an adaptive explanation. He was blunt in his criticism of this approach: "This procedure is wrong. They have not proved drift to be acting, but have failed to prove that selection is acting, and have invoked drift to cover the failure. An explanation which depends for its success on the failure of the investigator cannot be regarded as satisfactory . . ." (Cain 1951a, p. 1049). The fundamental problem is that some biologists are far too hasty in explaining features of organisms as the result of drift, rather than pursuing the inquiry further: "This is the real basis for every postulate of random variation or (more recently) random drift. The investigator finds that he, personally, cannot see any correlations [with environmental factors] in a given example of variation, and concludes that, therefore, there are none" (Cain 1951b, p. 424). In view of the complexity of living things and their environments, however, a more cautious approach should be used. This is especially so, since "every supposed example of random variation that has been properly studied has been shown to be non-random . . . " (Cain 1951b, p. 424). Consequently, those characters or variations that have been described as nonadaptive or random should more properly be reclassified as *uninvestigated.* "One must not assume randomness (or selection) without proof" (Cain 1951b, p. 424). Indeed, Cain went even further: "[I]t is doubtful whether any example of variation in Nature can be so completely analysed that, after selective effects have been estimated, the residual variation can be ascribed with confidence to genetic drift. There is always the possibility, indeed the likelihood, that the analysis of selective effects was incomplete" (Cain 1951a, p. 1049). Wallace would be proud.

"The Perfection of Animals"

These themes were echoed by Cain a decade later in a long essay with the provocative title "The Perfection of Animals" (1964). In places the essay reads like a manifesto for adaptationism. Failure to see the adaptive or functional significance of some feature is far more likely to be due to our own abysmal ignorance, Cain argued, than to the feature being truly nonadaptive, selectively neutral, or functionless. He insisted that everything that is known about the power of natural selection and the nature of evolution suggests that organisms are as well adapted as they need to be for their distinctive modes of life (Cain 1964, p. 37).

Cain's work on land snails focused on the functional significance of seemingly trivial characteristics like slight differences in color and banding patterns. But what about the fundamental "body plans" of living things? Are the fundamental structural designs associated with vertebrates, or tetrapods, or radially symmetrical organisms, best understood as adaptations, or as primordial blueprints that, once initiated, cannot be altered except in the most superficial ways? Are the basic organizational patterns of living things themselves adaptations, or do they instead constitute the boundary conditions within which selection can play a significant, but less dramatic, role? Cain leaves the reader in no doubt that "the major plans of construction shown by the older groups are soundly functional and retained merely because of that. The phyla and classes are the main possible ways of living in the face of competition from each other. Their plans are adaptive for broad functional specializations; the particular features of lesser groups are . . . adaptive for more particular functions" (Cain 1964, p. 37). In other words, it is all adaptation from the most basic structural plans or organisms to the seemingly most insignificant.

Cain's position can be fruitfully compared with that of Darwin and his contemporary Richard Owen. Owen had asserted in no uncertain terms that the pentadactyl limbs of dugongs, moles, bats, monkeys, and humans could not be explained as expressly designed for their respective ways of life, since the ways of life in question are so very different. Instead, commonality of structure was evidence of the "unity of type." Recall that according to biological idealists such as Owen, organismic design is most evident, not in the intricate adaptive details of each living thing, but instead in the basic structural plans that underlie the diversity of organic forms. God created all living things according to a small number of basic blueprints, which then adapt (imperfectly) to each different environment. In this view, there is no reason to expect all structures to be perfectly adapted to their functions, and indeed one would expect, given the different environments occupied by organisms with essentially the same structural plans, significant imperfections in adaptive fit. This is, of course, precisely what we do find.

Whereas Owen appealed to "unity of type," Darwin invoked "common descent," arguing that common ancestry provided the key to understanding commonality of organismic structure: "If we suppose that the ancient progenitor, the archetype as it may be called, of all mammals, had its limbs constructed on the existing general plan, for whatever purpose they served, we can at once perceive the plain signification of the homologous construction of the limbs throughout the whole class" (Darwin 1859, p. 435). According to Darwin, marine mammals have pentadactyl flippers, not because of a mysterious "unity of type," but because they are descended from ancestors with pentadactyl limbs. The explanation for this follows directly from his theory of descent: "[T]he chief part of the organisation of every being is simply due to inheritance; and consequently, though each being assuredly is well fitted for its place in nature, many structures now have no direct relation to the habits of life of each species" (Darwin 1859, p. 199).

In the *Origin* Darwin had asserted that "Nothing can be more hopeless than to attempt to explain this similarity of pattern in the members of the same class, by utility . . ." (Darwin 1859, pp. 434–35). Hopeless or not, Cain didn't shy away from offering an adaptationist explanation of precisely this striking feature of living things:

> [W]here we are dealing with structures which have persisted for hundreds of millions of years in hundreds of billions of individual life-histories, and which are still so little understood from a functional point of view, it is a very rash assertion that they are merely ancestral. . . . The flipper of a seal, for example, is not used merely as a simple flat plane: it executes complicated movements during swimming involving bending both along and across the axis. It is still used to some extent for movement on land. The use of the ends of the digits, when bent, for scratching may be of great importance in dislodging settlers. (Cain 1964, p. 46)

Not surprisingly, Cain does not offer a similar adaptive story for the vestigial pelvic bones of whales, but instead optimistically maintains that: "Every fresh piece of work that bears on function at all shows us again and again functional significance where we might not have expected it and highlights our vast ignorance about almost all living things" (Cain 1964, p. 46).[5]

Critiquing "the Adaptationist Programme"

Cain's undiluted confidence that all characteristics of organisms can and should be given an adaptationist explanation represented a widespread trend for which he was just one of the most explicit proponents. Many biologists simply assumed that such a position was correct without bothering to provide explicit justification for it. In some respects the situation in late-twentieth-century Darwinism came to resemble that of Darwinism a century earlier in which (to recall Julian Huxley's memorable description, quoted earlier) "[T]he paper demonstration that such and such a character was or might be adaptive was regarded by many writers as sufficient proof that it must owe its existence to Natural Selection. . . . Paley *redivivus*, one might say, but philosophically upside down, with Natural Selection instead of a Divine Artificer as the *Deus ex machina*" (Huxley 1942, p. 23).

"The Panglossian Paradigm"

Extreme positions tend to provoke backlashes, and this one was no exception. The *locus classicus* for this round of attacks on adaptationism was an influential essay by Stephen Jay Gould and Richard Lewontin with the cryptic but intriguing title – "The Spandrels of San Marco and the Panglossian Paradigm: A Critique of the Adaptationist Programme" (1979). The first part of the title comes from a comparison they drew between certain architectural features of St. Mark's Cathedral in Venice, and the methodology of ultra-adaptationist evolutionary biologists. Spandrels are described as the tapering triangular spaces formed by the intersection of two rounded arches meeting at right angles, and as the necessary architectural byproducts of mounting a dome on rounded arches. Each of the spandrels in St. Mark's Cathedral is decorated with a Christian theme. If one was ignorant of the architectural necessity of spandrels for domed buildings, one might suppose that they existed in order to provide spaces for the depiction of religious themes. But this would be to reverse the order of explanation. The spandrels exist for architectural reasons, and are then pressed into service for decorative or other purposes. The fact that they provide suitable surfaces for Christian iconography is no part of the explanation for their existence. Gould and Lewontin claim that many evolutionary biologists make an analogous mistake in their analysis of organisms. They uncritically assume that every feature of an organism exists because it serves some adaptive purpose, thereby ignoring the "architectural constraints" that delimit the structures of organisms.

The second part of the title of their essay refers to "Dr. Pangloss" in Voltaire's satire *Candide*, who assumed that whatever exists (including earthquakes and all the rest), does so because it is for the best (rather than that such things had to exist for some other reasons). So, too, Gould and Lewontin maintain, evolutionary biologists exhibit unlimited "faith in natural selection as an optimizing agent" (Gould and Lewontin 1979, p. 147). They too often act like Dr. Pangloss when they assume that every characteristic of living things exists because it is best for them to have precisely those characteristics. Needless to say, comparing many of their colleagues to Dr. Pangloss was not intended as a compliment.

The "adaptationist programme" that was the object of their attack had, they claimed, committed a number of scientific sins during the forty years it had dominated Anglo-American evolutionary biology. Three objectionable features of the approach, in particular, are singled out for critique. First, adaptationists begin by atomizing organisms into discrete "traits," and then devise an adaptationist story for each trait considered separately, as if organisms consisted of discrete characteristics that could be mixed and recombined in any possible combination, without regard for the range of constraints that limit phenotypes. For adaptationists, the only brake on the perfection of each trait considered separately are tradeoffs among competing selection pressures. "Any suboptimality of a part is explained as its contribution to the best possible design for the whole. The notion that suboptimality might represent anything other than the immediate work of natural selection is usually not entertained" (Gould and Lewontin 1979, p. 151). Overall nonoptimality is thereby also accounted for in terms of adaptation.

Second, by simply assuming that all characteristics are adaptive, adaptationists fail to distinguish between the current utility of a characteristic and the evolutionary reasons for that characteristic's existence. "This program regards natural selection as so powerful and the constraints upon it so few that direct production of adaptation through its operation becomes the primary cause of nearly all organic form, function, and behavior" (Gould and Lewontin 1979, pp. 150–51). Constraints upon the power of selection are either ignored or, if acknowledged, are just as quickly dismissed as unimportant. A telltale symptom of this deficiency is the failure to consider various nonadaptationist explanations.

Finally, Gould and Lewontin charge that adaptationists rely upon plausibility alone as a criterion for accepting speculative adaptationist explanations. The standard adaptationist methodology is to offer an adaptive story for a given trait. If one adaptive explanation fails (e.g., if it can be shown that the supposed adaptive benefit is not really one), then try another adaptive explanation, and so on, until one succeeds (or until one hits upon an explanation that can't be shown to be false, thus rendering adaptationist explanations unfalsifiable). There is an obvious problem with this strategy: "Since the range of adaptive stories is as wide as our minds are fertile, new stories can always be postulated" (Gould and Lewontin 1979, p. 153). As a weaker version of this, even if no plausible adaptive purpose for a given trait can be discovered, assume that there is such an adaptive purpose, which perhaps eludes us because of an imperfect understanding of the environmental conditions under which an organism lives. As a last resort, speculate about how a given structural feature might now be used, even if that is not the adaptive explanation for how that feature arose in the first place. Thus, the sole criterion when constructing an adaptationist explanation becomes *consistency* with natural selection. But this makes it far too easy to concoct adaptationist stories. Plausible adaptationist stories can *always* be told, but in the absence of clear, agreed-upon criteria for discriminating between explanations (adaptationist or otherwise), there is no way to know whether a given explanation is the correct explanation for a given feature. Like the "Just So" stories Rudyard Kipling wrote to explain the elephant's trunk or other characteristics of animals, adaptationist stories might be entertaining to concoct or to read, but they should not be mistaken for scientifically credible explanations.[6]

Critiquing the Critique

Gould and Lewontin's critique has generally been acknowledged as a much-needed corrective to uncritical adaptationist thinking. It has certainly made biologists more self-conscious about what they are doing when they propose adaptationist explanations. But even critics have their critics. For example, rather than conceding in the wake of Gould and Lewontin's critique that adaptationism should be abandoned as a methodology in evolutionary biology, Ernst Mayr offered biologists advice on how best to *carry out* "the adaptationist programme." According to Mayr, thanks to Darwin's demonstration of the efficacy of natural selection, we have perfectly good reasons for assuming (in the absence of positive evidence to the contrary in particular situations) that organisms are very well adapted to their environments. In a sense, natural selection becomes the null hypothesis for any given characteristic biologists wish to explain. Consequently, the evolutionist is warranted in beginning with the assumption that natural selection has been operative, and that the traits being considered are adaptive, until forced by the evidence to relinquish this hypothesis: "He [the evolutionist] must first attempt to explain biological phenomena and processes as the product of natural selection. Only after all attempts to do so have failed, is he justified in designating the unexplained residue tentatively as the product of chance" (Mayr 1983, p. 326).

More recently, Daniel Dennett (1995) has taken Gould and Lewontin to task for what he sees as exaggerations and misrepresentations in their essay. For example, Gould and Lewontin emphasize constraints that limit selection's ability to forge adaptations. Dennett agrees that awareness of constraints is important, but points out that, ironically, "Good adaptationist thinking is always on the lookout for hidden constraints, and in fact is the best method for uncovering them" (Dennett 1995, p. 261). The procedure is familiar to most biologists. Imagine what might be an optimal solution to a particular biological problem, examine the actual (nonoptimal) solution nature has hit upon, and then start asking questions about why the optimal solution was *not* achieved. Rather than being an alternative to adaptationism, recognition of constraints is part and parcel of sophisticated adaptationist thinking.

Dennett also takes on Gould and Lewontin's talk of "spandrels." In their account, the spandrels of St. Mark's Cathedral exist only as architectural necessities, because there is no other way to mount a domed roof on arches. But as Dennett points out, there are in fact any number of other ways of doing this. For example, the architect could use "squinches" instead of spandrels. Even if he does use spandrels, there are several different sorts to choose from. As it turns out, the sort of spandrel chosen for St. Mark's – having a curved, smooth surface – possesses characteristics of more than merely aesthetic significance. Being curved, it is the structure having close to the "minimal-energy" surface needed to fill the space between the arches. Being smooth, it is ideally suited as a surface upon which mosaic images can be displayed, which is after all one of the reasons the cathedral was built in the place. The fact that the architects of St. Mark's Cathedral chose spandrels over squinches, and that they chose one sort of spandrel over the others available to them, rather than demonstrating the necessity of using precisely that architectural structure (as Gould and Lewontin contended), instead demonstrates that the architects had options amongst which they chose, and chose the one that best suited their plans.

Accordingly, Gould and Lewontin's critique of "the adaptationist programme" fails. The essential problem with their argument is that "from the outset Gould and Lewontin invite us to contrast adaptationism with a concern for architectural "necessity" or "constraint" – as if the discovery of such constraints weren't an integral part of (good) adaptationist reasoning" (Dennett 1995, pp. 269–70). But this is simply false. "The conclusion is inescapable: the spandrels of San Marco aren't spandrels [in the sense of architectural constraints that exist as they do just because they have to]. . . . They are adaptations, chosen from a set of equipossible alternatives for largely aesthetic reasons. They were designed to have the shape they have precisely in order to provide suitable surfaces for the display of Christian iconography" (Dennett 1995, p. 274). As with the analysis of church architecture, so too with the analysis of organismal design. Beginning with the assumption that organisms are well designed is fundamental for research in evolutionary biology: "Adaptationist reasoning is not optional; it is the heart and soul of evolutionary biology. Although it may be supplemented, and its flaws repaired, to think of *displacing* it from central position in biology is to imagine not just the downfall of Darwinism but the collapse of modern biochemistry and all the life sciences and medicine" (Dennett 1995, p. 238). According to Dennett (and Mayr), adaptationism is, and ought to be, here to stay.

Summary: Adaptation After Darwin

When Darwin wrote that "[Discovering] the use of each trifling detail of structure is far from a barren search to those who believe in Natural Selection" (Darwin 1862, pp. 351–52), he was both expressing confidence in the power of natural selection and endorsing a certain *methodological strategy* for studying living things, one that essentially presupposes the belief that many, if not most, features of organisms are adaptations explainable with reference to natural selection. The popularity of adaptationist thinking has alternately waxed and waned since Darwin. In the late nineteenth and early twentieth centuries, Wallace and other Darwinians proposed adaptationist explanations for a wide range of biological phenomena. However, many biologists remained unconvinced, and proposed various non-Darwinian evolutionary theories to account for phenomena (e.g., nonadaptive characteristics and trends leading to extinction) that seemed difficult or impossible to explain in terms of natural selection. Darwinism emerged from this period of "eclipse" in the early decades of the twentieth century as Mendelian genetics was fused with the theory of natural selection to form the neo-Darwinian theory of evolution. Sewall Wright's work, in particular, proved seminal for understanding how nonselective forces could actually enhance the power of natural selection to produce rapid adaptive evolutionary change.

By mid-century, adaptationism was regnant once again, so much so that, looking back, some commentators have detected a "hardening" of the once-pluralistic character of Darwinism into a rigid orthodoxy. By the late 1970s, the pendulum was due to swing back in the other direction – or so some critics of "the adaptationist programme" hoped. However, adaptationism proved to be more resilient (or from the perspective of its critics, more recalcitrant) than its critics had foreseen. Nonetheless, the critique of "the adaptationist programme" succeeded in "problematizing" adaptationism and thereby making biologists more circumspect, and reflective, in framing evolutionary explanations. Contemporary controversies about the meaning and status of adaptationism are the heirs of these earlier developments. It is to these that we turn next.

6

Adaptation(ism) and Its Limits

> Look round the world: Contemplate the whole and every part of it: You will find it to be nothing but one great machine, subdivided into an infinite number of lesser machines. All these various machines, and even their most minute parts, are adjusted to each other with an accuracy which ravishes into admiration all men who have ever contemplated them.
>
> (David Hume, 1779)

Introduction

Writing in the eighteenth century, the Scottish philosopher David Hume was duly impressed by the order, harmony, and apparent design of the natural world. It seemed to him, as it did to the vast majority of his contemporaries, to be a world that bespoke the activity of a wise Deity who arranged its various parts to function together with awe-inspiring precision.[1] Then, as now, the most impressive instances of nested sets of machines within machines were living things, in comparison with which whatever other "machines" the universe consists of pale in comparison.

Hume could not have foreseen how much more detailed our knowledge of living things would become in the following two centuries. We now understand, in ways Hume could have only dimly imagined, how intricately adapted these living machines are to their environments. With such marvelous adaptations in abundance, it is hard not to be impressed. But exactly how impressed should we be? Reflecting on Hume's remark, Cronin (1991, p. 23) asks exactly the right question: "Adaptations, in Hume's delightful phrase, 'ravish into admiration all men who have ever contemplated them'. But how ravishing, how perfect should we expect them to be?"

That living things are often superbly designed is beyond doubt. That they are "perfect" (or in the parlance of contemporary evolutionary biology, "optimal"), and what it might *mean* to claim that living things are perfect (or optimal), is far from clear, and raises a range of interesting questions. Can the idea of a "perfect organism" be made coherent? In what sense does selection *explain* adaptation? What are the strengths, and weaknesses, of adaptationist thinking in evolutionary biology? Finally, how might these considerations bear on Darwin's claim that, thanks to the operation of natural selection, "all corporeal and mental endowments will tend to progress towards perfection" (Darwin 1859, p. 489; 1959, p. 758)? All this will take some time to sort out. First, however, it is important to be clear about the meaning of the central idea(s) at issue. "Adaptation" is often discussed as if its meaning were self-evident and unproblematic. In fact, its meaning (and that of associated terms like "adaptationism") has been controversial. In order to arrive at a clear understanding of the status of "adaptation" in contemporary evolutionary biology, it is essential to begin by disambiguating this term. How should adaptations be characterized?

"Adaptation"

To begin with a fundamental (and unproblematic) distinction, "adaptation" refers to both the *process* of becoming better adapted (i.e., fitted to the environment) as well as to the *product* of this process (i.e., organisms with particular characteristics). As already noted, the key innovation of Darwin's theory was to conceive of adaptation in *phylogenetic* rather than in *ontogenetic* terms. Whereas in Lamarck's theory individual organisms within each generation adapt to their environments and then pass on beneficial changes to their offspring, in Darwin's theory the changes that matter for evolution are introduced in the transition from one generation to the next in the form of random variations. Those variations that prove to be more beneficial tend to be passed on differentially to the next generation, leading to a process of adaptation from one generation to the next. "Adaptation" can therefore refer either to a process of change associated with the operation of natural selection, or to the product of that process. Context is usually sufficient to indicate which of these meanings is intended. In the following, "adaptation" should be understood in the sense of *product*, unless otherwise specified.

Historical versus Engineering Definitions

This much is unproblematic. Problems arise once we consider more closely the meaning of adaptation as a product of evolutionary forces. Two quite different conceptions characterize discussions of "adaptation." According to the historical (or "selection-product") view, adaptations are characteristics that exist because they conferred an advantage on the ancestors of their possessors. For example, to say that wings are adaptations is to say that wings exist (now) because having wings conferred an advantage on the ancestors of today's winged creatures. Today's winged creatures have wings *because* having wings benefited their ancestors, and this advantageous trait was passed on to them. Because natural selection is held to be the evolutionary force most directly responsible for such adaptations, some taking this approach have defined "adaptation" more narrowly as any trait that has been produced by natural selection.[2] For example, according to Elliott Sober, "Characteristic ***c*** is an adaptation for doing task ***t*** in a population if and only if members of the population now have ***c*** because, ancestrally, there was selection for having ***c*** and ***c*** conferred a fitness advantage because it performed task ***t***" (Sober 2000, p. 85). In this view, there is an *essential* connection between adaptation and selection, since adaptations just *are* those phenotypic characteristics caused by selection.

By contrast, there is what might be called the engineering (or "good design") view, according to which adaptations are simply traits that are good for their possessors, that is, that provide them with a better fit with their environments, and thus confer a fitness benefit, regardless of how they arose in the first place.[3] The main idea is that adaptations are those characteristics of organisms that convince us that they are well-designed for survival and reproduction. Michael Ruse provides a good statement of the "engineering" model of adaptation: "[W]hat is adaptation all about? What is the essence of an adaptive characteristic? Simply that we have before us something which is design-like" (Ruse 1988, p. 121). Similarly, in George C. Williams's formulation, an adaptation is a characteristic that conforms to "*a priori* design specifications" (Williams 1992, p. 40).

Adaptations versus Adaptive Traits

Clearly these two conceptions will agree much of the time in identifying a given characteristic as an adaptation. Wings, for example, may be viewed as adaptations either because they conferred a fitness advantage on the ancestors of today's winged creatures, or because for their current possessors they provide an excellent design feature. Like many characteristics, wings are adaptations in both the historical and the engineering senses. Other characteristics, however, demonstrate that the two conceptions of "adaptation" can sometimes diverge. For example, consider the human appendix. Notoriously, the appendix can rupture, spewing bacteria into the body, and possibly killing its possessor. The appendix seems to be entirely useless, while imposing a significant risk. Why, then, does it exist? According to the standard account, the appendix exists because at one time it was an organ that housed bacteria necessary for the digestion of cellulose, a substance responsible for the rigidity of cell walls in plants. It would therefore be an adaptation in the historical sense. But, because the appendix no longer serves this function, and sometimes compromises the functional integrity of the human body, it would not be an adaptation in the engineering sense. Likewise, those regions of the human brain that ground the ability to read (or to use a computer) might be considered a characteristic that enhances the fitness of humans in their current environments, and hence would be an adaptation in the engineering sense. But since it is not the case that those regions originally evolved because the ability to read (or to use a computer) was selectively advantageous in preliterate Paleolithic environments, they would not be adaptations in the historical sense. Such examples can be multiplied. Does this mean that "adaptation" must be an equivocal term?

When faced with a problem of this sort, biologists and philosophers instinctively make a further distinction, applying different terms to capture the different sorts of phenomena under consideration. In this case, one can resolve the disagreement by separating the two issues of historical origin and current function, and then distinguishing between "adaptations" (*per se*) and "adaptive traits" (Burian 1983). Along with historical (selection-product) proponents, one can say that a trait is an "adaptation" if it exists because of its causal history; for example, because it was favored in the past by natural selection. A trait could properly be said to be "adaptive," by contrast, if it satisfies criteria for good engineering design by increasing the relative fitness of its present possessor. Calling a trait "adaptive" would thus be to make a claim about its *current functional utility*, regardless of how it came about.

Such a distinction helps to resolve the disagreement between the different senses of "adaptation" by reserving that term for traits that have arisen because of certain processes occurring in the past, while still permitting traits that enhance the fitness of their possessors to be identified as "adaptive." Thus, appendices are adaptations (*per se*), but having an appendix is not (so far as we know) an adaptive trait. It is a classic instance of a vestigial organ, which is another way of saying that it once was, but is no longer, *adaptive*. The ability to read (or to use a computer), on the other hand, is an adaptive trait but not an adaptation. The evolution of those parts of the brain that make such activities possible may be viewed as adaptations, but not as adaptations for *those* particular activities. The abilities to read and to use a computer are by-products of natural selection for other cognitive abilities that are themselves adaptations in the proper sense.

Clearly, being an adaptive trait is neither a necessary nor a sufficient condition for a trait to be an adaptation. Nor does being an adaptation entail that the trait in question is adaptive. The fact that there are so many characteristics that are both adaptations and adaptive (e.g., opposable thumbs, four-chambered hearts, upright posture, etc.) makes it understandable that "adaptations" and "adaptive traits" should sometimes be conflated. But the distinction is real, and it is important to distinguish adaptations from adaptive traits if we are to properly explain phenotypic traits. This becomes evident when one notices what happens when this distinction is not accepted. Some biologists attempt to combine the two conceptions. For example, according to Lauder (1996), "a component of design is an adaptation only if it enhances fitness *and* arose historically as a result of natural selection on that trait for its current function" (Lauder 1996, p. 79; emphasis added). The problem with this is that it restricts adaptations to just those characteristics having current adaptive value. Consequently, if the thermoregulatory theory of feather origination (Ostrom 1974, 1979) is correct, then feathers *qua* flight devices cannot be considered adaptations in this account. A different problem arises if the concepts of adaptation and (optimal) adaptive trait are treated as synonymous. For example, Reeve and Sherman (1993) present a sustained critique of the selection-product view, which they classify, somewhat pejoratively, as a "history-laden definition." According to their preferred "nonhistorical definition . . . an adaptation is a phenotypic variant that results in the highest fitness among a specified set of variants in a given environment" (Reeve and Sherman 1993, p. 1). This definition is explicitly designed to "decouple" (the definition of) adaptation from (identification of) the evolutionary mechanisms that produce them. An odd, and perhaps fatal, consequence of their view is that only optimal phenotypes can qualify as adaptations. Vertebrate eyes, for example, would presumably not be adaptations on this view, because they are not optimal from an engineering-design perspective. The simplest, and most useful, way of resolving this difficulty is to make a fourfold distinction between (i) adaptation as the *process* of becoming better adapted, (ii) an adaptation as the *product* of selection, (iii) an adaptive trait as a phenotypic characteristic that provides a fitness advantage for its possessor, and (iv) an optimal adaptive trait as a phenotypic characteristic that provides a superior fitness advantage for its possessor relative to whatever other phenotypic characteristics are available. This fourfold set of distinctions is presupposed in the discussions which follow.

Exaptations and Evolutionary Tinkering

The distinctions between adaptation as process and adaptation as product, and between adaptation and (optimal) adaptive trait, are important. To further clarify (or complicate?) matters, Gould and Vrba (1982) introduced another term into the evolutionary lexicon: "exaptation." Exaptations are "features that now enhance fitness but were not built by natural selection for their current role" (Gould and Vrba 1982, p. 4). That is, a trait is an exaptation when its *current* adaptive significance differs from its *original* adaptive significance. The classic example of an exaptation is feathers. According to a widely accepted account, feathers originally evolved as modified reptilian scales, and functioned as thermoregulatory devices. But over time they evolved into feathers, whose primary purpose (now) is to assist flight.

Such functional conversions are probably common in evolution. Jacob (1977) contrasts two images of evolution. Evolution has sometimes been compared to an *engineer* who designs organisms of unsurpassable perfection. The problem with this image, Jacob notes, is that it completely misunderstands the nature of the evolutionary process. Whereas an engineer works according to a preconceived plan, and selects materials precisely with an intended end in mind, evolution has no preconceived plan, no preset goal, and can only use whatever materials happen to be available. Evolution is more like a *tinkerer* who does not know ahead of time what it is he will produce, and in working makes use of whatever materials happen to be at hand, sometimes taking objects with one function and pressing them into service to perform a different function. Familiar examples would include the vertebrate ear, whose minute bones were formed out of parts of the jaw, and the lungs of terrestrial vertebrates, which are enlargements of the esophagus. Darwin recognized this feature of evolution quite clearly when he wrote:

> Although an organ may not have been originally formed for some special purpose, if it now serves for this end, we are justified in saying that it is specially adapted

for it. On the same principle, if a man were to make a machine for some special purpose, but were to use old wheels, springs, and pulleys, only slightly altered, the whole machine, with all its parts, might be said to be specially contrived for its present purpose. Thus throughout nature almost every part of every living being has probably served, in a slightly modified condition, for diverse purposes, and has acted in the machinery of many ancient and distinct specific forms. (Darwin 1877, p. 284)

It is, of course, no coincidence that this sounds remarkably like Jacob's description of evolution: "Evolution behaves like a tinkerer who, during eons and eons, would slowly modify his work, unceasingly retouching it, cutting here, lengthening there, seizing the opportunities to adapt it progressively to its new use" (Jacob 1977, p. 1164).

The concept of "exaptation" captures this important "tinkering" feature of evolution. However, it is not without its problems. One sort of problem is epistemic. Often it will be impossible to identify the original adaptive function of some characteristic. As Reeve and Sherman note, "It is virtually impossible to identify the original roles of many traits that are of interest to behavioral ecologists (e.g., complex behavioral traits like mating or social behaviors) owing both to their poor representation in the fossil record and to their plasticity and variability among individuals now and (presumably) in the past" (Reeve and Sherman 1993, p. 3). Gould and Vrba's account also faces the difficulty of discerning precisely where an adaptation ends and the exaptation begins since it is not clear how much the current function of a given trait must differ from its original function to be considered an exaptation. Are human ear bones exaptations because, in addition to their original function of serving as devices for sensing acoustic stimuli, they now also serve to mediate social interactions via telephone conversations? Finally, the very concept of "exaptation" is in danger of becoming vacuous because if one goes back far enough in evolutionary time, nearly every characteristic at one time served a function other than its current one. But if so, then the advantages of distinguishing between adaptations and exaptations becomes unclear.

In principle, Gould and Vrba could simply dismiss epistemic critiques of the adaptation/exaptation distinction as trading on a confusion between ontology and epistemology, that is, between the way something *is* and our ability to *know* how something is. But because the distinction in question is supposed to be a contribution to *science*, which is concerned with improving our knowledge of the nature of the world, such a response would entirely miss the point. Besides, as other critics have noted, the concept of "exaptation" as Gould and Vrba define it, leads to some very odd consequences. In their account, a trait is an exaptation if its *current* adaptive significance differs from its *original* adaptive significance. Thus a trait is an adaptation *only* for the function for which it was *first* selected. For example, suppose that feathers originally evolved for thermoregulation. Feathers are thus adaptations for thermoregulation. But therefore (beliefs to the contrary notwithstanding) they are not adaptations for flight, which only appeared later. Because this counterintuitive consequence arises from the stipulation that only the *first* function served by a trait determines its status as an adaptation, Sterelny and Griffiths (1999) rightly question this restriction:

> [W]hat justifies this special status for the first of many selection pressures? The importance of the concept of adaptation in biology is that it explains the existence of many traits of the organisms we see around us. This explanation is not just a matter of how traits first arose, but of why they persisted and why they are still here today. If we want to understand why there are so many feathers in the world, their later use in flight is as relevant as their earlier use in thermoregulation. (Sterelny and Griffiths 1999, p. 219)

They conclude that "The adaptation/exaptation distinction is not very useful except as an indication of the succession of evolutionary events. A trait is an adaptation for all the purposes it has served and which help to explain why it still exists" (Sterelny and Griffiths 1999, p. 220).

Their point is well taken. However, the concept of "exaptation" is useful for another reason. A common objection to Darwin's theory is that many of the most impressive organismic characteristics only serve their current function in fully developed form. Consequently, when they (hypothetically) first arose they could not have served this function, and hence could not have provided any selective advantage to their possessors. Hence they could not have evolved by natural selection. For example, of what use would be 1 percent of a feather (or a wing), as such a trait would be entirely useless for flight? As an adaptation for flight, wings could never have gotten off the ground, so to speak.

The Darwinian response is twofold. First (as Richard Dawkins has argued repeatedly), in many cases (e.g., vision, cryptic coloration, etc.) 1 percent of a trait may well be advantageous over 0 percent of a trait, and that is sufficient for it to evolve by natural selection. Second, and more to the present point, the concept of "exaptation" makes it clear how a trait could be selectively beneficial for one function, develop more fully as a result of the relevant selection pressures for that function, and then be coopted into use for a quite different function. The concept of exaptation helps to make clear how natural selection can get a jump-start in a certain direction, even when it seems that a small change in a given trait would not be useful for the function which that trait would later serve in the organism's evolutionary history.[4]

Adaptationism

We have been exploring the concept of "adaptation" (and related concepts) in some detail because it is a fundamental organizing concept for Darwinism – so fundamental, that some biologists embrace a doctrine (or collection of doctrines) that places "adaptation" at the center of their evolutionary theorizing. As we saw in the previous chapter, "the adaptationist programme" has come in for a good deal of criticism (and countercriticism). However, questions about the status of "adaptationism" continue. In the remainder of this chapter, we will examine "adaptationism" with the aim of identifying and evaluating its various manifestations.

What Is "Adaptationism"?

A reader delving into the literature on "adaptationism" hoping to find a simple, straightforward account of its central idea is quickly confronted by a number of different claims about what this term *really* refers to. For example, Richard Lewontin (1979a) once defined "adaptationism" as "that approach to evolutionary studies which assumes without further proof that all aspects of the morphology, physiology and behavior of organisms are adaptive optimal solutions to problems" (Lewontin 1979a, p. 6).[5] By contrast, according to Elliott Sober, "Adaptationism, as a claim about nature, is a thesis about the 'power' of natural selection" (Sober 2000, p. 121). More specifically, adaptationism asserts that "Most phenotypic traits in most populations can be explained by a model in which selection is described and nonselective processes are ignored" (Sober 2000, p. 124).[6] Finally, recall that when Ernst Mayr undertook a defense of "the adaptationist program" in response to Gould and Lewontin's critique, he construed it as "a program of research devoted to demonstrate [*sic*] the adaptedness of individuals and their characteristics" (Mayr 1983, p. 325).

Three Kinds of Adaptationism

Clearly, "adaptationism" has meant different things to different people. Disagreements about "adaptationism" are therefore hardly surprising. In attempting to sort out the debates about adaptationism, authors have distinguished several distinct theses that are often conflated. For example, according to Philip Kitcher, "Three separate issues are involved: the possibility of confirming explanations of the presence of traits that appeal to natural selection, the issue of whether evolution inevitably produces the best available phenotype, and the question of the reliability of our guesses about best available phenotypes" (Kitcher 1987, p. 98). Peter Godfrey-Smith (1999, 2001) presents a somewhat different, more elaborated tripartite set of distinctions. "Empirical Adaptationism" is the claim that natural selection is a uniquely powerful and ubiquitous force evolutionary force, fueled by abundant biological variation, such that to a large degree it is possible to predict and explain the outcome of evolutionary processes by attending only to the role played by selection. No other evolutionary factor has this degree of causal importance. "Explanatory Adaptationism," by contrast, is the claim that the apparent design of organisms, and the relations of "fit" between organisms and their environments, are the really "big questions" in biology, and that natural selection is the "big answer" to such questions. Natural selection, therefore, has unique explanatory importance among evolutionary factors. Finally, "Methodological Adaptationism" is the claim that "The best way for scientists to approach biological systems is to look for features of adaptation and good design. Adaptation is a good 'organizing concept' for evolutionary research" (Godfrey-Smith 2001, pp. 336–37). Whereas some writers have argued that "adaptationism should be regarded as a heuristic, not as an hypothesis" (Resnik 1997, p. 48), thus presupposing that if it is the former, then it can't also be the latter, Godfrey-Smith shows that various meanings of "adaptationism" can be found in the literature, and rightly argues that distinguishing them is essential for conceptual clarification.

The Key Issues

The distinctions between Empirical, Explanatory, and Methodological Adaptationism are helpful, and go a long way toward clarifying the otherwise puzzling plethora of "adaptationisms" found in the literature. For the discussion that follows, I want to change the emphases just a bit, focusing on what I take to be the central *controversial* elements of the views identified (but not necessarily endorsed) by Lewontin, Sober, and Mayr (above), namely: (1) the ubiquity of organismic traits as optimal solutions to (past or present) biological problems, (2) the unique explanatory power of natural selection in accounting for adaptive traits, and (3) the scientific fruitfulness of assuming, as a methodological approach, that all or most organismic characteristics are adaptations resulting from natural selection, rather than the products of other evolutionary processes. In the following three sections I will examine these issues in turn, for each considering how the claim in question could be further disambiguated or qualified, and whether, or in what sense(s), the claim in question might be true. Because it bears most closely on the evaluation of Darwin's claim that under the influence of natural selection the characteristics of organisms are "tending to progress toward perfection," I will devote comparatively more space to discussing the first of these adaptationist claims.

Empirical Adaptationism

Are All Phenotypic Characteristics Adaptations?

Organisms can be viewed as suites of phenotypic characteristics. Are all phenotypic characteristics adaptations? Recall that according to the historical (selection-product) view, adaptations are those phenotypic characteristics that result from selection processes. Note that it is not enough that a phenotypic character simply be a "product" of natural selection. In a trivial sense, *all* phenotypic characteristics are products of selection, as from the origin of life forward, selection has (presumably) played *some* role in the evolution of life. The issue is not whether phenotypic characteristics are the products of selection in this general sense but, rather, whether there has been selection *for* that characteristic, not just selection for other characteristics causally connected in some way with the characteristic in question.

As already noted, there seem to be clear examples of phenotypic properties that are *not* adaptations in this sense. The abilities to read and to use a computer provided no selective advantage in the ancestral environments in which the bulk of human evolution occurred, and hence would not be adaptations. But couldn't it be argued that while there was no selection for reading or computer use *per se*, there *was* selection for the cognitive abilities that underlie these activities, and that therefore in a deep sense they either are, or represent, genuine adaptations? Couldn't a distinction be made between "superficial" characteristics like the ability to use a computer, and the really "deep," fundamental phenotypic characteristics that constitute organisms, with the claim being that while the superficial characteristics cannot always be identified as adaptations, the fundamental characteristics always can be so identified?

Unfortunately, this move faces serious problems. If anything is "fundamental" in the constitution of organisms, it is their basic structural plan, for example, radial or bilateral symmetry, number of limbs, types of organ systems, and so on (i.e., *baupläne*, in the sense of Gould and Lewontin 1979). As Sterelny and Griffiths point out, the persistence of basic structural plans in related but ecologically diverse organisms (e.g., moles and bats) may have little or no adaptive significance *per se*. The conservation of these patterns is to be explained in terms of developmental and historical constraints in evolution, not in terms of natural selection. For example, consider the tetrapod structural plan common to many terrestrial animals. Why do most land vertebrates have four legs? It might be tempting to conclude that this structural plan is obviously an adaptation. But this would be to ignore the fact that the fish that were ancestral to terrestrial animals also have four limbs (i.e., fins). Possessing four limbs may be suitable for locomotion on dry land, but the correct explanation lies not in selection for being tetrapodal, but in the fact that the evolutionary predecessors to terrestrial vertebrates possessed the same pattern (Lewin 1980, p. 886). In sum, "The persistence of such similarities over hundreds of millions of years is as striking as the existence of complex adaptations, and it is *not* explained by natural selection" (Sterelny and Griffiths 1999, p. 227).

Against this it could be argued that, rather than settling the issue, appealing to the basic structural plan of ancestors simply pushes the question back to an earlier stage. Given that these structural plans were ancestral to later organisms, why did those plans, and not others, come to be? The answer might well be that the basic structural plan ancestral to land vertebrate structural design was itself the direct product of natural selection designing an organism able to move effectively through, and change direction in, a liquid environment, so therefore the tetrapodal structural plan of land vertebrates, despite being an evolutionary inheritance from fishy ancestors, is nonetheless a *bona fide* adaptation. Recall (from the previous chapter) A. J. Cain's assessment that even the deepest homologies of the vertebrate archetype are adaptations for existing vertebrate species. In other words, the basic structural plans, no less than feathers and coloration, are adaptations. Darwin himself, it is worth noting, was inclined to subsume inheritance of structural plan under the rubric of adaptation. Although he recognized both factors in evolution ("Unity of Type" and "Conditions of Existence," respectively), he thought that the effects of conditions of existence *explained* unity of type: "Hence, in fact, the law of the Conditions of Existence is the higher law; as it includes, through the inheritance of former adaptations, that of Unity of Type" (Darwin 1859, p. 206).

So, is talk of the inheritance of basic structural plans really an alternative to an adaptationist explanation of those same characteristics? Clearly there is a relationship between basic structural plans and selection, in the sense that if a structural plan imposed a sufficiently high fitness cost on its possessors, selection would ensure that species with that plan would go extinct. Structural plans are certainly subject to the operation of natural selection. But is this enough to make them adaptations? I think it is. Describing such characteristics as adaptations is entirely consistent with the claim that many of the most basic features of living things exist now because these features were present in their ancestors, and such features have simply been conserved. So long as they originally arose because they were favored by natural selection, they would merit the appellation "adaptation," even if a (large) part of the explanation for why they exist now appeals to inheritance of structures, and that such structures constrain subsequent evolution. One could contrast constraints with adaptive characteristics, or one could subsume constraints within the domain of adaptive characteristics. Perhaps at least some of the constraints exist precisely because they are adaptations.

It would still not follow, of course, that all phenotypic characteristics are adaptations. Four-chambered hearts are adaptations for pumping blood throughout the body. But the "thumping" sound the heart makes as it beats is not itself an adaptation, because (presumably) there was never any selection for that (or any) cardiac sound. Likewise, blood's ability to deliver oxygen to cells is an adaptation, because (presumably) there was selection for oxygen-carrying capacity. But the redness of oxygenated blood is not an adaptation, because (presumably) there was never any selection for that color rather than for some other color of blood. Consequently, not all phenotypic characteristics are adaptations in the strict sense. But many are.

Are All Adaptive Characteristics Optimal?

Consider next the claim that every *adaptive* phenotypic characteristic is an optimal solution to some current biological problem. As already noted, many phenotypic characteristics will be neither adaptations nor currently adaptive traits (e.g., the redness of oxygenated blood). Even those phenotypic characteristics that are adaptations may not be currently adaptive for their possessors (e.g., the human appendix). But for those phenotypic characteristics that *are* currently adaptive, we can intelligibly ask how perfect they are.

Talk of "perfection" with respect to biological properties might seem entirely out of place. How *could* they (or anything, for that matter) be perfect? Yet such talk is not as odd as it might seem. As already noted, references to the "perfection" of living things abound in Darwin's writings, as when he informs readers of the *Origin* that one of his chief aims in the book is to explain "how the innumerable species inhabiting this world have been modified, so as to acquire that perfection of structure and coadaptation which most justly excites our admiration" (Darwin 1859, p. 3). Still, evolutionary biology has undergone significant changes since Darwin wrote. Given our current understanding of evolution, does it make sense to describe any phenotypic characteristic as "perfect"? Is the idea of a "perfect" phenotypic trait even a coherent idea?[7]

It does seem to make sense to describe some features of organisms as "perfect" (in a carefully specified sense). "Perfect design" is simply the limit notion of *good* design, and would refer to cases where it is difficult or impossible to imagine a *superior* solution to a particular problem. Darwin considered the hexagonal structure of the storage cells in a bee-hive to be a striking example of perfect design, because it appears to be mathematically impossible to find a better solution to the problem of maximizing storage space while minimizing building materials: "Beyond this stage of perfection in architecture, natural selection could not lead; for the comb of the hive-bee, as far as we can see, is absolutely perfect in economising wax" (Darwin 1859, p. 235). Obviously, some qualifications are important. A trait can be considered "perfect" only insofar as no better solution to the problem it solves is conceivable. Hexagonal storage cells in a beehive appear to be a perfect solution to the problem of maximizing storage space while minimizing building materials. Hexagonal cells *that utilize no material at all*, on the other hand, while even better in a sense, are not even conceivable.

Another example of a "perfect" solution to a specified problem might be some cases of mimicry, in which one species copies the appearance and/or behavior of another species, or of some aspect of the abiotic environment. Examples of protective mimicry include the (tasty) Viceroy butterfly that mimics the (toxic) Monarch butterfly; nonvenomous snakes that mimic in their coloration highly venomous snakes; and (nonstinging) flies that closely resemble honeybees. Other examples come from protective camouflage, such as stick and leaf insects that closely resemble the foliage they live on; larva of swallowtail butterflies that resemble bird droppings, and so on. In each case there is a "model" and a "mimic." To the extent that the mimic is indistinguishable *to predators* from the model, to that extent the mimic is "perfect." Such features could not be changed without making the mimic appear less like the model. In these cases we seem to have a clear-cut and even operationally useful notion of "biological perfection."

Whereas Darwin routinely used the word "perfect" to describe some biological characteristics, it is now standard to use the word "optimal," with talk of "perfection" being replaced by talk of "optimality." What does it mean for a trait to be optimal? According to one commonly accepted definition, "an optimal trait is defined as one that maximizes individual fitness relative to other possible variants in the population" (Abrams 2001, p. 274). To be optimally designed is thus to be structured in such a way that no better solution to the set of problems being faced, weighted according to the relative importance of each problem, is possible. Thinking in terms of optimal rather than perfect design entails explicit recognition that adaptations involve costs as well as benefits.[8]

We have already encountered this issue in a different context in Chapter 2. Against Wynne-Edwards's claim that organisms modulate their reproductive output for the good of the group, David Lack drew upon his own studies of clutch size in various species of birds to argue that organisms, sensitive to resource availability, modulate their reproductive output in order to maximize individual fitness in their current environment. It might seem that each bird should lay as many eggs as physiologically possible, in order to maximize the total number of offspring produced. In fact, however, other factors enter into the equation, resulting in a characteristic number of eggs per breeding cycle for each species. Writing before the terminology of "optimality" came into common usage, Lack simply concluded that species-specific clutch size "is an adaptation due to natural selection" and that the corresponding reproductive rates "are as efficient as possible" (Lack 1954, pp. 28, 276). In contemporary terminology, the birds Lack studied optimized their egg-laying behavior in order to maximize successful reproduction. Such examples could be multiplied indefinitely.

Some characteristics can therefore be considered optimal to solutions to specific biological problems. Clearly, however, not all phenotypic characteristics can be optimal, because optimality in one characteristic frequently entails suboptimality in others. Phenotypic traits never come in isolation, and the fate of individual traits is tightly bound with the fate of the entire organism. Is an optimal (or "perfect") *organism* possible? Using standard measures of adaptedness, a perfect organism would be one that lives (and remains reproductively active) forever, converts all (or exactly the right amount) of its energy consumption into reproductive activities, produces viable offspring at an infinite rate, moves through the environment with zero friction, is impervious to enemies or predators, is able to hear (and usefully process) all frequencies of sound waves, see (and usefully process) all wavelengths of electromagnetic radiation, etc. The idea of there actually being such an organism is, of course, absurd. Optimal (or "perfect") organisms, in this sense, are not even possible.[9]

Factors Limiting Optimal Design

The phenotypes of actual organisms always represent compromises and tradeoffs among functional tasks. Optimizing one function entails suboptimal design of others, all within the limits set by various "constraints on perfection" (Dawkins 1982b). Constraints are boundary conditions that circumscribe the range of possible (and hence actual) phenotypes. At perhaps the most basic level are constraints imposed by the properties of the materials available for constructing phenotypes, and the laws of physics (Vogel 1988). The tubular construction of limb bones in land vertebrates is apparently the best possible design for maximizing strength relative to surface area, and in the case of the sauropods, mammoths, and present-day elephants, has proven capable of supporting a tremendous amount of weight. But given the materials composing bone, there are limits to how much weight bones made of that material can support. It is therefore no accident that the largest animal ever to have lived (the blue whale) supports its enormous bulk, not on vertically oriented limbs, but by taking advantage of the natural buoyancy provided by seawater.

Closely related to such physical constraints are what might be called architectural constraints. For example, while it might be advantageous in some respects for insects to be as large as buses, as a matter of fact there are no insects larger than a baseball. The reason apparently has to do with the insects' mode of oxygen transport. Lacking lungs, insects acquire oxygen through openings in their exoskeletons (malphigian tubules). Because this is an inefficient mode of oxygen transport (relative to the use of lungs), there is insufficient oxygen to support a larger body size. (When O_2 levels in the atmosphere were higher, larger insects evolved.) Of course, in a sense this just raises a further question. Assuming that a larger body size might be advantageous for insects, why didn't they evolve the sort of oxygen transport system that would have permitted this additional growth?

This is where the terminology of "design space" becomes particularly useful (Dennett 1995). Think of each organism as occupying a position in a multi-dimensional morphological space. Organismic phenotypes (actual and possible) represent particular sets of values in this hyperspace. To move from point **A** to point **B** in hyperspace is to define a trajectory. Some trajectories are open whereas others (at a particular time) are closed. If point **B** represents a better-adapted state than point **A**, and the trajectory from **A** to **B** is open, and if the necessary variations occur at **A**, then an evolving lineage (i.e., a species) may move from **A** to **B**, thereby assuming that better-adapted state. In the case of the insects, it may be that there was no open trajectory from a design involving malphigian tubules to lungs; that is, a trajectory every stage of which is more adaptive than the preceding one. Switching metaphors, recall Wright's metaphor of the "adaptive landscape" consisting of adaptive peaks and valleys. Because adaptation is always a local affair, unresponsive to either the future or to some global optimum, species will tend to occupy relatively high peaks, but not necessarily the highest peak each is capable of occupying. Since natural selection can never lead species toward locally less advantageous positions on the adaptive landscape, if these intermediary positions are adaptively disadvantageous then insects can never move from point **A** (obtaining oxygen via malphigian tubules) to point **B** (obtaining oxygen by the use of lungs). The intermediate territory between adaptive peaks is untrespassable (so long as natural selection is the sole force operative on the phenotypes in question).[10]

Of course, this raises yet another question: Why couldn't there be intermediate steps between **A** (tubules but no lungs) and **B** (lungs but no tubules) that were more advantageous than **A**? At this point the notion of "developmental constraints" makes an appearance. Organisms succeed (to the extent that they do) as integrated wholes, not as mere collections of individual traits. No matter how advantageous a given trait might be for an organism, taken in isolation, unless its development is consistent with other traits the organism has it will not increase the adaptedness of the organism as a whole. For example, having stronger bones, a layer of insulating fat, and a much bigger brain might all be thought of as advantageous for an organism living in a dangerous, cold, and complex environment. But if the organism is a bird that depends for its livelihood on flight, then there will obviously be limits to how far selection takes such traits before they have a negative impact on what might be considered the bird's primary adaptation(s). Once a certain way of life is acquired by an organism, there will be strong pressure to refine this way of life more (i.e., to specialize) rather than to abandon it for another way of life perhaps already being successfully pursued by other species. Thus phenotypic traits might fail to be optimal because the overall design of the organism limits their further development.[11]

Yet another constraint comes from the limited supply of genetic variability. So long as there is an ample supply of variation for selection to work upon, then given the right variations, strong selective pressure, and sufficient time, almost any phenotype may be produced. In fact, however, genetic variability is never unlimited. Because selection is always an opportunistic affair, making use of what is available, but unable to create the variations it might need to create more perfect organisms and unable to plan for the future, it is hardly surprising that organisms fall short of optimal design. When Darwin remarked in the *Origin* that "natural selection is... silently and insensibly working, whenever and wherever opportunity offers, at the improvement of each organic being in relation to its organic and inorganic conditions of life" (Darwin 1859, p. 84), he understood clearly that selection can only act upon, and thus is limited by, the *opportunities* presented to it in the form of organic variations, however they arise.

Another reason organisms may not be optimally designed is "time lags." Organisms may be thought of "tracking" their environments in evolutionary time. As environments change, species may adapt to the new conditions. But such adaptation is not instantaneous, especially for organisms with long generations. Unless a species' environment remains stable for a considerable time, to some extent organisms will always lag behind the environments they occupy. As Dawkins puts it: "The animal we are looking at is very probably out of date, built under the influence of genes that were selected in some earlier era when conditions were different" (Dawkins 1982b, p. 35). He cites the hedgehog antipredator response of rolling up into a ball as a sadly inadequate defense against motor cars. There are no hedgehogs where I live, but the same is true for the opossums that populate the outlying areas and (too often) the freeways and surface streets of Los Angeles.

Another important kind of constraint might be termed "historical." It has often been noted that some phenotypic properties are clearly far from optimal from a design point of view, and are not the sort of thing one would expect if natural selection acted without constraints in molding phenotypes. For example, the retinas in the eyes of vertebrates are covered with "photocells" (rods and cones) leading to "wires" (nerves) that converge in the optic nerve carrying signals to the visual processing centers in the brain, making vision possible. So far so good. However, the way in which the retina is constructed is puzzling from a design point of view. Rather than the photocells pointing toward the light with the wires leading backwards away from the retina toward the brain, the photocells point away from the light source, with the wires protruding on the side of incoming light. One consequence of this is that the light entering the eye has to pass through a thicket of nerves before encountering the photocells themselves, thus causing at least some attenuation and distortion. Another consequence is that in order for the wires to reach the brain, they must pass through an opening in the retina, a spot at which an image cannot be focused. This is the cause of the "blind spot" present in all vertebrate eyes, clearly a less than optimal solution to the problem of enabling vision. (See Dawkins 1986, p. 93 for discussion.)

If it is easy to imagine a better design than the one characterizing vertebrate eyes, why hasn't selection forged that (better) design? The answer involves considering the trajectory through genetic hyperspace that would have to be traversed in order to turn the retina the right way around, once it had *already* started off in the wrong direction. Suppose that some primitive ancestor to contemporary vertebrates acquired a light-sensitive photocell, and the "wires" from it just happened to be coming out the wrong side. Because this proved more advantageous than not having a functioning photocell at all, it provided some survival advantage for its possessor. Once this advantage was in place, any step backwards towards, say, no functioning photocell at all, would have been selected against. So the process continued to build on its initial relatively advantageous but deeply flawed beginning by adding additionally backwards-wired photocells, eventually resulting in the highly useful but functionally suboptimal vertebrate eyes of today. Initial contingency coupled with selective pressures blind to the future drove the process of eye-building in the direction further along the path to contemporary vertebrate eyes. With each step along the way it became progressively more difficult to go back and rewire eyes in the functionally superior way. Selection can continue to improve the vertebrate eye in the future, but it is unlikely to undertake a fundamental overhaul of its basic design features, flawed though they are. Historical constraints prevented the vertebrate eye from achieving optimal design. As always, Dawkins provides a vivid image: "Like a river, natural selection blindly meliorizes its way down successive lines of immediately available least resistance. The animal that results is not the most perfect design conceivable, nor is it merely good enough to scrape by. It is the product of a historical sequence of changes, each one of which represented, at best, the better of the alternatives that happened to be around at the time" (Dawkins 1982b, p. 46).

In summary, a range of factors limit the degree of perfection actually attained by living things. From the beginning of his evolutionary theorizing to the end, Darwin himself was acutely aware that selection-driven adaptation was but one force shaping organisms. In an early notebook entry he observed that: "The condition of every animal is partly due to direct adaptation and partly to hereditary taint" (Barrett et al. 1987, p. 182). Years later, in the *Origin*, after having thought about the issue for over twenty years, he stated the unavoidable consequence of this fact: "Natural selection will not produce absolute perfection, nor do we always meet, as far as we can judge, with this high standard under nature" (Darwin 1859, p. 202). Natural selection is capable of producing entities of astounding adaptive complexity. Yet in light of the factors operating to limit the products of selection, Darwin marveled that "The wonder indeed is, on the theory of natural selection, that more cases of the want of absolute perfection have not been detected" (Darwin 1859, p. 472).

Explanatory Adaptationism

There is nothing in Darwin's theory to predict that organisms will be optimally designed, and much to suggest that, for a range of reasons, organisms will fall short of optimality. Nonetheless, it is undeniable that one of the most striking features of living things that requires scientific explanation is their amazing adaptive complexity. Natural selection is the only known biological process that could produce such apparent design. Natural selection, therefore, plays a central role in explaining adaptations. This much is uncontroversial. What is controversial, however, is ascribing causal or explanatory sufficiency to natural selection. Is it true, as one version of Explanatory Adaptationism asserts, that adaptations can always be explained by a model in which selection is described and nonselective processes are ignored? The issue here is not whether organisms are, in fact, well adapted to their environments (clearly, often they are), but rather the question of how this adaptive fit (where it exists) is to be explained. The claim being considered is that selection in some sense *uniquely* explains adaptations. Clarifying the sense in which this might be true requires consideration of a broader question: What is the connection between selection and adaptations (and adaptedness)?

The Selection-Product View Revisited

Recall that according to one version of the Historical View, the connection between selection and adaptation couldn't be tighter. For proponents of the "Selection-Product" view, there is an *essential* connection between adaptation and selection, because adaptations just *are* (by definition) those phenotypic characteristics caused by natural selection. The claim is that if a trait is an adaptation, it is the product of natural selection. This view does not entail that all or even most features of organisms are adaptations. It might be true that all adaptations are the result of natural selection, without it being the case that all traits are adaptations. This view also does not entail that any trait that arises directly from a selection process is an adaptation. All sorts of traits may result from the operation of natural selection, including what Darwin referred to as "correlated characters," that is, traits that are causally connected with traits for which there has been selection, without themselves being adaptations. But if a characteristic is an adaptation then, according to this view, it is the product of a selection process.

Fisher (1985) raises the most obvious objection to this view: If we *define* adaptations as products of natural selection, then it becomes impossible to noncircularly *explain* adaptations by citing their selective causes. The question of whether adaptation is caused by selection – rather than by some other process – becomes trivially true, a matter of definition rather than an important empirical fact about the natural world in need of demonstration. (Compare: Does being an "unmarried male of marriageable age" *explain* why someone is a bachelor, or is it constitutive of *being* a bachelor?) Besides, weren't adaptations correctly identified long before natural selection was known? Didn't Darwin set out in the *Origin* to *show* that adaptations are caused by selection? Wasn't this his *achievement*? For this reason (and others), Fisher prefers the engineering ("good design") over the historical (selection-product) account of adaptation.

Despite their initial plausibility, there are problems with Fisher's arguments. First, and most fundamentally, his argument that the selection-product view begs the question about the causes of adaptation is itself in danger of begging the question about the correct view of the relationship between selection and adaptation. Fisher assumes that it is possible to identify adaptations independently of knowledge of their causal origins. But while it may be possible to identify *adaptive traits* (i.e., "features and relationships that can be observed in the world today" manifesting some "current function"), it is another matter to correctly identify phenotypic traits as *adaptations*. This is precisely what the dispute about the correct account of adaptations is all about, so it cannot simply be asserted that it is possible to identify traits as adaptations in the absence of knowledge of their causal origins without begging the question against the view being critiqued.

Second, the fact that adaptations (more precisely, adaptive traits) were recognized before Darwin explained them in terms of natural selection no more counts against the selection-product view than does the fact that lightning was recognized before Franklin countered against the claim that "Lightning is (caused by) an electrical discharge in the atmosphere." It would be odd to argue that since lightning (as the bright, jagged light that sometimes fractures the night sky) was known before Franklin, that therefore the "electrical discharge" view of lightning must be mistaken. The very definition of lightning includes within it a fact that represents an important scientific discovery. Recognizing a phenomenon is one thing. Understanding its nature and relationship to other factors is another. Rather than detracting from his achievement, it was a significant part of Darwin's achievement to demonstrate that adaptations are the products of natural selection. The selection-product view of adaptation cannot be disposed of so easily.

This does not, however, mean that the selection-product view is without difficulties. Another difficulty of an epistemic sort stems from the specific requirement that adaptations be the products of natural selection. In the case of living organisms, we have direct access to the traits themselves. In the case of extinct organisms, some of whose traits are recorded in their fossilized remains, we have indirect access to their traits. But in no case do we have direct (or even indirect) access to the *causes* of their traits. One can, of course, assume that many if not most traits of most organisms are the products of natural selection, but often this can at best be a plausible conjecture. As Leroi et al. (1994) argue, it will seldom be the case that we can be justified in identifying a trait as an adaptation for a specific function in the historical sense, because actual information underdetermines a trait's status as an adaptation for a specific function. The evidence available will never be able to distinguish between adaptation and constraint. One could perhaps avoid this difficulty by distinguishing between something *being* an adaptation and our *knowing it to be* an adaptation. But, because what we are concerned with here is science, which involves our ability to know the natural world, being told that there *are* adaptations, but that we *cannot know* which traits are adaptations, is not very satisfying. As Amundson (1996) puts it, "[I]f we require a strong epistemic warrant for ascriptions of historical adaptation there may be very few cases to discuss" (Amundson 1996, p. 48). Ironically, therefore, the selection-product view of adaptations, while avoiding the charge of explanatory circularity, threatens to make identification of adaptations difficult or impossible.

Summary: Explanatory Adaptationism

In what sense, then, does natural selection explain adaptations? To explain adaptations is to identify their causes. Sober (2000, p. 124) usefully distinguishes three versions (of increasing strength) of the claim that natural selection is the (or a) cause of adaptations:

1. Natural selection played *some* causal role in the evolution of the trait in question.
2. Natural selection was an *important* cause of the evolution of the trait in question.
3. Natural selection was the *only important* cause of the evolution of the trait in question.

We may designate these as Weak, Moderate, and Strong Explanatory Adaptationism, respectively. Weak Explanatory Adaptationism is true, but trivially so, because if one goes back far enough in the evolutionary process, selection will always be implicated in some way. Strong Explanatory Adaptationism is likewise uncontroversial, but for a different reason: No one holds that selection is the only important cause of evolutionary products, including adaptation. Selection could not even operate were it not for the mutations that introduce phenotypic variation into a population in the first place. And it is acknowledged on all sides that drift plays a role in every actual population. So we are left with Moderate Explanatory Adaptationism. Given the ambiguity of the phrase "important cause," it is understandable why there might be disagreement about whether selection "explains" evolutionary phenomena. Critics of adaptationism (rightly) complain that there is more to evolution than selection. Friends of adaptationism insist that without selection, some of the most interesting and important aspects of evolution (e.g., complex adaptive characteristics) would not arise. They are both right. Selection plays a fundamental explanatory role in accounting for adaptations and for many other evolutionary phenomena as well. But selection always operates in conjunction with other forces, some of which enhance, some of which limit, the power of selection. Identifying and studying how these factors interact is a fundamental task of evolutionary biology.

Methodological Adaptationism

We turn, finally, to Methodological Adaptationism, which maintains that the most fruitful research strategy in evolutionary biology is to proceed as if organisms, in their parts and taken as a whole, are well adapted to their environments. An extreme version of this would be to follow Kant (who was not, of course, thinking of organisms in an evolutionary context): "In the natural constitution of an organized being, i.e., one suitably adapted to life, we assume as an axiom that no organ will be found for any purpose which is not the fittest and best adapted to that purpose" (Kant 1785, p. 11). A weaker version would simply proceed *as if* no organ will be found for any purpose which is not the fittest and best adapted to that purpose, but then be prepared to modify this view in light of empirical findings to the contrary. This seems to capture Mayr's insistence that "[The evolutionary biologist] must first attempt to explain biological phenomena and processes as the product of natural selection. Only after all attempts to do so have failed, is he justified in designating the unexplained residue tentatively as a product of chance" (Mayr 1983, p. 326).

The Virtues of Adaptationist Thinking

Methodological Adaptationism presupposes that natural selection is a potent evolutionary force, and therefore that assuming that many, or even most, features of organisms are adaptations is a (or perhaps the most) fruitful scientific strategy to pursue. The distinction between Empirical and Methodological adaptationism corresponds to the distinction between the *acceptance* and the *pursuit* of a hypothesis. As John Beatty notes,

> To pursue (or entertain) a hypothesis is to seek evidence for or against it. To accept a hypothesis is to assert its truth on the basis of evidence already gathered in its behalf. Clearly, what makes a hypothesis worthy of pursuit is not the same as what makes it worthy of acceptance. What makes any hypothesis worthy of acceptance (if anything does) is a high degree of evidential support. A hypothesis need not, however, have a high degree of evidential support in order to be worthy of pursuit. Presumably, for instance, it would be rational to pursue a hypothesis in order to determine *just how much* evidential support it has. (Beatty 1987, pp. 54–55)

Methodological Adaptationism directs one to determine, if possible, the optimal suite of characteristics for any biological entity for its environment, and to then compare this ideal to the actual situation. Optimality modeling is a paradigm example of Methodological Adaptationism in practice. An optimality model tells us how an organism *should* be designed. Analysis of actual organisms can then tell us how far the organism is from the optimal state. Thinking in terms of optimal adaptation gives the biologist a standard by which to judge the actual adaptiveness of organisms (Beatty 1980). As Holcomb notes, "Optimality modeling does not affirm or deny that organisms are optimally adapted, or nearly so. It merely provides an ideal standard of comparison in order to discover the extent to which various traits are optimized according to various criteria within constraints given by conditions of the organism and its environment" (Holcomb 1989, p. 206). In other words, "The method by itself does not deliver any factual claims about how organisms do act. But when factual claims about how organisms do act are made, the method delivers a way of assessing the kind and degree of deviation of fact from norm" (Holcomb 1989, pp. 206–7).

As a heuristic, methodological adaptationism would be warranted to the extent that it leads to fruitful biological research. Organisms do for the most part seem to be well adapted, and looking for adaptations does seem to have been a fruitful research strategy in evolutionary biology from Darwin to the present. How, then, could one possibly object to Methodological Adaptationism as a heuristic to guide scientific research?

The Perils of Adaptationist Thinking

It is a truism that one often finds what one expects to find. The problem with Methodological Adaptationism, from the point of view of its critics, is that it can easily lead biologists to "find" adaptations everywhere, leading them to generate unfalsifiable "just-so stories," rendering evolutionary theory nonempirical. As Ahouse notes, "That selectionism tied to optimality arguments might serve the function of generating scenarios in which selection is the primary explanation for the distribution of characters in a population is not controversial.... That we can redescribe everything around us in purely adaptationist terms is not controversial, whether we should, is" (Ahouse 1998, pp. 360, 364).

A fundamental problem with optimality modeling is that we generally don't have any independent insight into what the "problems" might be with respect to a trait whose status as optimal we are trying to assess. If a given trait appears to be an optimal solution to a given problem, how do we know that that is indeed the problem that the trait in question evolved to solve?

The problem is acute. Many phenotypic traits have multiple functions, and organisms are highly complex, interconnected systems. It is therefore unlikely that any biologist will be able to describe all the selective costs and benefits of different values for any given trait in an organism. A related problem with establishing the optimality of a trait consists in part in our inability to know all the ramifications of the presence of a given trait for an organism. As Abrams observes, "It is hard to see how one could prove that all the potential constraints, costs, and benefits have been correctly represented in a model" (Abrams 2001, p. 283). For example, it is well-known that mammals living at high latitudes tend to have thicker or more effective layers of insulation in the form of fur or subcutaneous fat. This is generally regarded as an adaptive response to maintain a stable body temperature in cold climates. How would one construct an optimality model to predict the thickness of fat in a particular species? One would have to have an enormous amount of exact information concerning the genetic, physiological, and developmental constraints on, and costs of, producing and bearing fatty tissues in this species, as well as knowledge of past climates. Additionally, one would have to know that heat conservation was the primary function of increased fat storage. There are, of course, other possibilities. Perhaps food supplies are more variable at higher latitudes, placing a greater premium on energy storage. Or perhaps predation is less intense at higher latitudes, permitting species to increase fat thickness without paying an inordinate cost in decreased ability to escape from predators. And so on. In order to determine that fat thickness was optimal, one would also therefore need to consider these alternative explanations, and assign each a weight in the optimality model. Even if a quantitative optimality model predicted exactly the fat thickness measured, it would still not settle the issue of the reason for the observed fat thickness. It could be because of its insulating properties, but the same thickness might also be optimal for energy storage, or as a compromise between two or more possible functions. By itself, an optimality model delivering an accurate prediction (assuming one could be constructed) would not by itself reveal the function of the adaptation in question.

Summary: Adaptation(ism) and Its Limits

Rose and Lauder (1996) use the apt expression "Post-Spandrel Adaptationism" to refer to the study of adaptation after Gould and Lewontin's influential paper. No longer can biologists simply assume *a priori* that all features of organisms are optimal features produced by natural selection specifically for current function. Nonetheless, approaching the study of organisms with optimality considerations in mind can be a particularly fruitful methodology for understanding the extent to which organisms achieve, and fall short of, optimal design. As Abrams rightly concludes, "Exact optimality seems unlikely and may never occur. Even if it did occur, we would not be able to recognize it because of lack of knowledge of the full set of selective consequences of a given trait. Near-optimality (adaptation) probably occurs frequently, but it does not occur all the time, and we don't know the exact frequency with which it occurs" (Abrams 2001, p. 285). Nonetheless, optimality models are useful in helping us to understand the ways in which natural selection, operating in conjunction with other evolutionary factors, produces the astounding but flawed biological entities we observe.

Clearly one can be a methodological adaptationist without endorsing either empirical or explanatory adaptationism. Likewise, one could be an explanatory adaptationist without embracing empirical adaptationism. Ironically, perhaps, embracing methodological adaptationism might be the best way to reveal the limitations of empirical adaptationism. Look for adaptations and assume that they are there to be found; but do it honestly, so that cases where a phenotypic feature cannot be accounted for as a result of natural selection stand out, and require some other sort of account. Rather than lamenting the fact that adaptations are constrained by various factors, biologists can immerse themselves in the richness of the living world as it actually exists. As Ahouse remarks with respect to a particularly enthusiastic proponent of adaptationism (in each of its three forms), "While Dennett may believe that biologists long for a crystal palace of pristine adaptationism, many biologists glory in the exquisite mix of contingency, adaptation and constraint that is in evidence in the uncountable compromises that result in a particular ecosystem on a particular day" (Ahouse 1998, p. 371).

Darwin described all corporeal and mental endowments as tending to "progress toward perfection." But he understood that "perfection" remains an elusive destination, like the horizon that maintains a fixed distance away as one approaches it. Absolute perfection as an attainable state would preclude continued progress. In some ways, it was evolutionary progress, rather than biological perfection, that most intrigued Darwin and those who followed him. It is to this difficult but fundamental topic that we turn next.

PART III

PROGRESS

7

Darwin on Evolutionary Progress

> It may be said that natural selection is daily and hourly scrutinising, throughout the world, every variation, even the slightest; rejecting that which is bad, preserving and adding up all that is good; silently and insensibly working, whenever and wherever opportunity offers, at the improvement of each organic being in relation to its organic and inorganic conditions of life.
>
> (Darwin 1859, p. 84)

Introduction

Considered as a whole, the two most striking aspects of the evolution of life on earth are the staggering diversity of living forms that have come into existence, and the fact that older forms have given way to new and improved forms that seem (for the most part) to be admirably adapted for their respective ways of life. Darwin captured both aspects of evolution in the closing words of the *Origin*, where he remarked that "from so simple a beginning, endless forms most beautiful and most wonderful have been, and are being, evolved" (Darwin 1859, p. 490). Although there are fascinating problems associated with the evolution of diversity (for example, why are there so many different kinds of living things? How do new species come into existence? What *are* "species," anyway?), it is the second aspect of the evolutionary process that is at issue here. Life has not only diversified from its initial humble beginnings, it has also advanced. Multicellular organisms ("metazoa") arose from unicellular organisms; mammals arose from earlier, nonmammalian ancestors; and in general larger and more complex organisms arose from smaller and simpler ones. There is a temporal order in the appearance of the planet's biota. Life on earth has a *history*. Once there were only the simplest sorts of living things – replicating molecules, perhaps. Now the world contains creatures of amazing adaptive complexity that surpass anything the best human engineers could design, let alone construct.

That there has been an overall *direction* in the evolution of life seems obvious. That the history of life on earth also manifests some sort of *progress*, that is, that living things have *improved* in some sense, has to many biologists seemed equally obvious. How could anyone who accepts an evolutionary view of life *deny* that progress has occurred, that some organisms are simply more "advanced," are "higher" in the scale of nature than are representatives of "lower" more primitive classes? E. O. Wilson captures nicely this intuitive view about the living world:

> Many reversals have occurred along the way, but the overall average across the history of life has moved from the simple and few to the more complex and numerous. During the past billion years, animals as a whole evolved upward in body size, feeding and defensive techniques, brain and behavioral complexity, social organization, and precision of environmental control – in each case farther from the nonliving state than their simpler antecedents did. More precisely, the overall averages of these traits and their upper extremes went up. (Wilson 1992, p. 187)

Wilson then draws the conclusion that would also be shared by many biologists:

> Progress, then, is a property of the evolution of life as a whole by almost any conceivable intuitive standard, including the acquisition of goals and intentions in the behavior of animals. It makes little sense to judge it irrelevant. Attentive to the adjuration of C. S. Peirce, let us not pretend to deny in our philosophy what we know in our hearts to be true. (Wilson 1992, p. 187)

Ernst Mayr apparently concurs: "On almost any measure one can think of, a squid, a social bee, or a primate, is more progressive than a prokaryote" (Mayr 1982, p. 532). To such biologists, progress in evolution is as obvious, and as undeniable, as the manifest progress in the development of the automobile (Mayr 1994, p. 40).

Unfortunately for such sentiments, Darwin's theory is *also* widely understood as having banished for all time from scientific discourse the idea of "evolutionary progress," and along with it the cogency of describing organisms as "higher" or "lower." Stephen Jay Gould's assertion is perhaps the most sweeping: "Progress is a noxious, culturally embedded, untestable, nonoperational, intractable idea that must be replaced if we wish to understand the patterns of history" (Gould 1988a p. 319).

Other critics are somewhat less contemptuous but equally dismissive of the idea of evolutionary progress. William Provine speaks for many (perhaps most) biologists when he insists that "The problem is that there is no ultimate basis in the evolutionary process from which to judge true progress" (Provine 1988 p. 63). In a classic case of British understatement, John Maynard Smith remarks that "The concept of progress has a bad name in evolutionary biology" (Maynard Smith 1988, p. 219). That's putting it mildly.[1]

Given these widely divergent positions, it is hardly surprising that the notion of "evolutionary progress" remains a hotly contested idea in evolutionary biology. Like contemporary life forms, contemporary ideas owe their existence to their ancestral precursors. Consequently, in order to understand, clarify, and possibly resolve contemporary debates about evolutionary progress, a critical examination of this concept as it has been understood in the historical development of evolutionary biology is necessary. Michael Ruse (1996) has examined the reciprocal relationship between ideas of biological and cultural progress (or Progress, as Ruse designates it), showing how to some extent the two have often developed in tandem. Such studies are essential for placing the idea of progress in its wider cultural contexts. My aim here, however, is to understand the idea of evolutionary progress itself, as it has functioned within evolutionary biology, in order to determine whether the idea, in some form, merits assent or rejection. Such an aim precludes an exhaustive history of the idea, but instead requires a careful examination of those "moments" in the development of evolutionary biology in which the issue first took shape, acquired its classic interpretation, and then entered into contemporary debate.

Although there is much more to this story, I will focus on three particularly illustrative episodes that prefigure and capture the main points of contention in current debates about evolutionary progress. In this chapter, we will examine Darwin's extended struggle to come to grips with the idea of evolutionary progress, especially in the *Origin of Species.* The following chapter looks at Julian Huxley's enthusiastic endorsement of evolutionary progress in the first half of this century, along with the responses it provoked from George Gaylord Simpson, and then considers the contemporary debate between Stephen Jay Gould and Richard Dawkins. Drawing upon this background, I will argue (in Chapter 9) that although some ideas of evolutionary progress are best rejected as hopelessly flawed, not all are, and that the seeds of a solution to the problem are to be found in the history about to be surveyed. More specifically, I will argue that there is a sense in which evolution *is* genuinely progressive (and, perhaps more importantly, directional), but a significant amount of disambiguation of the idea of "evolutionary progress" is necessary before this fact can be properly understood.

Darwin's Evolving View of Progress

We must begin, however, at the beginning, to see how the problem took shape in the thought of evolution's first effective advocate. This is not as simple as one might hope, because Darwin's view of progress in evolution has presented something of a puzzle to historians. Consider two statements, both made by Darwin, albeit at different times in his life. The first is an entry in his "B Notebook," composed in 1838/9: "It is absurd to talk of one animal being higher than another. – *We* consider those, where the cerebral structure/intellectual faculties most developed, as highest. – A bee doubtless would where the instincts were" (Darwin, B Notebook, p. 74; in Barrett et al. 1987, p. 189). The second statement is from the *Origin of Species*, published twenty years later: "The inhabitants of each successive period in the world's history have beaten their predecessors in the race for life, and are, in so far, *higher* in the scale of nature" (Darwin 1859, p. 345; emphasis added). Taken together, these statements illustrate perfectly the two distinct (but related) problems for understanding Darwin's account of evolutionary progress. The first problem is simply to understand what Darwin himself believed about evolutionary progress. While at times he seems to flatly *reject* as meaningless the idea that organisms appearing later in the history of life are in some sense "higher" than earlier organisms, at other times he appears to *embrace* this view. Consequently, Darwin's view has been something of an enigma, and scholarly opinion has been divided on the question of whether or not he was a "progressionist." (The scholarly dispute over Darwin's "genuine" beliefs about progress is examined in the Appendix to this chapter.)

The second problem concerns whether Darwin's view of evolutionary progress was consistent with the basic principles of his theory of natural selection. This problem is especially pertinent on the view that Darwin *did* believe in some sort of overall progress in the history of life. According to the standard account of the operation of natural selection, all adaptation is to *local* conditions, and species either evolve along with changing conditions or fail to evolve and go extinct. There is apparently no way that a series of randomly changing local environments can elicit cumulative progressive advance. So if Darwin did believe in some sort of overall progressive advance, that view seems to be at odds with the basic principles of his theory.

I will argue that Darwin's apparently equivocal remarks about evolutionary progress can be rendered consistent by clearly distinguishing the senses in which he believed evolution is progressive, from those senses in which he believed it is not. There is also a developmental aspect to Darwin's thinking about evolutionary progress that it is critical to recognize. Darwin carefully crafted his remarks on progress in his published works in order to commit himself to the idea only to the extent that the idea could be made conceptually respectable. As his confidence grew that the idea made sense, so, too, did his boldness in endorsing it in his writings. At first he puts forth the view tentatively, with plenty of qualifications. By the end, he shows no hesitation in describing the evolutionary progress as progressive. This trend is already evident in his writings before the publication of the *Origin of Species*.

Notebooks (1837–1839)

Darwin's circumnavigation of the globe as naturalist aboard the H.M.S. *Beagle* ended in October 1836, and there immediately followed what he later described as the two most productive years of his life (Bowler 1989, p. 164). He began a series of private notebooks to record his speculations concerning transmutation and its implications. The Notebooks have become an invaluable tool for reconstructing Darwin's thought at that time. One thing is quite clear: he was already speculating about the relationship between evolutionary change, complexity, and progress. In his "B Notebook" (July 1837–February 1838) he wrote, rather cryptically: "Each species changes. Does it progress . . . the simplest cannot help. – becoming more complicated; & if we look to first origin there must be progress" (B Notebook, 18; in Barrett et al. 1987, p. 175). One can only guess at what Darwin might have had in mind. A reasonable conjecture is that if organisms began in simple form and then changed, they could only change in the direction of greater complexity, because that was the only direction of change open to them. This would be a form of progress in the sense that life on earth thereby moved a step closer to the level of complexity it would latter attain. In his "E Notebook" (October 1838–July 10, 1839) Darwin expands upon the simple idea that complexity must increase by speculating about the ecological and evolutionary *causes* of the tendency toward increased complexity, and by clarifying the sense in which this "must" occur:

> The enormous *number* of animals in the world depends, on their varied structure & complexity. – hence as the forms became complicated, they opened *fresh* means of adding to their complexity.– but yet there is no NECESSARY tendency in the simple animals to become complicated although all perhaps will have done so from the new relations caused by the advancing complexity of others. (Darwin, E Notebook, p. 95; in Barrett et al. 1987, pp. 422–23)

There are hints here of a self-propelling positive feedback loop driving increasing complexity. As organisms become more complex, this causes (in some unspecified manner) other organisms to become more complex, and so on. As stations in the economy of nature become filled, there will be increasing pressure for organisms to exploit previously vacant niches. Exploiting these niches effectively requires that structural changes occur. Since the *simplest* structural organizations have appeared *first*, the only structural organizations remaining will be more complex ones. As the previously vacant niches become occupied by these structurally more complex organisms, the same process is repeated at *this* level, thus driving up the overall level of structural complexity.

The "necessity" here is "extrinsic": it is external conditions ("the advancing complexity of others"), rather than some intrinsic necessitating tendency, that accounts for the overall trend. This reading is corroborated when we turn to his "N Notebook," written at about the same time (2 October 1838 – 1 August 1839), where we find Darwin exclaiming to himself: "In my theory there is no absolute tendency to progression, excepting from favourable circumstances!" (N Notebook, p. 47; in Barrett et al. 1987, p. 576). In one sense, in his notebook entries Darwin was simply reflecting on what *had* to be the case. If one begins by assuming that the first life forms were simple, what else could life do as it evolves but become more complex? But something is still missing from this account. Why couldn't life have simply stayed *relatively* simple, instead of resulting in the highly complex organisms that exist today? Conceivably, organisms could vary in *structure* without this variance representing greater or lesser *complexity*. Even if the increasing complexity of some organisms makes *possible* the increasing complexity of others, such increasing complexity does not necessarily *follow*. What is still left unclear, therefore, is a plausible dynamic that might drive the increasing complexity found in nature. In summary, in the Notebooks we find Darwin toying with the idea of evolutionary progress, clearly sympathetic toward the idea, but at the same time distancing himself from the idea that such progress is somehow intrinsically "necessary."

Essay of 1844

Despite the suggestive ideas Darwin entertained in his notebooks, a clear theoretical justification for supposing that natural processes would account for increasing organic complexity was still missing. By October 1838, thanks to his reading of Malthus, Darwin already had "a theory by which to work" (Darwin 1958, p, 120), that is, the theory of natural selection. But the connection between natural selection and increasing complexity is far from straightforward. If increased complexity was always selectively advantageous, then of course natural selection could explain the increasing complexity found in the history of life. But as Darwin recognized, there are circumstances in which simplification, rather than complexification, will be selectively advantageous. In the "Essay of 1844," his first attempt to explain the theory of natural selection in detail, he noted:

> A long course of selection might cause a form to become more simple, as well as more complicated. . . . According to our theory, there is obviously no power tending constantly to exalt species, except the mutual struggle between the different individuals and classes; but from the strong and general hereditary tendency we might expect to find some tendency to progressive complication in the successive production of new organic forms. (in F. Darwin 1909, p. 227)

The introduction of the principle of natural selection both solved one problem and raised others. Natural selection can explain, in principle, why there might be some tendency toward increasing complexity, but it can also explain why some forms become simpler, and why yet others do not change at all. By itself, it gives no reason to conclude that any tendency in one direction would be more significant than one in the opposite direction. Darwin's remark that "from the strong and general hereditary tendency we might expect to find some tendency to progressive complication in the successive production of new organic forms" is left vague and undeveloped, and runs counter to his desire to find reasons for supposing that *external* (i.e., environmental) conditions are sufficient to account for increasing organic complexity. As it turns out, Darwin subsequently had a lot to say on this topic. For such reasons we have to turn to the *Origin of Species.*

Evolutionary Progress in the *Origin of Species* (1859–1872)

After composing the "Essay of 1844" (lest he die prematurely and his theory with him), Darwin devoted the next eight years (1846–54) to morphological and taxonomic work on barnacles, culminating in his two volume *Monograph on the Sub-Class Cirripedia* (1854). From September 1854 until June 1858 he was occupied full-time with work on his "species theory." In 1856 he began writing his "Big Species Book," documenting in overwhelming detail the evidence supporting his theory (Stauffer 1975). But when he received that fateful letter from Alfred Russel Wallace in 1858 detailing a theory remarkably like his own, he abandoned work on the larger treatment and quickly composed the *Origin of Species* as an "abstract" of his theory. When we turn to this work – the most widely read, and most carefully crafted, of all his published works – we find Darwin increasingly preoccupied with the issue of evolutionary progress as each new edition appeared.

"Competitive Highness"

In the first edition of the *Origin* (1859), after noting that there has been much discussion of whether recent forms are "higher" than more ancient forms, Darwin informs the reader that he will not discuss this issue in any detail because naturalists have not yet adequately defined what is meant by "high" and "low" forms. Nonetheless, he claims that

> in one particular sense the more recent forms must, on my theory, be higher than the more ancient; for each new species is formed by having had some advantage in the struggle for life over other and preceding forms. If under a nearly similar climate, the eocene inhabitants of one quarter of the world were put into competition with the existing inhabitants of the same or some other quarter, the eocene fauna or flora would certainly be beaten and exterminated. . . . I do not doubt that this process of improvement has affected in a marked and sensible manner the organisation of the more recent and victorious forms of life, in comparison with the ancient and beaten forms. (Darwin 1859, pp. 336–337)[2]

The "one particular sense" is what he came to call "competitive highness."[3] More recent forms are "higher" than earlier forms because they have beaten them in direct competition. "As natural selection acts solely by the preservation of profitable modifications, each new form will tend in a fully-stocked country to take the place of, and finally to exterminate, its own less improved parent or other less-favoured forms with which it comes into competition" (Darwin 1859, p. 172). As he says later in the *Origin*, "The inhabitants of each successive period in the world's history have beaten their predecessors in the race for life, and are, in so far, higher in the scale of nature" (Darwin 1859, p. 345).

It is important to be as clear as possible about what Darwin meant by "competitive highness." The "highness" in question is essentially a matter of how well adapted to their environment organisms are compared to their *immediate* predecessors. Left unclear is the extent to which later organisms should be considered "higher" than *all* their predecessors. Because adaptedness is a function of the properties of an organism in a specific environment, it is conceivable that in a lineage of organisms **A**, **B**, and **C**, that **B** is better adapted than **A**, and **C** is better adapted than **B**, but that **C** is not better adapted than **A**. In other words, being better adapted might not be a transitive relation. Consequently, this notion of "highness" will only be applicable to organisms successively occupying the *same* (or a very similar) environment. Darwin's specification that the competition take place "under a nearly similar climate" is important, because if the climate changed dramatically over evolutionary time, then earlier forms might be just as well adapted to *their* environments as later forms are to *theirs*, and it would then be the climate in which the competition takes place, not the "highness" of the organisms involved, that would determine the winners. Finally, the notion of "competitive highness" in the sense in which Darwin describes it is only applicable to organisms that are, or would be, truly in competition with one another. This might include organisms in the same lineage (i.e., a group of organisms and their descendants), or it could include organisms in entirely different lineages that nonetheless compete for the same resources (e.g., reptilian and mammalian carnivores). But it would not include organisms that are not in competition for the same resources, or that occupy different environments (e.g., benthic scavengers and savanna grazers). It would not, that is, warrant any global judgments about how the organization of life "as a whole" has progressed. Nonetheless, finding a warrant for making such a claim was Darwin's ultimate aim, because he immediately added to the foregoing remarks that "this may account for that vague yet ill-defined sentiment, felt by many palæontologists, that organization on the whole has progressed" (Darwin 1859, p. 345). Nonetheless, to move from "competitive highness" to progress in the organization "on the whole," something else was needed.

"Specialisation" and "Division of Physiological Labour"

The "something else" appears in the second (and all subsequent) editions of the *Origin.* By 1860 Darwin was willing to go public with a definition of "highness" that he declined to give in 1859:

> The best definition probably is, that the higher forms have their organs more distinctly specialised for different functions; as such division of physiological labour seems to be an advantage to each being, natural selection will constantly tend in so far to make the later and more modified forms higher than their early progenitors, or than the slightly modified descendants of such progenitors. (Darwin 1959, p. 547)

The notions of "specialisation" and "division of physiological labour" as criteria of "highness" that appear in the second edition of the *Origin* can be traced directly to Darwin's reading of Karl Ernst von Baer and Henri Milne Edwards, respectively. Von Baer had defined the "grade of development" of an animal as:

> the greater or less heterogeneity of its elementary parts of the separate divisions of a complex apparatus; in a word, its greater histological and morphological differentiation. The more homogeneous the whole mass of the body is, so much the lower is the grade of its development. The grade is higher when nerves and muscles, blood and cell-substance, are sharply distinguished. (von Baer 1828, pp. 207–8; quoted in Ospovat 1981, p. 119)

Darwin had read von Baer several years earlier (in Huxley's 1853 translation), and had discussed the applicability of von Baer's criteria to the barnacles he was studying at the time (Darwin 1854, vol. 2, pp. 19–20). He also was much impressed with the work of the Belgian/French biologist Henre Milne Edwards, whose work he had read in 1846, and to whom he dedicated the second volume of his work on barnacles. Milne Edwards had argued that specialization of parts with an attendant "division of physiological labour" makes organisms more efficient:

> When . . . life begins to manifest more complicated phenomena, and the final result produced by the interplay of the different parts of the body becomes more perfect, . . . the life of the individual, instead of being the sum of a larger or smaller number of identical elements, results from essentially different acts produced by distinct organs. (Milne Edwards 1827; quoted in Ruse 1996, p. 159)

With the definitions of "organisation" articulated by von Baer and Milne Edwards in hand, Darwin had criteria that could be connected to the operation of natural selection to explain why evolution will tend toward advancement. In the third edition of the *Origin* (1861) he cites with approval von Baer's standard of "the amount of differentiation of the different parts . . . and their specialisation for different functions; or, as Milne Edwards would express it, the completeness of the division of physiological labour" (Darwin 1959, p. 221). In all subsequent editions, Darwin identifies "the great Von Baer" and Milne Edwards by name, treating their definitions of "specialisation" and "division of physiological labour" as essentially equivalent (Darwin 1959, p. 221). These biologists supplied precisely the concepts Darwin needed in order to link together natural selection and evolutionary improvement.

Nonetheless, one aspect of Darwin's use of the ideas of von Baer and Milne Edwards is initially puzzling. If he was aware of and in agreement with their definitions of "highness" when he was composing the first edition of the *Origin*, why did postpone their use until the second edition? It is at this point that the identification of Darwin's primary audience becomes crucial. According to Ospovat (1981), Darwin intentionally downplayed the inevitability of progress in the first edition of the *Origin* because he wanted to avoid for his book the punishment that Robert Chambers's *Vestiges* (1844) had received at the hands of T. H. Huxley, and because he wanted the support of Charles Lyell, who in his *Principles of Geology* (1830–33) had argued against progressionism. After gauging reactions to the first edition of the *Origin*, however, these concerns lost much of their strength. First, from the very beginning the *Origin*, in contrast to *Vestiges*, was treated as a serious contribution to science. Second, Darwin had in "morphological differentiation" and "division of physiological labour" a conception of "highness" that was generally accepted among professional zoologists. Third, by 1859 Lyell had abandoned his nonprogressionism, so Darwin felt freer in taking a progressionist stand in later editions of the *Origin*. Although Lyell had abandoned his nonprogressionism, he was still not convinced that natural selection could account for evolutionary progress, believing instead that a continued intervention of creative power or some "principle of improvement" was necessary to produce successively higher levels of organization. In a letter to Lyell dated 25 October 1859, Darwin responded by emphasizing the power of natural selection:

> When you contrast natural selection and "improvement," you seem always to overlook... that every step in the natural selection of each species implies improvement in that species in relation to its conditions of life.... Improvement implies, I suppose, each form obtaining many parts or organs, all excellently adapted for their function. As each species is improved, and as the number of forms will have increased, if we look to the whole course of time, the organic condition of life for other forms will become more complex, and there will be a necessity for other forms to become improved, or they will be exterminated: and I see no limit to this process of improvement, without the intervention of any other and direct principle of improvement. All this seems to me quite compatible with certain forms fitted for simple conditions, remaining unaltered, or

> being degraded. If I have a second edition [of the *Origin*], I will reiterate "Natural Selection," and, as a general consequence, "Natural Improvement." (in F. Darwin 1888, vol. 2, p. 177)

True to his word, in the second edition of the *Origin*, in the summary for Chapter IV, Darwin added that natural selection "leads to the improvement of each creature in relation to its organic and inorganic conditions of life (Darwin 1959, p. 271).

"Highness" and Natural Selection

As we have seen, Darwin employed two distinct notions of "highness" – "competitive highness" and "specialisation and division of physiological labour" – and implied that when applied to the actual history of life, the two standards yield the same results. On the one hand, "highness" is represented by competitively superior organisms that beat their predecessors in the struggle for existence. On the other hand, "highness" is represented by organisms displaying greater "specialisation" and "division of physiological labour," that is, greater complexity.[4] Unless Darwin could *connect* the two notions in some theoretically justified way, he would be left with two distinct, and not obviously related, conceptions of biological "highness." How did he attempt to resolve this problem?

Unsurprisingly given its centrality in his theorizing, the idea of natural selection is used to bridge the two conceptions of "highness." There are hints of Darwin's solution in the second edition of the *Origin* (1860). There he remarks that since "division of physiological labour seems to be an advantage to each being, natural selection will constantly tend in so far as to make the later and more modified forms higher than their early progenitors, or than the slightly modified descendants of such progenitors" (Darwin 1959, p. 547). But the connection between the two conceptions of "highness" is still not very clear. In the third edition of the *Origin* (1861) he added to the chapter on Natural Selection an entirely new section entitled "On the degree to which Organisation tends to advance." There his view is more developed and spelled out in a bit more detail. In a passage in the third edition of the *Origin* that nearly (but not quite) equates the two conceptions, Darwin wrote:

> [A]s the specialisation of parts and organs is an advantage to each being, so natural selection will constantly tend thus to render the organisation of each being more specialised and perfect, and in this sense higher. . . . In another and more general manner we can see that on the theory of natural selection the more recent forms will tend to be higher than their progenitors, for they will in the

struggle for life have to beat all the older forms with which they come into close competition. (Darwin 1959, pp. 547–548)

Elsewhere in the third edition of the *Origin* he is even more explicit on the linkage:

If we look at the differentiation and specialisation of the several organs of each being . . . as the best standard of highness of organisation, natural selection clearly leads toward highness; for all physiologists admit that the specialisation of organs, inasmuch as they perform in this state their functions better, is an advantage to each being; and hence the accumulation of variations tending towards specialisation is within the scope of natural selection. (Darwin 1959, p. 222)

Darwin's mature view is now beginning to come into focus. An increased specialization of parts supporting a division of physiological labor, by rendering organisms competitively superior, would be favored by natural selection. Hence, natural selection produces (and explains) "highness" in both senses, and explains as well why evolution will tend toward advancement. In a later section we will return to this account in order to explore in more detail the connection between the two sorts of "highness."

On the "Necessity" of Evolutionary Progress

Although Darwin connects the two conceptions of "highness," and sees both as consequences of natural selection, he continues to treat them as distinct standards. The reason why he could not simply *equate* them is that he was well aware that in some cases achieving greater competitive highness involves a simplification, rather than a complexification, of structure. Given the struggle for existence, organisms will seize on every underexploited niche in the economy of nature, with the result that "it is quite possible for natural selection gradually to fit an organic being to a situation in which several organs would be superfluous and useless: in such cases there might be retrogression in the scale of organisation" (Darwin 1959, p. 222). Cases of "retrogression of organisation" may be expected "under very simple conditions of life [in which] a high organisation would be of no service [and] possibly would be of actual disservice, as being of a more delicate nature, and more liable to be put out of order and thus injured" (Darwin 1959, p. 225). Darwin gives no examples here, but one thinks of examples like cave fish, living in complete darkness, that have lost their eyes. Given the vulnerability of eyes to infections, and the energetic costs of maintaining such organs, loss of eyes in such an environment might well render such fishes competitively higher than their competitors with eyes, despite (or rather because of) their simplified structure.

So on Darwin's view, natural selection may simplify, as well as complexify, organization. A different and potentially more difficult issue concerns Darwin's claim that the ultimate result of natural selection is that "each creature will tend to become more and more improved in relation to its conditions of life," inevitably leading to "the gradual advancement of the organisation of the *greater number* of living things throughout the world" (Darwin 1959, p. 221; emphasis added). The claim seems to be that although *some* organisms will not manifest "advancement in organisation," *most* will. If this is so, then two objections arise, both of which Darwin anticipates: First, "How is it that throughout the world a multitude of the lowest forms still exist"? After all, simpler forms far outnumber more complex forms in terms of practically every measure one can think of: in absolute numbers, in number of species, and even in total biomass. So why haven't these forms been eliminated, or at least reduced to an insignificant proportion of living things on earth? Second, "How is it that in each great class some forms are more highly developed than others? Why have not the more highly developed forms everywhere supplanted and exterminated the lower?" (Darwin 1959, pp. 222–23). This problem is particularly acute since Darwin suggested that natural selection operates most intensely among closely related organisms, that is, among organisms competing for essentially the same resources. If increased complexity is usually an advantage, then one might expect that natural selection would have eliminated the lower, less developed forms in favor of the more highly developed forms.

In response to the first objection, Darwin mentions four possibilities. Simple forms may continue to exist because in some cases the favorable variations for natural selection to act upon have never arisen. In other cases perhaps not enough time has been available for "the utmost possible amount of development" (Darwin 1959, p. 225). In yet others the aforementioned "retrogression of organisation" may be responsible. But the main explanation is simply that an advance in organization may not *always* be a definite advantage, and thus ought not always be expected. This point permits Darwin to sharply distinguish his theory from Lamarck's and all similar theories.[5] Lamarck had posited "an innate and inevitable tendency toward perfection in all organic beings," and was consequently forced to posit the continual production of new and simple forms by spontaneous generation in order to explain why there were still any such forms in existence. But Darwin insisted that his own theory was in no need of such a questionable belief. "On my theory the present existence of lowly organised productions offers no difficulty; for natural selection includes no necessary and universal law of advancement or development – it only takes advantage of such variations as arise and are beneficial to each creature under its complex relations of life" (Darwin 1959, p. 223).[6] To make sure that this point was clearly understood, in the fifth edition of the *Origin* he changed this to read: "On our theory the continued existence of lowly organisms offers no difficulty; for natural selection, or the survival of the fittest, does not necessarily include progressive development" (Darwin 1959, p. 223).[7] Darwin returned to this point in the last chapter of the *Origin*, where he emphasized that the sort of "advancement" he was advocating "is perfectly compatible with numerous beings still retaining a simple and little improved organisation fitted for simple conditions of life; it is likewise compatible with some forms having retrograded in organisation, though becoming under each grade of descent better fitted for their changed and degraded habits of life" (Darwin 1959, p. 742). This interpretation is further strengthened in the sixth edition by the inclusion of a new chapter entitled "Miscellaneous Objections to the Theory of Natural Selection." In the section entitled "Progressive Development" Darwin explicitly contrasts belief in "an innate and necessary law of development" with "the doctrine of natural selection or the survival of the fittest, which implies that when variations or individual differences of a beneficial nature happen to arise, these will be preserved; but this will be effected only under certain favourable circumstances" (Darwin 1959, p. 228). Consequently, "there is no need . . . to invoke any internal force beyond the tendency to ordinary variability . . . which through the aid of natural selection would . . . well give rise by graduated steps to natural races or species. The final result will generally have been, as already explained, an advance, but in some few cases a retrogression, in organisation" (Darwin 1959, p. 264).

Clearly Darwin did not believe in any sort of law that would invariably cause all organic beings to progress together up a ladder of being. As he makes clear in every edition of the *Origin*: "I believe in no fixed law of development, causing all the inhabitants of a country to change abruptly, or simultaneously, or to an equal degree. . . . Hence it is by no means surprising that one species should retain the same identical form much longer than others; or, if changing, that it should change less" (Darwin 1859, p. 314; Darwin 1959, p. 523).[8] The fact of organisms unchanged through time *would* be fatal to his theory, Darwin says in a peculiar turn of phrase, if his theory entailed "advance in organisation as a necessary contingent." But there is no necessity on the theory of natural selection for all organisms to experience continual advance, even in the face of relatively unchanging environmental conditions (Darwin 1959, pp. 549–50). Since "natural selection acts . . . exclusively by the preservation and accumulation of variations, which are beneficial under the organic and inorganic conditions of life to which each creature is at each successive period exposed" (Darwin 1959, p. 221), progress in the sense of advancement in organization will occur only when increased complexity would be an advantage. How would it benefit an earthworm, Darwin asks, to be more highly organized? It not at all clear that it would. Consequently, "If it were no advantage, these forms would be left by natural selection unimproved or but little improved; and might remain for indefinite ages in their present little advanced condition" (Darwin 1959, p. 223). Thus, while he rejected any notion of evolutionary progress as determined by a necessary law of progression, Darwin nonetheless accepted evolutionary progress as a contingent general consequence of natural selection.[9]

Because Darwin's answer to the question of whether natural selection necessarily leads to progressive development is so liable to misunderstanding, it is worth trying to make it as clear as possible. Unfortunately, Darwin himself did not always express himself on this issue as clearly as one might like. For example, in a remarkable passage that appears in the sixth edition of the *Origin*, he writes:

> Although we have no good evidence of the existence in organic beings of an innate tendency towards progressive development, yet this necessarily follows . . . through the continued action of natural selection. For the best definition which has ever been given of a high standard of organisation is the degree to which the parts have been specialised or differentiated; and natural selection tends towards this end, inasmuch as the parts are thus enabled to perform their functions more efficiently. (Darwin 1959, p. 241)

There are three different ways of reading this passage, only one of which coheres with Darwin's other statements. Given his concern to distance his view from Lamarck's, he cannot mean that despite any evidence for it, an *innate* tendency in organic beings towards progressive development nonetheless exists. Likewise, he cannot mean that progressive development necessarily exists in *every* case, as his earlier remarks make clear. Instead, what Darwin claims is that natural selection necessarily *tends toward*, rather than *necessarily results in*, progressive development. The distinction is crucial, because the former construction makes progress contingent in a way that the latter does not. In a sense, Darwin's theory of natural selection embodies what might be called a "perturbationist model" of evolutionary progress according to which deviations from the expected outcome are explained as arising from numerous contingencies which may deflect the actual course of evolution from exemplifying its central causal tendency. In this sense it is akin to physical theories like Newton's theory of motion. Objects moving in a given direction at a given velocity will continue to do so unless acted upon by some additional force. All objects in the real world are, of course, acted upon by numerous forces, but one's analysis starts from the idealized case, and complications are introduced only as necessary. In the present case, Darwin is asserting that natural selection will inevitably produce more advanced, "higher" organisms, except in those cases where other factors overrule this tendency. The result is that progress will characterize the evolution of life "on the whole," but not necessarily every part of it. From the theory of natural selection one can deduce that there will be a progressive tendency in the evolution of life, but this will be consistent with empirical data from paleontology that reveal many cases of stasis and retrogression. As Darwin wrote to the botanist W. H. Harvey, "There is nothing in my theory necessitating in each case progression of organisation, though Natural Selection tends in this line, and has generally thus acted" (in Darwin and Seward 1903, vol. I, p. 164).

Finally, recall the second objection to his general thesis that Darwin identified: "[H]ow is it that in each great class some forms are more highly developed than others? Why have not the more highly developed forms everywhere supplanted and exterminated the lower?" Darwin does not dispute the crucial presupposition of the objection, namely, that within each class some forms *are* more "highly developed" than others. He notes that among vertebrates, mammals and fish coexist, as do sharks and amphioxus, "which latter fish in the extreme simplicity of its structure closely approaches the invertebrate classes" (Darwin 1959, p. 223). According to Darwin, the reason why "higher" organisms like mammals and sharks do not supplant "lower" organisms like fish and amphioxus, is that they are generally not in direct competition with one another. "[M]ammals and fish hardly come into competition with each other; the advancement of certain mammals or of the whole class to the highest grade of organisation would not lead to their taking the place of, and thus exterminating, fishes" (Darwin 1959, p. 224). Likewise, "members of the shark family would not, it is probable, tend to supplant the amphioxus; the struggle for existence in the case of the amphioxus apparently will lie with members of the invertebrate classes" (Darwin 1959, p. 224).

In other words, Darwin affirmed competition, and hence the possibility for "advancement to the highest grade of organisation" *within* each class while denying such competition *between* major classes. Although mammals are higher than fish, one would not expect mammals to supplant fish in the struggle for existence. Fish will presumably continue to exist even as at least some mammals advance in organization. Consequently, "Although organisation, on the whole, may have advanced and be advancing throughout the world, yet the scale will still present all degrees of perfection; for the high advancement of certain whole classes, or of certain members of each class, does not at all necessarily lead to the extinction of those groups with which they do not enter into close competition" (Darwin 1959, p. 224).

"Grade of Development" versus "Type of Organisation"

I have been exploring Darwin's view of evolutionary progress in some detail in order to answer the first of the two problems concerning Darwin's view identified earlier, viz.: What precisely did Darwin believe about evolutionary progress? Although I have been striving to present a consistent interpretation of Darwin's position, it has to be admitted that Darwin himself makes this a difficult task. The difficulty arises from taking into consideration *other* remarks Darwin makes about judging "higher" and "lower." Consider, for example, the first of the two quotes I used to introduce the "puzzle" of Darwin's view: "It is absurd to talk of one animal being higher than another. – *We* consider those, where the cerebral structure/intellectual faculties most developed, as highest. – A bee doubtless would where the instincts were" (Darwin, B Notebook, p. 74; in Barrett et al. 1987, p. 189). It would be tempting to just write off this remark as a view held by Darwin very early on in his theorizing that he eventually abandoned in favor of whole-hearted progressionism. But the problem with this move is that he repeats, and even emphasizes the very same point, many years later. In the third edition of the *Origin* (1861), he asks: "[W]ho will decide whether a cuttle-fish be higher than a bee?" (Darwin 1959, p. 550). By the sixth edition (1872), he was prepared to answer his own question with a degree of confidence that seems to leave no doubt about his position: "To attempt to compare members of distinct types in the scale of highness seems hopeless; who will decide whether a cuttle-fish be higher than a bee, that insect which the great Von Baer believed to be 'in fact more highly organised than a fish, although upon another type'?" (Darwin 1959, p. 550).

The problem is how to square Darwin's *rejection* of comparing distinct types of organisms in the scale of highness in this quote with his willingness to make such comparisons elsewhere in the *Origin*. One possibility is to take seriously Darwin's reference to the problems inherent in comparing "distinct types" in the scale of highness. As we have seen, Darwin made use of von Baer's concept of "grade of development" as a general standard in terms of which to distinguish between higher and lower organisms. Von Baer (1828) also distinguished between "grade of development" and "type of organization." Whereas the former notion refers to the perfection of an animal's structure, the latter refers to "the relative position of the organic elements and of the organs." The idea of "types of organization" came from Cuvier who, in *The Animal Kingdom* (1817), divined that there exist "four principal forms, four general plans . . . on which all animals appear to have been modeled," viz., Vertebrata (vertebrates), Mollusca (molluscs), Articulata (arthropods), and Radiata (radially shaped animals). Each one represents a different fundamental body plan. An implication of this classification for Cuvier was that the traditional chain of being was broken. No linear series of animals from simplest to most complex could be constructed. Only within each major group (or "embranchement," as Cuvier termed it) could comparisons be made, and even here forms could not be classified in a straight line from simplest to most complex.

According to von Baer, the type of organization characterizing a given organism is totally distinct from the "grade of development" of that animal. Two important consequences follow from this: (1) The same "grade of development" may be attained in many different "types"; (2) a single linear arrangement of all animals in terms of "grade of development," irrespective of "type", is impossible.[10] Consequently, cuttle-fish and bees cannot properly be compared in the scale of highness, because they belong to two different "types of organization." But a modern bee (or cuttle-fish) and other organisms of the same "type" may be properly compared in terms of "highness" because they have the same basic structural organization.

Employing this distinction, it could be argued that each of the two different notions of "highness" that Darwin worked with has its own proper domain of application. The notion of "competitive highness" in which "the inhabitants of each successive period in the world's history have beaten their predecessors in the race for life" is applicable to organisms of the same "type" that come into direct competition. A present-day predator and the direct ancestor it supplanted can be compared in this way. But two organisms of the same "type" that do not compete (e.g., mammals and birds) cannot be compared on this standard. They can, however, be compared in terms of "grade of organisation" since they belong to the same "type" (both are vertebrates). Finally, some organisms (e.g., insects and molluscs) cannot be compared in terms of either standard of highness because they belong to entirely different "types." So perhaps the *apparent* ambiguity in Darwin's talk of "higher" and "lower" can be resolved by paying careful attention to the distinctions he accepted, and the proper domains of applicability of the different concepts of "highness" he deployed. The "cost" of this interpretive strategy, however, is that it would prevent Darwin from comparing organisms of different types in terms of "degree of organisation," with the result that he would be unable to say that a monkey is "higher" in this sense than a polyp. This seems like a high price to pay in order to render a single anomalous remark consistent with Darwin's other, more numerous claims about evolutionary progress, so I am inclined to look for another way to square Darwin's apparently anomalous remark with his overall view of evolutionary progress.

A more attractive strategy is to interpret Darwin as making an *epistemological* point. Whereas in his early notebook entry he says flatly that it is *absurd* to talk of one animal being higher than another, in the *Origin* he merely emphasizes the *difficulty* of comparing organisms of distinct types. Given his criterion of "specialisation" and "division of physiological labour," it will often be the case that one organism is "higher" than another in this sense. But it is also entirely possible, and perhaps indeed inevitable, that there will be "tie." This will *obviously* be true for two members of the same species, but can also be true for two members of widely divergent types (e.g., cuttle-fish and bees). What follows is that although it makes sense to distinguish organisms as "higher" or "lower" in terms of structural organization, it will not always be possible in *practice* to make such determinations, and that the actual ranking of organisms as "higher" or "lower" (should one wish to undertake such a task) need not correspond *exactly* to any traditional "scale of nature." On purely structural grounds, it *may* turn out that a bee is "in fact more highly organised than a fish." There is nothing obviously inconsistent with such a view.[11]

Progress in *The Descent of Man* (1871)

We have been considering Darwin's remarks about progress in the *Origin of Species* at some length, attempting to work out some of the subtleties of Darwin's view. But we should not lose sight of the central point that

Darwin was, or rather became, a proponent of evolutionary progress. If there is still any doubt that Darwin believed that the evolution of life manifests advancement, consider this passage from the *Descent of Man* (1871), which may also serve as a fine summary of his mature view:

> The best definition of advancement or progress in the organic scale ever given, is that of Von Baer; and this rests on the amount of differentiation and specialisation of the several parts of the same being, when arrived, as I should be inclined to add, at maturity. Now as organisms have become slowly adapted by means of natural selection for diversified lines of life, their parts will have become, from the advantage gained by the division of physiological labour, more and more differentiated and specialised for various functions... and thus all the parts are rendered more and more complex. But each organism will still retain the general type of structure of the progenitor from which it was aboriginally derived. In accordance with this view it seems, if we turn to geological evidence, that organisation on the whole has advanced throughout the world by slow and interrupted steps. (Darwin 1871, vol. 1, p. 211)

This passage is remarkable because it sums up Darwin's conception of evolutionary progress so well. The specialization of parts with its associated division of physiological labor provides a competitive advantage upon which natural selection acts to increase the degree of complexity of organisms within each structural and organizational type. Although this process will not necessarily result in advance in every instance, nonetheless the general theoretical claim finds empirical support in the geological evidence (e.g., the fossil record), which records a definite tendency for the organization of life on the whole to advance. According to Darwin, evolutionary progress is both a well-grounded theoretical prediction derived from the theory of natural selection, and an established empirical fact confirmed by the geological evidence.

Was Darwin's View Cogent?

With this account as background, we can now return to the two problems for understanding Darwin's view. The first is simply to understand what he believed about evolutionary progress. If the interpretation above is correct, he was a committed progressionist who nonetheless was well aware of the difficulties *in practice* of making determinations of "higher" and "lower." The second problem concerns determining whether Darwin's view of evolutionary progress was consistent with the basic principles of his theory of natural selection. Later (in Chapter 9) we will consider contemporary arguments that attempt to show that greater complexity is often of selective advantage, but at this point we can note that one way in which this problem could be resolved would be to simply admit that whereas the theory of natural selection does not *predict* that increased complexity will always or usually go hand in hand with increased adaptation, nonetheless as a matter of *empirical fact* these two have often been conjoined, and there is a good Darwinian explanation for why this has happened. Life necessarily began in a simple form. If greater specialization of parts and division of physiological labor was *sometimes* adaptively advantageous, then the overall level of organic complexity would increase. What is true at this early stage in the evolution of life would be true as well at later stages. So long as some lineages manifest increasing complexity, the overall level of complexity may increase, even if some organisms do not undergo complexification, or even undergo simplification of structure. There has been an increase in complexity in the history of life, and Darwin provided a reasonable *explanation* for this phenomenon, even if his theory cannot *predict* in any precise manner the details of this process. But as has been clear for a long time, in evolutionary biology explanation can proceed in the absence of predictability (Scriven 1959).

Summary: Darwin on Evolutionary Progress

According to David Hull, what is remarkable about Darwin is that "at a time when a belief in progress was pandemic, he had so little to say about it, and when he did, expressed himself so equivocally" (Hull 1988b, p. 30). According to Richardson and Kane, "Divergence and specialization, rather than progress, were the hallmarks of *The Origin of Species*" (Richardson and Kane 1988, p. 149). Such claims need to be reassessed in light of the discussion presented above. Darwin had plenty to say about evolutionary progress; his views on the issue developed throughout his writings, especially in the successive editions of the *Origin*; and when considered carefully, a clear, unequivocal thesis emerges. To talk of divergence and specialization *rather than* progress as the motif of the *Origin* is to anachronistically draw a false contrast. For Darwin, an important form of evolutionary progress consists precisely in advancement in the organization of living things, where the latter is marked by increasing specialization of parts and division of labor. When his scattered remarks are taken into account and interpreted carefully, Darwin emerges as both fascinated with the issue of evolutionary progress and as deeply sympathetic with this idea.

As Darwin himself noted in the *Origin*, "I am well aware that scarcely a single point is discussed in this volume on which facts cannot be adduced, often apparently leading to conclusions directly opposite to those at which I have arrived. A fair result can obtained only by fully stating and balancing the facts and arguments on both sides of each question" (Darwin 1859, p. 2). This is sound advice not just in science, but also in the study of science. In seeking to understand Darwin's views on evolutionary progress, we have considered a wide range of his remarks on this topic. Darwin emerges from this study, not just as a proponent of evolutionary progress, but as a careful and subtle thinker whose ideas and arguments, while developed in relation to the cultural values and scientific thought of his day, also display remarkable originality and insight. Certainly the notion that some organisms are "higher" than others, and that there has been direction and progress in the evolution of life, had currency before Darwin. But such ideas did not become serious *problems* inspiring debate until, and because of, Darwin's work. Darwin's vision of the evolution of life as an ascent driven by replacement of less fit organisms by superior forms seemed to reinforce the traditional view of the hierarchy of nature, while his conception of the history of life as a branching treelike structure (rather than a strictly linear ladder of creation à la Lamarck), a true "descent with modification," seemed to call it into question. Darwin's own writings embody this dual conception of life. Despite Darwin's endorsement of a qualified form of evolutionary progress, he never felt entirely comfortable with the idea, calling it "vague" and "ill-defined" on more than one occasion. The challenge for an evolutionary progressionist is to specify more exactly just what this progress consists in, and to provide compelling theoretical arguments in its behalf. As we shall see in the chapter, the question of "evolutionary progress" continued to intrigue, vex, and sometimes infuriate biologists right through the twentieth century.

8

Evolutionary Progress from Darwin to Dawkins

> Evolutionary biologists, it seems, can neither live with nor live without the idea of progress.
>
> (Greene 1990, p. 55)

Introduction

Darwin viewed the evolutionary process as contingently but nonetheless significantly progressive. Under the influence of natural selection organisms become not only better adapted to their conditions of life, but also tend to become more complex and specialized – more improved, in a sense. The synthesis of Darwin's ideas with Mendelian genetics in the first half of the twentieth century resolved many of the problems that led some biologists immediately after Darwin to embrace non-Darwinian evolutionary theories. Ironically, such developments also exacerbated the problems posed for the idea of evolutionary progress. If evolution simply consists of shifts in gene frequencies resulting from selection operating on randomly generated mutations in fluctuating environments, in what sense could the evolutionary process as a whole be considered "progressive"? Different biologists responded to this problem in different ways. The result was a sustained controversy over the meaning and reality of evolutionary progress, the terms of which continue to inform contemporary debates on this issue.

In considering the controversy over evolutionary progress in the twentieth century, I will focus on two debates that serve to highlight the critical biological and philosophical issues involved. The first occurred in the middle decades of the twentieth century and pitted against one another two of the most prominent evolutionists of the day. In writings spanning some four decades of professional activity, Sir Julian Huxley (1887–1975) enthusiastically defended the idea of evolutionary progress as central to the evolutionary process, at one point even declaring that "[G]ranted the existence of Variation and Natural Selection, then biological progress ... must come about" (Huxley 1928, p. 337). By contrast, George Gaylord Simpson (1902–84) argued that "evolution is not invariably accompanied by progress, nor does it really seem to be characterized by progress as an essential feature" (Simpson 1949, p. 262). As it turns out, however, the difference between their views is both less dramatic, and more interesting, than these quotes suggest.

The second debate is more recent, and bears an eerie (but in retrospect hardly surprising) resemblance to the Huxley/Simpson debate (Shanahan 2001). Stephen Jay Gould has been an outspoken opponent of the idea of evolutionary progress in all its forms. Richard Dawkins, by contrast, has argued that the evolutionary process is, in an essential sense, fundamentally progressive. A critical examination of their respective views will bring us as far as we can go in surveying the historical development of this issue, and will provide many of the conceptual resources necessary to tackle the issue philosophically in the chapter.

Julian Huxley's Progressive Evolutionism

Although the details of his views developed during his long career, Julian Huxley's core conviction that the evolutionary process has been characterized by progress never wavered. It was Huxley who christened the coming together of disparate fields of biology under the Darwinian umbrella of natural selection the "modern synthesis" in his *magnum opus*, *Evolution: The Modern Synthesis* (1942), a book that stands as one of the central documents in twentieth-century evolutionary biology. The book is an expanded treatment of his 1936 presidential address to the Zoology Section of the British Association for the Advancement of Science entitled "Natural Selection and Evolutionary Progress" (Huxley 1936), so it is hardly surprising that the issue of evolutionary progress receives special attention. Through it and his other writings, Huxley disseminated the message that the evolutionary process has been characterized by progress as perhaps its most important feature. His impassioned advocacy of evolutionary progress makes him both the outstanding proponent of evolutionary progress in the twentieth century, as well as the favorite whipping-boy for opponents of evolutionary progress (see, for example, Provine 1988).

Because Huxley has long been a lightning rod for critics of evolutionary progress, it is important to accurately understand his view before considering the merits of these criticisms. Fortunately, Huxley left us plenty of material with which to work.[1]

Directional Trends

That the issue of evolutionary progress continued to be problematic and in need of defense in the first decades of the twentieth century is evident from the way in which Huxley frames his earliest essay on the topic. He notes that "To the average man it will appear indisputable that a man is *higher* than a worm or a polyp, an insect *higher* than a protozoan, even if he cannot exactly define in what resides this highness or lowness of organic types" (Huxley 1923, p. 10). It is primarily among professional biologists that doubts about the progressiveness of evolution with its corollary, the distinction of higher and lower forms of life, arise. Their reservations stem from a set of fundamental objections. Huxley considers, and then dispatches, these objections as the prerequisite for proposing his own account of evolutionary progress.

First, there is the issue of adaptation. According to critics, we cannot say that the adaptations characterizing the so-called lower organisms are inferior to those characterizing the so-called higher organisms, and therefore that the processes leading from the former to the latter represent progress. A man is not better adapted to his environment than is a flea which lives upon him as a parasite. Likewise there is little sense to be attached to the notion that a bird is better adapted to life in the air than is a jellyfish to life in the sea. We have, therefore, according to the critics of progress, "no right to speak of one as higher than the other, or to regard the transition from one type to another as involving progress" (Huxley 1923, p. 11). Second, there is the issue of complexity. While some biologists will admit that in the course of evolution there has been an increase in complexity and in the degree of organization, they will nonetheless refuse to regard such increases as having value in themselves, or as constituting biological progress. Being more complex is not necessarily better than being simple. Thirdly, there is the issue of "living fossils," that is, life forms that have remained virtually unchanged for millions of years – like the lamp-shell Lingula, for example. If there is a "Law of Progress," the critics ask, how is it that such creatures are exempt from its operations? The existence of such creatures seems inconsistent with a supposed "Law of Progress." Finally, there is the issue of degeneration, that is, cases where structures are eliminated, reduced, or simplified over the course of evolutionary time. For example, in sedentary or parasitic forms locomotor organs may disappear, sensory and nervous systems may be reduced, and digestive systems may become simplified. How can such "degeneration" be considered progressive?

Huxley had little trouble disposing of the last two objections. No one should deny that there exist life forms that have remained virtually unchanged for long periods of time, or that degeneration is a common biological phenomenon. However, these facts only count against biological progress if one assumes that such progress requires that *all* life forms in *all* lineages must be progressing in order for evolution as a whole to manifest progress. But this is plainly false: "To deny progress because of degeneration is really no more legitimate than to assert that, because each wave runs back after it has broken, therefore the tide can never rise" (Huxley 1923, p. 13). What is important is the *overall* direction or tendency of evolution, not every individual instance.

Huxley handles the first two objections in similar fashion. Granting that the degree of adaptedness has not increased in evolution, all that follows is that progress does not consist in greater adaptedness. Likewise, if complexity has increased in evolution, but the objectors refuse to recognize such increasing complexity as progress, then the basis for judgments of progress must be sought elsewhere. The onus is then on defenders of the notion of progress to identify that quality or quantity that has increased in evolution and that constitutes evolutionary progress.

In this early essay, and again in another one published five years later (Huxley 1928), Huxley enthusiastically takes up the challenge. He begins by asking two key questions: (1) Is there a *direction* discernible in the general evolution of life? (2) If there is, can we say that this direction is *progressive*? To Huxley the answer to the first question was obvious from an examination of the fossil record. There are at least *six* directional trends discernible in the history of life. First, in the course of evolution there has been an increase in *size*, both in the "units of life" themselves (i.e., cells), and in their aggregations (single-celled organisms to metazoan individuals and communities). Second, there has been an increase in *complexity* or the division of labor amongst the parts of organisms. More "efficient" (e.g., faster, more sensitive, more intelligent) organisms appeared after less efficient ones. Each of the improvements better fits its possessor to a certain way of life and renders them less fit for others. They are therefore termed "specializations." As Huxley notes, "Biological specialization moves always in one direction only and is achieved at the expense of improvements in other directions" (Huxley

1928, p. 330).[2] Third, there has been an increase in the *harmony* of these parts, leading to greater unification within organisms. Fourth, there has been an increase in *self-regulation*, according to which organisms become more independent of changes in their external environments. Becoming homeothermic ("warm-blooded") would be an example. Organisms that can maintain a constant body temperature in the face of fluctuating temperatures are more independent of the vagaries of the environment than those lacking this feature. Fifth, there has been an increase in the possibility of bringing the past to bear on present problems, in the form of greater memory, then in rationality, and finally in tradition; in short, in *learning*. Finally, the *psychical* faculties of knowing, feeling, and willing have increased, as has their relative importance for the life of the individual organism. Acknowledging that increases in each of these dimensions has not been universal, Huxley writes that "It is to this increase, continuous during evolutionary time, in the average and especially in the *upper level* of these properties that, I venture to think, the term biological progress can be properly applied" (Huxley 1923, p. 31; emphasis added).

Recognition of these directional trends, Huxley went on to argue, demonstrates that evolution is progressive as well, because it is precisely to the increase in the average and especially in the "upper level" of these properties that the term "biological progress" is properly applied: "Such changes, involving the improvement of the all-round achievements of the organism without depriving it of valuable possibilities, may properly be called biological progress" (Huxley 1928, p. 332). According to Huxley, biological progress is no accident: "[G]ranted the existence of Variation and Natural Selection, then biological progress as well as adaptation (which is the product of specialization) must come about" (Huxley 1928, p. 337).

Control and Independence

In his early writings Huxley emphasized that there are a number of important directional trends in evolution that together constitute evolutionary progress. In subsequent writings he condensed this list to identify what he took to be the single uniquely correct standard of evolutionary progress, viz., increase in those properties facilitating greater control over and independence from the environment, stating that "advance in these respects may provisionally be taken as the criterion of biological progress" (Huxley 1942, p. 562).[3] Huxley did not mean to suggest that higher organisms are in fact completely independent of the environments. Obviously all organisms depend directly on the environments in which they exist. Rather what Huxley was pointing to is the greater ability to cope with *changing* environmental conditions.[4] Control over the environment is simply the ability to alter aspects of the environment to suit one's needs, and thus contributes to independence.

Progress, thus defined, is not the same as specialization. "Specialization . . . is an improvement in efficiency of adaptation for a particular mode of life: progress is an improvement in efficiency of living in general. The latter is an all-round, the former a one-sided advance" (Huxley 1942, p. 562). The more specialized the life form, the more likely it is to go extinct when the conditions to which it is so finely adapted change. Becoming more efficient and independent of environmental vicissitudes, by contrast, are always evolutionary assets.

Biological Advancement

Unsurprisingly, given the contentiousness of the issue, not everyone found Huxley's view convincing. First, there was the charge of anthropocentrism. The American paleontologist George Gaylord Simpson argued that, despite Huxley's disclaimer, his concept of progress was in fact anthropocentric. Simpson was skeptical about those defenses of evolutionary progress that seemed to him to uncritically read back into the evolutionary process a story that eventuates in our own peculiar human adaptations. Gaining greater control over and independence from the environment is precisely the *human* evolutionary specialization, and Huxley (Simpson claimed) had merely read these characteristics back into the history of life (Simpson 1949, p. 251).[5] Second, there is what we might call the "problem of the criterion." Simpson pointed out that it is far from clear that the various criteria Huxley proposed all yield the same conclusions about the evolution of life. For example, "an increase in the psychical powers of organisms, an increase of willing, of feeling, and of knowing" is not the same thing as "increased control over and independence of the environment" (Simpson 1949, p. 258). An organism could be more aware of its environment and have a richer mental life without necessarily having any greater control over or independence from the environment. Pit vipers, for example, have heat sensitive pits located between their eyes and nostrils which enable them to detect the body heat of the small mammals upon which they prey – a mechanism found in no other animals. "This peculiar development certainly counts as progress in perception . . . , but the absence of such an apparatus has no relevance to progress in animals that do not live on warm-blooded prey or that have other adequate means of locating such prey" (Simpson 1949, p. 260).

In writings toward the end of his life, Huxley revised his description of progress in order to respond to these criticisms. In *Evolution in Action* (1953) he reflected that on rereading his youthful essay on "Progress, Biological and Other" (1923), he found that his ideas on the subject were still much the same, with one exception. "The only difference is that I then thought that biological progress could be wholly defined by its results; I now realize that any definition must also take into account the path that it has followed" (Huxley 1953, p. 113). He also distinguishes biological *improvement* from biological *progress*. Improvements are "adaptations which benefit individual organisms at the expense of the species; minor adjustments of the species; specializations of a type for a particular way of life; and advances in the general efficiency of biological machinery" (Huxley 1953, p. 113). Most of these improvements eventually come to a stop, but occasionally a line of advance continues on resulting in continuity of improvement between one group and its successor, for example, between reptiles and mammals. The word "progress," by contrast, denotes the sum of these continuities over the whole of evolutionary time. In light of this later definition of progress, it becomes clear that what had earlier made control over and independence from environmental changes constitutive of biological progress for Huxley is that such innovations make possible further evolutionary breakthroughs. It is from life forms characterized by greater efficiency and independence that later dominant life forms develop. For example, developing a moisture-conserving outer covering, a shelled egg, and lungs, are all biological improvements that permitted invasion of the land by reptiles. Survival in terrestrial environments of fluctuating temperatures placed a premium on the evolution of internal temperature regulation, a property that makes mammals more advanced than reptiles. Maintaining a constant internal environment, in turn, made possible the evolution of complex nervous systems in which experience may be stored, and transmitted to future generations. Each innovation opened up additional evolutionary trajectories. Consequently, "There have been many attempts to define biological progress, or advance in organization. It would seem that the most satisfactory definition is as follows: biological progress consists in biological improvements which permit or facilitate further improvements" (Huxley 1954a, p. 11). Biological progress is thus identified with "improvement which permits or facilitates further improvement; or, if you prefer, as a series of advances which do not stand in the way of further advances" (Huxley 1953, p. 86).[6] Biological progress is thus identified with those adaptive breakthroughs that make possible *further* biological progress.[7]

In one sense, Huxley noted, given the operation of natural selection, such progress is inevitable: "[N]atural selection plus time produces the various degrees of biological improvement that we find in nature" (Huxley 1953, p. 38). Take any world in which natural selection operates and progress as defined above will occur in some of its lines of life. But such progress need not occur in *every* evolutionary lineage. Evolution is always conditioned by accidents that may impede or facilitate progress in particular lineages. In other words, "Progress is inevitable as a general fact; but it is unpredictable in its particulars" (Huxley 1953, p. 115). Still, given that progress is occurring in at least some lineages, one should expect an *overall* progressive tendency in the history of life. Huxley quotes with approval Darwin's remark in the *Origin* that: "The ultimate result [of natural selection] will be that each creature will tend to become more and more improved in relation to its conditions of life. This improvement will, I think, inevitably lead to the gradual advancement of the organization of the greater number of living things throughout the world" (Darwin 1959, p. 221; quoted in Huxley 1953, pp. 39–40; also in Huxley 1954, p. 6). Huxley notes that Darwin never pursued this part of his argument to its logical conclusion, but he did realize that natural selection must in the long-run result in something that deserves to be called "improvement."

Simpson's Pluralistic Conception of Progress

The casual reader might conclude from Simpson's criticisms of Huxley's progressive evolutionism (mentioned in passing above) that he was completely opposed to the idea of evolutionary progress. This would be a mistake. Simpson's view did differ from that of Huxley in significant ways, to be explained below, but to anticipate the discussion that follows he was not opposed to the notion of progress in evolution *per se,* but only to particular criteria of progress that seemed to him to *uncritically* reflect merely a human point of view. Simpson's view can be gleaned from three sources (spanning some twenty-five years) in which he was especially explicit about his view of progress: *The Meaning of Evolution* (1949), an essay on "The History of Life" (1960), and his final statement, "The Concept of Progress in Organic Evolution" (1974).

The Meaning of Evolution

As his example of the pit viper suggests, Simpson thought that increases in perceptual capacities represent a widespread and more fundamental criterion of progress than the "control and independence" criteria proposed by Huxley:

> It is, indeed, progress in perception of and reaction to environment that underlies and makes possible such quite limited degrees of independence and control as have been achieved. It is also evident that this concept of progress is not man-centered but provides a general criterion applicable without necessary reference to the human condition. (Simpson 1949, p. 260)

However, rather than propose *this* as the ultimate criterion of evolutionary progress, Simpson wished to emphasize that there are multiple valid criteria of evolutionary progress. According to Simpson, any progress that we "detect" in evolution will be according to some criterion or another, and there is no criterion that is the sole correct one. "Progress can be identified and studied in the history of life only if we first postulate a criterion of progress or can find such a criterion in that history itself" (Simpson 1949, p. 241). Unlike Huxley, who sought for a single defining characteristic of evolutionary progress, Simpson embraced a thoroughly pluralistic conception according to which there are many forms of evolutionary progress. Nonetheless, "there is no criterion of progress by which progress can be considered a *universal* phenomenon of evolution" (Simpson 1949, p. 243). It is a mistake to assume that there is a single standard of progress that has general validity in evolution, rather than "a multitude of possible points of reference" (Simpson 1949, p. 242). In *The Meaning of Evolution* Simpson catalogues at least a dozen different, alternative, equally valid, criteria of progress. Among them are: (1) "a tendency for life to expand, to fill in all the available spaces in the livable environments, including those created by the process of that expansion itself" (p. 243); (2) increase in variety and abundance, not in life as a whole, but within a given group of organisms; that is, "successive dominance" (p. 245); (3) successive invasion and development of new adaptive zones (e.g., the land, the air, etc.); (4) replacement *within* adaptive zones, connected with "better adaptation" (pp. 248–49); (5) increasing specialization (p. 250); (6) change that broadens the chances of further change (p. 250); (7) increased ability to cope with a greater variety of environments (p. 251); (8) greater control over the environment (p. 252); (9) increasing complexity (pp. 252–53); (10) increase in the general energy or maintained level of vital processes (p. 256); (11) reproduction and care of young (p. 257); and (12) increased awareness and perception of the environment (p. 258). In short, "there is no sense in which it can be said that evolution *is* progress. Within the framework of the evolutionary history of life there have been not one but many different sorts of progress . . ." (Simpson 1949, pp. 261–62). He goes on to make clear that in his view, "Among many possible definitions of progress, and many corresponding sorts of progress in evolution, that of change toward a particular sort of organism is as valid as any other as long as it is clearly understood to be specific with respect to a selected point and subjective in this sense" (Simpson 1949, p. 262). The contrast here with Huxley's view is striking. For Huxley, there is objective progress (however this is to be defined). For Simpson, progress is imposed upon biology by humans and is a matter of interpretation. ("Progress is not an intrinsic quality, that exists independently of human thought.") Since we can postulate different criteria of progress, we can view evolution as progressive in a number of different ways.[8]

To summarize, Simpson was not really opposed to the notion of progress in evolution at all. His objections, rather, were with *particular* criteria of progress that had been proposed that seemed to him to uncritically reflect a human point of view. More importantly, his target was not so much particular criteria of progress that might be proposed from an inductive scrutiny of the history of life on earth, as it was the idea that however progress is defined, such progress was in some sense *necessary* or *inevitable*: "Progress has occurred within it but is not of its essence" (Simpson 1949, p. 261). On this point, Simpson and Darwin were in complete agreement.

The History of Life

In his 1960 essay, Simpson further clarified his view. He pointed out that the history of life is obviously "progressive" by one definition. Insofar as the history of life proceeds by successive stages, each one derived from and differing from those preceding it, then it is undergoing "progressive change." But all this means is that the change is in a certain direction. Asserting that this change constitutes overall *improvement* or change for the *better* is another, much stronger, claim. "Better" is an evaluative term that is meaningful only in light of some standard for rendering such judgments. One such standard does make biological sense:

> If a useful function comes to be performed more effectively or if a new function adds to utility, then there has been improvement. It is meaningful to consider such improvement as progress. . . . The limb of a mammal cannot be considered an improvement over the fin of a fish, because neither performs more effectively the functions of the other or adds (without equivalent loss) to the functions of the other. But some fins function better as fins than others, and some legs as legs. There is improvement among fins and among legs, but not between the two. (Simpson 1960, p. 176)

It might be tempting to consider some sorts of improvement to be more general, and to thus transcend the functional needs of particular lineages. Take eyes, for example. It might be thought better to have eyes than to lack them, and surely the vertebrate eye is an improvement over a protozoan's simple light-sensitive pigment spot. But even this example fails as an instance of a general improvement because the vertebrate eye is only an improvement if accompanied by a nervous system that can process visual information. A plant would not obviously be better off with eyes, even if accompanied by a nervous system. In the end, Simpson concludes, the only good candidates for genuinely universal improvements are those features common to *all* organisms and involved in the origin of life.

The Concept of Progress in Organic Evolution

In his 1974 essay Simpson distinguishes between "succession," "progression," and "progress": "Succession" means no more than change in time; progression indicates, or should indicate, no more than that the changes are continuous and connected . . . more or less gradual; progress means, or should mean, that the changes are for the better" (Simpson 1974, p. 330). Clearly there can be succession without progression, and progression without progress. While there is no doubt that a succession of organic entities has occurred (i.e., organic evolution), questions of both progression and progress are different matters. Directly after these remarks, however, Simpson seems to qualify his notion of progress. He writes that,

> Some organisms *are* better than their ancestors or than some of their relatives at doing certain things in certain ways. Some oysters are better at being oysters than their ancestors. Some trees are better at living on mountain tops than others. . . . With such examples it is perfectly reasonable to say that improvement has factually occurred and that there is therefore evolutionary progress. The progress is, however, ad hoc in every case. Our ancestors' progress was not the oysters', the trees', or the monkeys', nor was theirs ours. (Simpson 1974, pp. 50–51)

In other words, recognition of adaptive improvement within specific lineages does not warrant comparisons between lineages nor claims about the overall progressiveness of evolution.

Summary: Huxlean versus Simpsonian Evolutionary Progress

Simpson's general approach suggests an important alternative to the Huxlean conception of evolutionary progress.[9] According to Huxley, evolutionary progress is a natural fact about the living world that exists independently of human thought. "Progress" is no mere imposition on the history of life by human beings. Rather, "there was progress before man ever appeared on earth, and its reality would have been in no way impaired even if he had never come into being. His rise only continued, modified, and accelerated a process that had been in operation since the dawn of life" (Huxley 1923, p. 40). On Simpson's view, by contrast, every claim that the history of life manifests progress requires the postulation of a standard which we select in light of our interests. Despite the fact that such standards need not correspond to some *uniquely correct* fact about evolution, claims that evolution manifests long-term progress can nonetheless be *true.* Taking Simpson's approach permits one to say, for example, there has been long-term directional change and improvement in the biologically-relevant property of seeing. From the origin of life until the present, the ability to sense the environment using photoreceptors has increased; and when it has, the organisms possessing greater visual abilities have benefited. This is so despite the fact that not all organisms would benefit from possessing eyes, or eyes of the most improved sort. Long-term evolutionary progress, on this view, simply requires that the maximum value of some biologically relevant property increases during the history of life. The more biologically relevant properties whose maxima have increased during the history of life, the more life on earth will be characterized by evolutionary progress.

Gould on Evolutionary Progress

Among twentieth-century evolutionists, Stephen Jay Gould was surely the most dogged opponent of the idea of evolutionary progress. As a recurring theme in a number of popular essays, in a professional paper devoted to this topic, and finally in a full book-length treatment, he attempts to quash the notion that the march of evolution is ever upward and onward. One essay begins as follows: "Progress is a noxious, culturally embedded, untestable, nonoperational, intractable idea that must be replaced if we wish to understand the patterns of history" (Gould 1988, p. 319). Writing eight years later, he adamantly denies "that progress characterizes the history of life as a whole, or even represents an orienting force in evolution at all" (Gould 1996, p. 3). Indeed, the overarching aim of his book *Full House* is to present "the general argument for denying that progress defines the history of life or even exists as a general trend at all" (Gould 1996, p. 4). Examining Gould's arguments is essential for understanding the contemporary status of evolutionary progress. The reasoning in *Full House* is diffuse and nonlinear, but at least five distinct arguments intended to impugn the notion of evolutionary progress can be isolated.[10] To anticipate the discussion that follows, I will argue that Gould's arguments fail to establish what they are intended to establish, and thus that Gould has not presented convincing arguments showing that evolutionary progress is an illusion.

The Anti-Egocentric Argument

It is often wise to be deeply skeptical of those claims that, were they believed to be true, would improve our individual or collective self-image. A belief in evolutionary progress that puts *Homo sapiens* at the very pinnacle of the history of life is, according to Gould, just such a belief. It may seem intuitively obvious that some organisms are "higher" than others, and that human beings are the "highest," but trusting one's intuitions is liable to be unreliable in this case because we have an insidious (but natural) tendency to place ourselves at the top, and to arrange all other living things in relation to us somewhere down the evolutionary ladder. We do this, says Gould, because "We crave progress as our best hope for retaining human arrogance in an evolutionary world. Only in these terms can I understand why such a poorly formulated and improbable argument maintains such a powerful hold over us today" (Gould 1996, p. 29). Gould's "Anti-Egocentric Argument" is simply this: The very fact that we are so predisposed to believe in evolutionary progress, and to place ourselves at evolution's pinnacle, should render this belief deeply suspect. "Our geological confinement to a moment at the very end of recorded time must engender suspicions that we are a lucky accident, an afterthought rather than the goal of all creation. Progress is the doctrine that dispels this chilling thought..." (Gould 1988a, p. 319).

As an argument against evolutionary progress, The Anti-Egocentric Argument is clearly unsound. Granting the reasonable point that we should be wary of uncritically accepting conclusions that we have good reason to think we are predisposed to accept, this provides no positive reason to *reject* those conclusions. After all, at least some of the things that it is in our collective self-interest to believe (e.g., that it is unlikely that our sun will explode tomorrow, destroying all life on earth) also happen to be *true*. Likewise, if there is evolutionary progress according to some standard, then it is certainly possible that *Homo sapiens* is the most progressive organism according to that standard. Just as we should be extremely wary of succumbing to an anthropocentric bias, so, too (as Simpson pointed out long ago), should we be wary of rejecting claims simply because they are *consistent* with a detested anthropocentric bias.[11]

The No Inherent "Thrust" Argument

A second Gouldian argument against belief in evolutionary progress equates progress with some sort of inherent, ineluctable "thrust" making progress inevitable and unavoidable. Gould points out that there is no empirical evidence that a "pervasive and predictable thrust toward progress permeates the history of life" (Gould 1996, p. 146). Instead, the history of life appears to be rife with contingency, making each stage in the process unpredictable given what came before. There is nothing about the evolutionary process that would make progress inevitable, or even likely. Gould has elsewhere explored in detail the themes of chance, contingency, and historicity in evolution (Gould 1986), and the present argument complements those reflections.

Obviously the key concept in this argument is that of "thrust." As noted in Chapter 5, at the beginning of the twentieth century various theories of "orthogenetic evolution" thrived, according to which there are inherent forces operating in living things that drive evolution along predetermined pathways, for example, toward greater size, or complexity, or specialization. Since the evidence for such forces is nonexistent, Gould is quite right to reject such reasons for supposing that there is evolutionary progress. But, because belief amongst evolutionary biologists in such forces essentially died out with the triumph of the modern synthesis, the argument understood in this sense only defeats a belief that no longer exists in the scientific community.

The Random Motion Argument

Gould summarizes a related but distinct argument for the nonreality of progress in evolution in the following words: "The vaunted progress of life is really *random motion away from simple beginnings*, not *directed impetus toward inherently advantageous complexity*" (Gould 1996, p. 173; emphasis in original). If one thinks of organisms as occupying an abstract multidimensional "morphological space," then evolution consists largely in the migration (or better yet, the drifting) of lineages into different regions of this space. Some regions represent organisms with greater complexity, others represent organisms of lesser complexity. Because life began in a simple, relatively uncomplicated form, the only regions of morphological space available for colonization were those for more complex organisms. Consequently, some organisms became more complex, not because increased complexity was "better," but just because there was nothing else to do *but* become more complex.[12] The crucial point for Gould is that even if lineages change *randomly* one would predict this result. If evolution occurs at all, then greater complexity will arise – not as a result of an internal "push" in this direction, but just in virtue of the limited range of possibilities open to it. Gould considers this observation to fatally undermine a belief in evolutionary progress.

It is important to be clear about what this argument does and does not establish. The most that this argument shows is that "directed impetus toward inherently advantageous complexity" is not *necessary* to give rise to greater complexity, because there are ways of explaining increased complexity without it. However, it does not and cannot show that trends toward increased complexity are illusory (indeed, the argument seems to presuppose that such trends are real). Nor can it show that there is no inherent evolutionary dynamic driving lineages toward greater complexity. Showing that a particular cause is not *necessary* is obviously not adequate to show that that cause is *non-existent.* Those biologists friendly to the idea of evolutionary progress believe that there are forces operating within evolution that make progress extremely likely, although not guaranteed. The most important of these forces is natural selection. As we have seen, Darwin believed that natural selection operating amongst predators and prey will contingently result in progressive evolution of a certain sort. Therefore, at most Gould has offered an alternative explanation of the tendency for life to become more complex as it evolves, not a refutation of this trend – one that, moreover, evolutionary progressionists can and do easily embrace alongside their own explanations.[13]

The Biotic Domination Argument

A fourth Gouldian argument against evolutionary progress focuses on the relative *numbers* of different kinds of living things. Despite the fact that the earliest fossils show only rather simple bacteria, we now have eagle-eyed eagles, sonar-equipped bats, electric fish, spitting cobras, bombardier beetles, and all the rest. According to Gould, however, this is no evidence of progress, because "the earth remains chock-full of bacteria, and insects surely dominate among multicellular animals – with about a million described species versus only four thousand or so for mammals. If progress is so damned obvious, how shall this elusive notion be defined when ants wreck our picnics and bacteria take our lives?" (Gould 1996, p. 145). Thus, evolutionary progress is an illusion because bacteria and insects far outnumber mammals.

The logic of this argument may not be apparent, so it is worth spelling it out more formally:

Premise: There are more bacteria (and species of insects) in the world than there are mammals (or species of mammals).
Conclusion: Hence, evolutionary progress is an illusion.

Stated in this way, the argument is clearly invalid. The missing premise needed to make this into a *valid* argument would be something like this: The claim that evolutionary progress is real and not just an illusion can only be justified if the organisms (or species) deemed to be more "advanced" also *outnumber* all other organisms (or kinds of organisms). But once the missing premise is stated so baldly its absurdity becomes evident. No one who has argued for the reality of some form of evolutionary progress has insisted that the most "advanced" forms also be the most *numerous*. What the most ardent defenders of the notion of evolutionary progress have always maintained is that the "upper limit" of structural and behavioral complexity amongst organisms constitutes "progress" relative to some set of earlier forms. There is nothing whatsoever in this claim about the relative numbers of the organisms of each type. To put the point in the simplest possible terms, Gould's central empirical argument against evolutionary progress in *Full House* trades on a confusion between *quality* and *quantity*. The issue of evolutionary progress concerns the "quality" of organisms that evolve (as measured, perhaps, in biomechanical advances), not how many of them there are (the "quantity"). Despite the fact that the phylum Arthropoda constitutes 80 percent of multicellular animals, yet "displays no trend to neurological complexity through time" (Gould 1996, p. 15), this has no bearing on the question of evolutionary progress. To answer Gould's rhetorical question in kind, Why should the fact that ants wreck our picnics be thought to have any relevance to the issue of evolutionary progress?

The "Full House" Argument

Finally, Gould identifies the aim of his book *Full House* as motivating "viewing a history of change as the increase or contraction of variation in an entire system (a 'full house'), rather than as a 'thing' moving somewhere." The contrast in question is between an appreciation of "life's infinite variety," and the view that a "pervasive and predictable thrust toward progress permeates the history of life" (Gould 1996, p. 146). Gould's sympathies are clearly with the former vision rather than the latter.

But it is not clear why increased biodiversity and evolutionary progress should be treated as mutually exclusive alternatives. Conceivably the history of life might show *both* increase (or contraction) in variation in an entire system *and* directionality, or even progress. Gould's discussion in *Full House* is largely unintelligible unless one accepts this questionable dichotomy.

Summary: Gould on Evolutionary Progress

The five Gouldian arguments against evolutionary progress identified above are apparently intended to refute the following five theses, respectively: (i) *Homo sapiens* is the predetermined *raison d'être* or *telos* of the evolutionary process; (ii) there is an inherent force in the evolutionary process "thrusting" it forward along a predetermined route; (iii) increasing complexity is *inherently* advantageous, irrespective of environment; (iv) in order for evolution to be progressive, the most advanced organisms must outnumber (in either species or individuals) less advanced forms; and (v) the history of life resembles an unbroken chain of ascent rather than a branching tree structure. The problem is that there are very few, if any, professional biologists who would be tempted to accept any of these theses. Why, then, does Gould devote an entire book to their critique?

Gould's arguments are puzzling if assumed to be directed to other evolutionary biologists. However, when understood as an exercise in educating a public that does harbor many such false ideas about evolution, the puzzle is resolved. By his own admission, his treatment is explicitly aimed at a popular audience:

> The truth of evolution and the power of Darwinian explanation have been so liberating and transforming in the history of Western thought – while the popular drive to oppose, or simple failure to understand, remain so entrenched (given the threat and uncongeniality of evolution to so much held dear both by psyche and society) – that writings for the general public must lean toward the hedgehog's task of documenting and defending the veracity and explanatory range of evolution itself. (Gould 1997a, p. 1020)

Not only is the nonscientific public antipathetic toward evolution, but their view of evolution is so distorted as to be almost unrecognizable by professional biologists, a point Gould was already making two decades before he wrote *Full House*: "[S]cientists . . . long ago abandoned the concept of necessary links between evolution and progress as the worst sort of anthropocentric bias. Yet most laymen still equate evolution with progress"

(Gould 1977, p. 37). This might not be so bad, Gould has argued, were it not for the dire consequences of this equation:

> This fallacious equation of organic evolution with progress continues to have unfortunate consequences. Historically, it engendered the abuses of Social Darwinism. . . . Today, it remains a primary component of our global arrogance, our belief in dominion over, rather than fellowship with, more than a million other species that inhabit our planet. (Gould 1977, pp. 37–38)

Gould's primary interest in his popular works is in educating the public about evolution by disabusing them of ideas which have no place in a Darwinian understanding of life. Despite the international appeal of his books, Gould writes for a predominantly American audience, many of whom still consider evolution a threat to traditional religious values. Gould thus feels acutely the need to distance Darwinian thinking as much as possible from popular misconceptions of it.[14]

Dawkins on Evolutionary Progress

Richard Dawkins has produced a stream of books on evolution acclaimed for their lucid prose and accessibility to the scientific layperson. Sometimes unjustly dismissed as a mere "popularizer," he also has made important original contributions to science, while successfully articulating a distinctive perspective in contemporary evolutionary biology. Dawkins and Gould have never been reticent to point out the perceived flaws in each other's understanding of evolution. The two-decades-old "warfare" between them has been waged on a number of distinct but related battlefields, but central to these skirmishes are fundamental theoretical and methodological issues concerning the units of selection (Dawkins 1989a; Gould 1980a), the cogency of adaptationism (Gould and Lewontin 1979; Dawkins 1982b) and, most recently, the reality of evolutionary progress (Dawkins 1997; Gould 1996a, 1997a; see Sterelny 2001 for an overview). Here I want to examine in some detail Dawkins's view of evolutionary progress, including his critique of Gould's view.

The "Adaptationist" Definition

In his review of *Full House* (which appeared in the journal *Evolution*) Dawkins writes: "Gould's definition of progress, calculated to deliver a negative answer to the question of whether evolution is progressive, is 'a tendency for life to increase in anatomical complexity, or neurological elaboration, or size and flexibility of behavioral repertoire, or any criterion obviously concocted... to place *Homo sapiens* atop a supposed heap'" (Dawkins 1997, p. 1016). While agreeing with Gould that "complexity, braininess and other particular qualities dear to the human ego should not necessarily be expected to increase progressively in a majority of lineages . . ." (Dawkins 1997, p. 1018), he nonetheless finds fault with Gould's broader critique of evolutionary progress:

> By Gould's enthusiastic account... there is no general evidence that a statistical majority of evolutionary lineages show driven trends in the direction of increased complexity.... Gould is sailing dangerously close to the windmill tilting he has previously made his personal art form. Why should any thoughtful Darwinian have expected a majority of lineages to increase in anatomical complexity? Certainly it is not clear that anybody inspired by adaptationist philosophy would. (Dawkins 1997, p. 1017)

In Dawkins's view, "Gould is wrong to say that the appearance of progress in evolution is a statistical illusion . . ." (Dawkins 1997, p. 1018), because there is an alternative, and more plausible, way of construing evolutionary progress than simply as "increased anatomical complexity."

Dawkins's alternative, "adaptationist" definition of progress is "a tendency for lineages to improve cumulatively their adaptive fit to their particular way of life, by increasing the numbers of features which combine together in adaptive complexes" (Dawkins 1997, p. 1016). This "adaptationist" definition of progress, Dawkins states,

> takes progress to mean an increase, not in complexity, intelligence or some other anthropocentric value, but in the accumulating number of features contributing towards whatever adaptation the lineage in question exemplifies. By this definition, adaptive evolution is not just incidentally progressive, it is deeply, dyed-in-the-wool, indispensably progressive. It is fundamentally necessary that it should be progressive if Darwinian natural selection is to perform the explanatory role in our world view that we require of it, and that it alone can perform. (Dawkins 1997, p. 1017)

It is the *necessity* of adaptive evolution being progressive that comes through most clearly in Dawkins's presentation. Natural selection is a *cumulative* process in which small gains in adaptive fit are saved, and become the foundation for further adaptive gains (Dawkins 1986, pp. 43–74). Consequently, "the evolution of complex, manyparted adaptations must be progressive" because "[l]ater descendants will have accumulated a larger number of components towards the adaptive combination than earlier ancestors" (Dawkins 1997, pp. 1017–18). Take, for example, that favorite example of natural theologians and natural selectionists alike, the vertebrate eye. Starting from ancient ancestors possessing a simple light-sensitive patch containing only a few features good for detecting light gradients, there has been a cumulative process of step-by-step "gradual, progressive increase in the number of features which an engineer would recognize as contributing towards optical quality," eventually resulting in "the modern, multifeatured descendent of that optical prototype" (Dawkins 1997, p. 1018). Consequently, "The evolution of the vertebrate eye *must* have been progressive.... Without stirring from our armchair, we can see that it must be so" (Dawkins 1997, p. 1018).

What is especially striking about this argument, of course, is the forthrightness with which Dawkins reveals its virtual *a priori* character. It is that we *must*, if we wish to solve the problem of how complex, functional systems come into existence without any appeal to intentional agency, attribute this to the power of natural selection, because it is natural selection, and natural selection alone, that can perform this essential explanatory role. But since evolutionary progress just is "a tendency for lineages to improve cumulatively their adaptive fit to their particular way of life, by increasing the numbers of features which combine together in adaptive complexes" it follows that evolution is and must be progressive. Thus, "[If] you define progress less chauvinistically – if you let the animals bring their own definitions – you will find progress, in a genuinely interesting sense of the word, nearly everywhere" (Dawkins, 1997, p. 1018).

Arms Races

What inspires Dawkins' confidence that this account of evolution is correct? Part of the answer, he tells us, concerns evolutionary "arms races" in which predator and prey are locked into a contest to achieve any slight superiority over the other, each driving the other to greater adaptive refinement (Dawkins and Krebs 1979). As predators become more efficient at capturing prey, prey in turn evolve greater efficiency at escaping, which in turn places greater selective pressure on the predators, and so on in an upward spiral of adaptive improvement. As Dawkins notes, "The resulting positive feedback loop is a good explanation for driven progressive evolution, and the drive may be sustained for many successive generations" (Dawkins 1997, p. 1018). He regards such a process "as of the utmost importance because it is largely arms races that have injected such 'progressiveness' as there is in evolution" (Dawkins 1986, p. 178).[15]

Dawkins is careful to offer two caveats to such claims. First, this does *not* mean that all living things will be progressing in all features, but only that a certain kind of progress is likely: "Adaptation to the weather, to the inanimate vicissitudes of ice ages and droughts, may well not be progressive: just an aimless tracking of unprogressively meandering climatic variables. But adaptation to the biotic environment is likely to be progressive because enemies, unlike the weather, themselves evolve (Vermeij 1987)" (Dawkins 1997, p. 1018; see also Dawkins 1986, pp. 178–179). Second, on the adaptationist view Dawkins favors, progressive evolution is not expected to continue indefinitely. A given coevolutionary arms race may last for millions of years but probably not for hundreds of millions of years. Other factors, such as physical constraints, will eventually play a limiting role (e.g., predators will not keep increasing in speed until they are running at Mach 2). In addition, a given kind of organism may be involved in *multiple* arms races simultaneously (e.g, with predators, with parasites, etc.) and an acceptable level of success in one race may entail compromises in another. Third, one should not expect the improvement to be continuous and "smooth." It is more likely to be "a fitful affair, stagnating or even sometimes going 'backwards', rather than moving solidly 'forwards' in the direction suggested by the arms-race idea.... There may well be long stretches of time in which no 'progress' in the arms race, and perhaps no evolutionary change at all, takes place" (Dawkins 1986, p. 181). Arms races will sometimes culminate in extinction, at which point a new arms race may begin. Consequently, it may be true that for a given lineage "there was no global progress over the hundreds of millions of years, only a sawtooth succession of small progresses terminated by extinctions. Nonetheless, the ramp of each sawtooth was properly and significantly progressive" (Dawkins 1997, pp. 1018–19).

Not surprisingly, Gould finds this sort of argument less than fully convincing. The problem has to do with the *sort* of progress such a process is capable of generating:

> Do we consider a poker game progressive when players up the ante? Not usually, I think. The stakes are higher, but the rules don't change; a full house still beats a flush.... Is a snail with a thick shell "better" than its thinner-shelled ancestor because an increase in the power of crushing predators requires this degree of strength to achieve the same adaptation that ancestors attained with thinner shells? The later world is different by virtue of such "arms races," but in what usual sense of the term can we proclaim it better? (Gould 1988a, p. 325)

In other words, a snail with a thick shell is no "better off" (i.e., in terms of *fitness*, the number of offspring it produces) in relation to a starfish with great crushing power than its thin-shelled ancestor was against its relatively weaker enemy.

Dawkins anticipates this objection, and his response helps to further clarify his view. He acknowledges that "The participants in the race do not necessarily survive more successfully as time goes by – their "partners" in the coevolutionary spiral see to that (the familiar Red Queen Effect).

But the *equipment* for survival, on both sides, is improving judged by engineering criteria" (Dawkins 1997, p. 1018; cf. Dawkins 1986, pp. 182–83). Although later organisms in an evolving lineage may be no fitter relative to their enemies than earlier organisms were relative to theirs, later organisms in an evolving lineage are progressive relative to earlier organisms in the sense that were one to pit a contemporary prey organism and its direct ancestor against a contemporary predator, the contemporary prey organism would stand a much better chance of surviving. For example, "[I]t does at first seem to be an expectation of the arms race idea that modern predators might massacre Eocene prey. And Eocene predators chasing modern prey might be in the same position as a Spitfire chasing a jet" (Dawkins and Krebs 1979, p. 506). Although the contemporary prey organism might not fare any better against a contemporary predator than the ancestor would against the ancestral predator (that is, it would not have greater "fitness" in relation to its environment), it would still be true that within the prey organism's lineage there has been a progressive shift toward greater anatomical effectiveness (Dawkins 1986, p. 183).

How "Pervasive" Is Evolutionary Progress?

A distinct but related issue dividing Gould and Dawkins concerns the *pervasiveness* of evolutionary progress. Recall that Dawkins claimed that no Darwinian inspired by adaptationist philosophy would expect a statistical majority of lineages to show an increase in anatomical complexity. Dawkins' preferred construal of evolutionary progress is "a tendency for lineages to improve cumulatively their adaptive fit to their particular way of life, by increasing the numbers of features which combine together in adaptive complexes" (Dawkins 1997, p. 1016). Later he writes: "Progressive increase in morphological complexity is to be expected only in taxa whose way of life benefits from morphological complexity. . . . But what I do insist on is that in a majority of evolutionary lineages there will be progressive evolution toward *something*. It won't, however, be the *same* thing in different lineages" (Dawkins 1997, p. 1018). Putting these two claims together, we see that according to Dawkins, "in a majority of evolutionary lineages there will be progressive evolution toward *something*," progressive evolution is just "a tendency for lineages to improve cumulatively their adaptive fit to their particular way of life, by increasing the numbers of features which combine together in adaptive complexes," from which it seems to follow that in a majority of evolutionary lineages there will be a tendency to increase the number of features that combine together in adaptive complexes. It is unclear how this can be squared with Dawkins's claim that no thoughtful Darwinian would expect a majority of lineages to increase in anatomical complexity. What would "an accumulating number of features contributing toward adaptive fit by increasing the numbers of features which combine together in adaptive complexes" *be* if not increasing complexity? Although the anatomical details will vary from lineage to lineage, anatomical complexity itself can be defined univocally as 'the information content of the description of that animal' (Dawkins 1992, p. 265). What is true *across* lineages should, *a fortiori*, also be true *within* a given lineage. It is therefore hard to avoid the conclusion that, despite his explicit disavowal of this notion, Dawkins really does see natural selection driving a majority of evolutionary lineages toward greater complexity. As he remarks at one point, "Directionalist common sense surely wins on the very long time scale: once there was only blue-green slime and now there are sharp-eyed metazoa" (Dawkins and Krebs 1979, p. 508).

The Evolution of Evolvability

Finally, according to Dawkins there is a sense in which evolution itself may evolve, progressively, over a longer timescale than the individual ramps of the arms race sawtooth. Major innovations in embryological technique represent "watershed events" in the history of life opening up new vistas of evolutionary possibility (Maynard Smith and Szathmáry 1995). Using language that sounds strikingly reminiscent of Huxley's final definition of genuine evolutionary progress, Dawkins writes that such events "constitute genuinely progressive improvements . . . [n]ot just in the normal Darwinian sense of assisting individuals to survive and reproduce, but . . . in the sense of boosting evolution itself in ways that seem entitled to the label progressive" (Dawkins 1997, pp. 1019–20). Elsewhere he has christened this phenomenon "the evolution of evolvability" (Dawkins 1989).

In summary, on Dawkins's view, progress is an important aspects of evolution. Rather than being merely a minor feature of evolution, it is one of its most distinctive features, arising in numerous lineages. Given the operation of natural selection, evolutionary progress is virtually inevitable. And far from characterizing just small atypical parts of the ramifying tree of life, it recurs repeatedly. The contrast with Gould apparently could not be more complete.

Summary: Evolutionary Progress from Darwin to Dawkins

The question of whether evolution is "progressive" has occupied (and sometimes preoccupied) evolutionists from Darwin to the present (Ruse

1996). As we saw in Chapter 7, Darwin viewed the evolutionary process as contingently but nonetheless significantly progressive. Under the influence of natural selection organisms become not only better adapted to their conditions of life but also tend to become more complex and specialized. But this view faces a serious problem: What reason is there to believe that there is anything inherent in the operation of natural selection that would lead one to expect such a result?

This issue first received the serious attention it deserves in the writings of Julian Huxley and George Gaylord Simpson. Huxley was a lifelong advocate of the idea of evolutionary progress. Although his view continued to develop throughout his long career, he consistently maintained that the evolutionary process is progressive, and that there were objective criteria that rendered this conclusion unavoidable. Like Huxley, Simpson, too, believed that the evolutionary process could be described as progressive, but unlike Huxley maintained that there are multiple ("subjective") criteria by which such a verdict could be rendered.

More recently, Stephen Jay Gould and Richard Dawkins have debated the nature and status of the claim that evolution is, in some sense, progressive. Whereas Gould presents a battery of arguments designed to undermine belief in evolutionary progress, Dawkins maintains that in an essential sense, evolution is fundamentally progressive. It is still too soon to determine which of these positions (if either) will ultimately prevail. But it is undeniable that among evolutionary biologists there is a strong tendency to dismiss the idea of evolutionary progress as "non-Darwinian" (e.g., Durant 1992), despite the fact that Darwin himself argued for the reality of evolutionary progress, and that many of the major evolutionists in the twentieth century believed that evolution does manifest progress in one sense or another.

The historian of science John C. Greene is evidently correct: "Evolutionary biologists, it seems, can neither live with nor live without the idea of progress" (Greene 1990, p. 55). Nonetheless, identifying the views of influential biologists at different periods in the history of evolutionary thought is one thing; identifying good *arguments* for each position in order to determine which is better justified, is another. In the next chapter, I examine some of the distinctions already introduced in passing in order to spell out more precisely a conception of evolutionary progress that even its staunchest critics might be able to live with.

9

Is Evolution Progressive?

> Each species changes[.] [D]oes it progress? . . . [T]he simplest cannot help becoming more complicated; & if we look to first origin[,] there must be progress.
>
> (Darwin, B Notebook, p. 18; in Barrett et al. 1987, p. 175)

Introduction

As we have seen, the idea that evolution manifests progress has long been problematic for Darwinians. On the one hand, the history of life appears to be the story of the gradual emergence of higher, more advanced life forms. Thus evolution appears to embody spectacular biological progress. On the other hand, every supposed example of evolutionary progress is confronted with a counterexample. Alongside masterpieces of biological engineering are organisms with flawed designs, creatures that have changed little in millions of years, and even creatures that have apparently regressed to structurally simpler forms, confounding the claim that evolution is ever onward and upward. Evolutionary progress is also suspect on theoretical grounds. Natural selection contains no inherent perfecting mechanism, but only insures that organisms will be relatively well adapted to whatever local environment their immediate ancestors occupied. As environments change, organisms will either track them or go extinct, but there is no reason to assume that this change represents consistent advance in any particular direction. Finally, the idea of evolutionary progress has been subjected to philosophical critique as well. In particular, "progress" has seemed too anthropocentric, value-laden, and subjective to be considered a respectable scientific concept, and has thus been dismissed as merely symptomatic of various historical or social factors lacking any epistemic force (Gould 1988a; Ruse 1996). In this view, claims that the history of life manifests progress are to be understood as thinly veiled attempts to provide a scientific basis for what is essentially faith that the present is an improvement on the past, and that the future promises even greater glories. If evolutionary progress once seemed obvious to biologists, the tide has now clearly turned. As Nitecki notes, "The concept of progress has been all but banned from evolutionary biology as being anthropomorphic or at best of limited and ambiguous usefulness" (Nitecki 1988, p. viii). Anyone wishing to endorse the idea of evolutionary progress must be prepared to confront this concern, offer positive arguments on its behalf, and rebut the slew of objections that have been leveled against it – no easy task.

Aims and Strategies

Reflection on the longevity of the debate over evolutionary progress might tempt one to conclude that it is inherently unresolvable, and to focus instead on explaining the debate in social or cultural terms. An examination of various "contextual factors" underlying debates about evolutionary progress is certainly necessary (Ruse 1996; Shanahan 2001). But an overemphasis on such factors can obscure the fundamental epistemic issues at stake. The discussion that follows approaches the question of evolutionary progress as a substantive issue requiring serious conceptual analysis and critical evaluation. My focus in the following, therefore, will be on understanding the idea of evolutionary progress itself, and on whether the history of life is (in some sense) properly described as manifesting progress. As I will try to show, the question "Is evolution progressive?," properly understood, admits of the same sort of answer as other questions in evolutionary biology – which is not to say that it can necessarily be definitively answered one way or the other, but that there are good reasons to conclude that some views on this issue are more satisfactory than others.

My strategy is as follows. Determining whether the history of life manifests progress requires first of all a clear understanding of what "evolutionary progress" means. Following a number of other writers, I will suggest that it includes both a descriptive and an evaluative component (direction and improvement, respectively). What it might *mean* for evolution to have a "direction," and to embody "improvement," are then explored in some detail. I then turn to the question of whether evolution *does* have a direction, and in what sense (if any) it might be said to embody improvement. Finally, I consider various objections to the claim of evolutionary progress, and respond to each. Whereas the reality of evolutionary progress could once be taken for granted, contemporary critics of that idea now almost take for granted that there are and can be no good arguments for evolutionary progress. On the basis of the present analysis, however, I conclude that there are several ways in which the claim that evolution is progressive can be justified, and that although evolutionary progress *may* be illusory, it is not *obviously* so.

What Is Evolutionary Progress?

The preceding chapters discussed evolutionary progress as it has been understood by various influential biologists. Here I want to identify those features of the idea with which virtually everyone, despite their other differences, would agree. For proponents and critics alike, evolutionary progress is "directional change towards the better" (Ayala 1988, p. 78). More exactly, "A lineage shows progressive change precisely when two conditions are satisfied: (1) there must be an increase in some quantitative characteristic (like complexity or size); and (2) the increase must be a change for the better. In short, Progress = Directional Change + Values" (Sober 1994, p. 20). There is, of course, no reason why directional change must be described as an *increase* in any characteristic. Any increase can also be described as a *decrease*: An increase in metabolic efficiency is also a decrease in metabolic inefficiency. The converse is also true. Perhaps the most famous directional evolutionary trend of all is the reduction of toes in horses, which presumably corresponds to an increased ability to efficiently traverse grassy plains. What is essential is that there be a directional trend in which some characteristic changes in a consistent direction.

Succession, Progression, and Progress

Understood in this way, progress must be distinguished from two closely related ideas: succession and progression. "Succession" is simply "the occurrence of different entities at different times," that is, mere change in the sense of replacement (Simpson 1974, p. 32). Succession need not manifest any consistent directional pattern at all. "Progression," by contrast, is "a continuous and connected series" (Simpson 1974, p. 34), that is, change in a particular direction. As Francisco Ayala puts it, "The concept of 'direction' implies that a series of changes have occurred which can be arranged in a linear sequence so that elements in the later part of the sequence are further from early elements of the sequence than intermediate elements are" (Ayala 1988, pp. 76–77). Progression entails succession, but the converse is not true. Finally, "progress" means directional change in which "the changes are for the better;" that is, with improvement. Progress entails both progression and succession, but neither succession nor progression by themselves entail progress. Succession, progression, and progress have often been conflated in the biological literature, but given their different meanings, it is essential to treat these as distinct concepts.

The following remarks by John T. Bonner (1988) illustrate well the sorts of confusions that are liable to occur if we are not careful about our use of terminology:

> There is an interesting blind spot among biologists. While we readily admit that the first organisms were bacteria-like and that the most complex organism of all is our own kind, it is considered bad form to take this as any kind of progression. . . . [O]ne is flirting with sin if one says that a worm is a lower animal and a vertebrate is a higher animal, even though their fossil origins will be found in lower and higher strata. (Bonner 1988, pp. 5–6)

While I am sympathetic to the general aims of Bonner's book, and will refer to it positively later, the above passage simply confuses the issues. What biologists generally object to is not the idea that there has been a *progression* (i.e., a series of directional changes) from bacteria to *Homo sapiens,* but rather the idea that this progression should be viewed as genuine *progress* (i.e., improvement) in some scientifically intelligible sense with *Homo sapiens* being the most perfect living thing. Likewise, biologists do frequently object to describing some organisms as "higher" and others as "lower" (in the sense of superior and inferior), but such designations are only contingently related to the fact that fossils are found in different locations in a sequence of strata. It is simply an unfortunate fact of the English language that many words do double-duty for entirely different ideas. Distinguishing succession, progression, and progress is essential.

The history of life on earth has embodied progress, then, just in case there has been directional evolutionary change for the better. Has evolution been progressive in this sense? I will approach this question in the next two sections by dividing it into two subquestions. First, are there any directional trends in evolution? Second, can any of these directional trends be considered "for the better," that is, as improvements in some sense?

Directional Evolutionary Change

A process has a direction just in case the value of some property involved increases during a specified time interval. Directional evolutionary change ("progression") merely requires that there be an increase in some quantitative characteristic (i.e., some property) among living things arranged in a temporal sequence from earlier to later. Understood in this way, the question of whether the history of life manifests directional change is entirely unproblematic. Not only does the history of life manifest directional change, it is ubiquitous. Every ancestor-descendent relationship in which some characteristic increases (i.e., every anagenetic trend) is by definition directional. No one doubts that evolution is directional in this "local," short-term sense. What is controversial is whether the history of life "as a whole" has a direction. Is there a long-term (cumulative) direction in evolution? Contributing to the disagreements is the fact that there are a variety of nonequivalent ways in which the history of life might be globally directional.

Universal versus Episodic Directional Change

One obvious way in which the history of life might manifest long-term directional change is if it is everywhere and at all times manifesting directional change, that is, if evolution is *universally directional*. The problem is that few biologists would suggest that evolution is always and everywhere *anything*, much less directional, in any interesting sense. Because evolutionary progress has often been dismissed simply because particular sorts of directional change have not been *universal* in the history of life, this point is worth dwelling on for a moment. Biological systems are distinguished as much by their failure to conform to the general rules humans devise as by the rules themselves. Females make a greater investment in offspring in the form of parental care than males – except when they don't (e.g., in seahorses). Offspring in sexually reproducing species are just as closely related to their parents as they are to their siblings – except when they aren't (e.g., in Hymenoptera with haplodiploidy). Bright, conspicuous colors indicate extreme toxicity – except when they don't (e.g., in harmless nonvenomous snakes that mimic poisonous coral snakes). Such exceptions to the general rules distinguish the history of life on earth from the subject matter of physics, in which apparent exceptions to the basic principles (e.g., the anomalous perihelion of Mercury in relation to Newton's laws) are rare and demand either a revision of the theory or replacement by a more adequate theory. In evolutionary biology, by contrast, we are intrigued, but hardly surprised, when we learn of some deviation from the normal patterns we have come to expect. Although we cannot always anticipate each particular biological novelty that appears, we nonetheless expect that there will be novelties of one sort or another. Darwin captured the distinction between the character of physics and biology beautifully in the final words of the *Origin*, in which he contrasted the earth, considered simply as another planet "cycling on according to the fixed law of gravity," with the "grandeur" of life on earth, in which "endless forms most beautiful and most wonderful have been, and are being, evolved" (Darwin 1859, p. 490).

Returning now to the issue of directional evolutionary change, it is clear that directional change need not be universal in order for the history of life as a whole to have a direction (or directions). A given property may be increasing in some lineages, decreasing in others, and remaining constant in the rest. Evolution might still have a long-term direction, however, even if directional change is merely episodic, so long as the episodes of directional change can be summed to provide an overall trend (more on this below). Episodic directional change is compatible with periods of stasis and even with periods of regression, in which some lineages revert to an earlier form. It follows that merely pointing out that some lineages have not undergone directional change for a long time does nothing to undermine the claim that evolution "as a whole" has a direction. By the same token, merely showing that there has been directional change within one or more lineages is clearly insufficient to demonstrate that the history of life as a whole has a direction. What is true of the whole need not be true of each part, and vice versa.

Uniform, Net, and Apex Directional Change

If the history of life "as a whole" is directional, it will have to be the cumulative result of episodic directional change. But there are three different ways in which episodic directional change could sum to give evolution "as a whole" a direction. First, whenever the value of some property changes (as frequent or infrequent as this might be), it might *always* change in the same direction (uniform directional change). Second, although the value of a property may sometimes increase and sometimes decrease, so long as the *average* value increases over time, a directional trend will be present (net directional change). Third, even if the average value of some property remains constant or decreases, the *maximum* value of that property may still increase over time (apex directional change).[1]

The significance of distinguishing between uniform, net, and apex directional change can be demonstrated by considering a specific (but highly idealized) example. Suppose (perhaps counterfactually) that "intelligence" is a quantifiable biological property. Suppose as well that it is possible to rank organisms on a scale from 1 to 10 with regard to this property, with organisms at Level 1 being exceedingly stupid and organisms at higher levels being increasingly more brilliant. Finally, suppose that at time T_1 there are organisms at Level 4 but no higher, whereas at some later time, T_2, there are organisms at Level 7. In this case, the upper level of intelligence would have increased from T_1 to T_2 (apex directional change). But this increase in the upper intelligence level could have occurred even if the average intelligence level dropped or if the sequence leading up to it changed in a nonuniform fashion. For example, the organisms occupying lower levels of the intelligence scale might be far more numerous than those occupying the higher levels, or the number of individuals within species at the lower levels might far outnumber those occupying the higher levels, dragging the average level of intelligence down. Conversely, the average level of intelligence could increase (net directional change) without a corresponding increase in the highest level of intelligence achieved (e.g., if the number of entities at an intermediate level of intelligence increase relative to the number of entities at lower levels). For example, suppose that there are numerous species (or organisms) at Levels 2 and 3, but just a few at Levels 7 and 8. Suppose further that those at Levels 7 and 8 fail to evolve any further in the direction of increased intelligence, while those at Level 3 increase substantially in number relative to those at Level 2. In this case there would be an increase in the average level of intelligence, with no change in the highest level of intelligence attained. Clearly, therefore, increase in the *average* of some property and increase in the *highest level* (i.e., maximum value) of some property are distinct and independent measures of directional change.

This is important, because disagreements concerning direction in evolution can sometimes be resolved simply by specifying explicitly the sort of directional change under consideration. Critics point out that there is no directional trend in evolution toward increased intelligence, because not all lineages show an increase in this property. Furthermore, it cannot even be said that the average level of intelligence has increased, because the vast majority of organisms (e.g., bacteria) occupy relatively low positions on the scale of intelligence. Proponents of directional evolutionary change regarding intelligence, however, focus on the fact that the upper level of intelligence (as manifested in the large brain volume and behavioral flexibility of primates, cetaceans, etc.) has increased over time. Clearly, an increase in the upper level of intelligence is consistent with only a minuscule fraction of all living things enjoying this higher level of intelligence. In this case, the apparent disagreement about the evolution of intelligence is based on a simple confusion of types of directional change.

What, If Anything, Has Increased in Evolution?

Like universal directional change, uniform directional change is unlikely. As environments change, properties will sometimes increase and sometimes decrease in value. Net directional change is a more promising candidate for identifying directional evolutionary change, but there are practical problems in determining the average value of some property at any one time, and in changes in the value of that property over time. Take an apparently simple property like size, for instance. An average is simply the total quantity of whatever is being measured divided by the total number of units. But what are the relevant units in this case? Should the average size of organisms be calculated on the basis of the number of individual organisms in existence at a given time, or on the basis of the number of species of a given size? As life diversifies, the average size of different kinds of organisms may increase (e.g., if individuals of new species tend to be larger than their predecessors), while the average size of all individual organisms may decrease (e.g., if the smallest organisms increase disproportionately in number). Given these problems, proponents of directional evolutionary change generally claim that there has been an increase in the maximum value of some property (apex directional change) when proposing the idea that life as a whole has advanced from its simple beginnings. That is, the claim is that the upper level of some property has increased in the history of life.[2]

But is the history of life directional in this sense? Many biologists have thought so. We have already seen numerous examples. Darwin toyed with the idea that perhaps "organization on the whole has progressed" (Darwin 1859, p. 345). Early in his career Julian Huxley proposed that no less than six major directional trends were discernible in the history of life (Huxley 1923, p. 31). E. O. Wilson takes a similarly expansive view, asserting that "[D]uring the past billion years, animals as a whole evolved upward in body size, feeding and defensive techniques, brain and behavioral complexity, social organization, and precision of environmental control – in each case farther from the nonliving state than their simpler antecedents did. More precisely, the overall averages of these traits and their upper extremes went up" (Wilson 1992, p. 187). Richard Dawkins argues that directional change in evolution is characterized by "a tendency for lineages to improve cumulatively their adaptive fit to their particular way of life, by increasing the numbers of features which combine together in adaptive complexes" (Dawkins 1997, p. 1016). More recently, Robert Wright has argued that biological evolution is characterized by the rise of what he calls "non-zero-sumness," that is, intra- or interspecific cooperation for mutual benefit. In Wright's view, "biological evolution . . . can be viewed as the ongoing elaboration of non-zero-sum dynamics" (Wright 2000, p. 252). Non-zero-sumness is the logic that organized genes into simple prokaryotic cells, simple prokaryotic cells into more complex eukaryotic cells, eukaryotic cells into organisms, and then organisms into societies (Wright 2000, p. 263). Other properties sometimes thought to increase in the history of life include evolutionary versatility, intelligence, developmental entrenchment, specialization, and ability to sense the environment (McShea 1998).

Intermission: Complexity

The above list is only partial, but provides a representative sample of the kinds of proposals that have been advanced. Although many of the claimed increases are impressionistic rather than based on carefully undertaken empirical studies, in principle a case could be made for each of these proposals. But by far the most popular candidate for what has increased in the history of life is "complexity" – a property that seems to be implicated in most if not all of the other properties identified above. Even those who are explicitly skeptical about the reality of evolutionary progress are often willing to endorse evolutionary progression with respect to complexity.[3] Unfortunately, it is also one of the most difficult properties to rigorously define.[4]

A notable attempt to render this idea more precise is found in John Tyler Bonner's book, *The Evolution of Complexity by Means of Natural Selection* (1988), where he offers the following definition of biological complexity: "The greater the number of cell types [in an organism], or the number of species [in a community], the greater the complexity" (Bonner 1988, p. 101). Bonner collects evidence to show that during the course of evolution there has been an increase in both the upper limit of the size of organisms and in both measures (organismic and ecological) of biological complexity.[5] Finally, the two trends are not unrelated: larger organisms tend to be more complex than smaller organisms, with greater size providing a necessary (but obviously not sufficient) condition for greater complexity.

Evolutionary Progression: Preliminary Conclusions

Although I have focused on complexity, a similar case for directional change could be made for several other biologically relevant properties. For example, if intelligence is positively correlated with brain size, then it is evident that the upper level of intelligence has increased from the origin of life to the present (Jerison 1973). None of the properties cited by proponents as examples of directional evolutionary change are claimed to increase in every lineage over every segment of its evolutionary history. But a good case can be made that there is a straightforward and unproblematic sense in which evolution "as a whole" is correctly described as directional in several important aspects. Despite the reservations of critics, evolution has been directional.

Improvement

A strong case can be made that the history of life "as a whole" has a direction (or rather, several directions). Can it also be described as manifesting progress? Although direction is necessary for progress, it is not sufficient. Directional change ("progression") must also be "for the better" (i.e., embody improvement) if it is to qualify as progress. But this is precisely where the most difficult philosophical issues concerning evolutionary progress arise. Critics argue that there is and can be no "ultimate" or "objective" basis for claims of evolutionary improvement. Such criticisms have convinced most biologists that the very idea of evolutionary progress is hopelessly flawed. This judgment may, however, be premature, because a common defect of *critiques* of evolutionary progress is that inappropriate or arbitrary standards of improvement are presupposed, and then applied to purported cases of evolutionary progress, with the predictable result that these cases are dismissed as unfounded. Before the issue of evolutionary progress can be resolved, therefore, a more charitable approach is needed. In particular, the idea of "improvement" itself must be examined, in order to clarify its meaning and requirements.

The "No Ultimate Basis" Objection

Later I will consider and respond to a range of objections to the idea of evolutionary progress. But one misunderstanding is so common, and the mistake it is based upon is so often considered fatal to the idea of evolutionary progress, that it needs to be treated at the outset. The idea of evolutionary progress is often rejected because "there is no ultimate basis in the evolutionary process from which to judge true progress" (Provine 1988, p. 63). Different critics flesh out this claim in different ways. According to Michael Ruse, for example, it is essential to distinguish between "absolute" and "relative" (or "comparative") evolutionary progress. Ruse defines absolute progress as "the climb up some objective scale," and makes it clear that only if there is some "objective scale" of improvement can judgments about absolute progress make sense. But, because there is no objective scale of biological improvement, the notion of absolute progress drops out. Comparative progress, by contrast, involves "competition between groups" (Ruse 1993, p. 55). Placed in a competitive situation, some groups do better than others. If group **A** succeeds in outcompeting group **B**, then group **A** can be considered superior to group **B**, and there has been comparative progress. In this view, whereas the fairly benign idea of comparative or relative progress is acceptable, the robust notion of absolute progress of the "onward and upward" sort envisioned by Huxley and many others has been discredited.

Despite its intuitive appeal, however, the distinction between "absolute" and "relative" progress is unhelpful for clarifying the issue of evolutionary progress, because upon closer examination it threatens to unravel. Begin with the conception of "absolute progress," as defined above. For any "objective scale" that organisms or species may be said to "climb," it would still be true that any judgment that organisms or species had progressed would have to be made *relative* to this particular scale. Even in the case where a given directional change in a species is thought to fulfill more completely God's will, improvement in this case would still be only relative to this divine standard. In relation to some other standard, it might not be a progressive shift at all. It is true that "[N]o satisfactory epistemic criterion of [absolute] progress has yet been given" (Ruse 1996, p. 534). However, this is not because the concept of progress itself is inherently problematic, but rather because the idea of "absolute progress" (i.e., progress that is not measured in relation to some standard) may well be incoherent.

On the other side of the distinction, to say that "comparative progress" involves "competition between groups" itself presupposes some standard independent of both groups, according to which one group can be judged competitively superior to another. For example, judgments of relative progress of the sort "group **A** succeeds in outcompeting group **B**" require some standard by which such success is determined. But there are an array of different ways in which competitive success can be judged. Is group **A** competitively superior to group **B** if (or only if) group **A** increases in numbers relative to group **B**? Or is it a relative increase in biomass that matters? Does group **A** outcompete group **B** only if group **A** drives group **B** to extinction? What if group **A** comes to occupy a greater geographical range than group **B**? The point is that none of these standards of success, *qua* standards, are simply given by nature. What is to count as competitive success depends on which biological factors interest us most. But if so, then even judgments of relative (comparative) progress, no less than judgments of absolute progress, presuppose some standard in terms of which such progress is assessed.

These observations help to pinpoint the critical issue. Judgments of improvement in any context presuppose some *standard* in terms of which a given directional change can be evaluated. In the broadest sense, two different sorts of standards may be distinguished: those that might be shared by the scientific community, and those that are entirely idiosyncratic to a single agent. Describing a standard as "subjective" implies that it is idiosyncratic and personal, simply a matter of individual taste or preference, incapable of any further justification, and perhaps even an experience that is incapable of being shared with others. When I claim that chocolate tastes better than vanilla ice cream, for example, I am merely expressing my subjective tastes and preferences, without for a moment suggesting that my preferences could or ought to be confirmed by other agents.

By contrast, standards might be based on some publicly accessible, biologically relevant property of living things. For example, saying that one measure of evolutionary progress consists in "improvements in sense perception" might be a standard that is invented by behavioral biologists simply because that property of living things interests them. Such a standard could be justified to the extent that it can be made clear and "enables us to say illuminating things about the evolution of life" (Ayala 1988, p. 84). But precisely because it is a shared standard for which good (i.e., biologically relevant) reasons might be offered, it would not be "subjective" in the sense defined above. Of course, one *could* invent a purely subjective standard of evolutionary progress. If I were to simply declare that evolutionary progress consists in "increasing blueness," just because blue happens to be my favorite color, that would indeed be a completely personal, subjective, idiosyncratic standard, and surely dismissed as such.

It is worth emphasizing that even if standards of evolutionary progress are constructed relative to our interests, it does not follow that they are subjective. Clearly, standards can be both constructed relative to our interests yet not subjective in the sense defined above. Constructed standards of measurement like those constituting the metric system can and often do become shared public property, understood and applied in exactly the same way by everyone who grasps the basic concepts. In like manner, although a standard of evolutionary progress might be invented by us relative to our own interests, this does not entail that it is "subjective" or idiosyncratic. The basis for a standard may be constructed by us in order to pick out properties of things that interest us, where nature itself does not provide this basis ready-made. But such standards are no less objective for having been selected by us, for our own purposes.

In summary, standards of evolutionary progress can be both relative and objective. The spirit of this point is captured by Ayala, who writes that although the choice of a standard by which to evaluate organisms or their features depends on decisions we make, based on our interests, "once a standard of progress has been chosen, decisions concerning whether progress has occurred in the living world, and what organisms are more or less progressive, can be made following the usual standards and methods of scientific discourse" (Ayala 1988, p. 90).

Still, claiming in abstract terms that evolutionary progress entails improvement relative to some standard is one thing; identifying and justifying that standard is quite another. Either way, proponents of evolutionary progress are obligated to provide some account of how improvement might be measured. Accordingly, the central problem in determining whether evolution manifests progress concerns the identification and justification of a *specific* standard according to which improvement can be measured.

Domain-Relevant versus Domain-Irrelevant Standards

In general there is nothing mysterious about judgments that one thing is objectively better than another. Often this requires little more than specifying a particular task or function in terms of which the entities can be compared. Entity **B** is better than entity **A** for a particular function, **F**, if **B** performs **F** better than **A**. Notice that in order to judge that **B** is better than **A** with regard to **F** it is not necessary that **B** and **A** were both designed or intended to perform **F**. A hammer is better for pounding nails than is a can opener, and this is so regardless of the fact that hammers but not can openers are designed to pound nails. Likewise, to the extent that can openers change in such a way as to become better at pounding nails, they have improved with respect to that function, even though they may never have been intended to be used for this purpose.

In principle, therefore, any two things can be compared with respect to their ability to perform some specific task or function. But only a small subset of these comparisons will be even remotely interesting. The most basic requirement that an *interesting* comparison of entities in any domain must satisfy is relevance to what being "better" in that domain consists in. This, in turn, requires specifying a particular task or function that entities in a particular domain are intended to perform. Thus, ball-peen and claw hammers can be compared for their ability to pound nails, just as manual and electric can openers can be compared for their ability to open cans. In each case it is possible to ask whether a change from the former to the latter constitutes improvement with respect to the domain-relevant task or function at issue. Sometimes it is even possible to give reasonable answers to such questions (Ayala 1999).

Biologically Relevant versus Biologically Irrelevant Standards

The distinction between domain-relevant and domain-irrelevant standards can be applied directly to the issue of evolutionary progress. Comparisons of biological entities will be most interesting when they are made with respect to specific biologically relevant properties, that is, properties that are directly related to the tasks or functions biological entities are designed to perform. In the most general sense, organisms are designed to propagate their genetic information (Williams 1966; Dawkins 1989). They do this by being successful at surviving and reproducing. Surviving and reproducing, in turn, are subserved by more specific biological functions, that is, acquiring energy, perceiving the environment, avoiding predators, securing a mate (for sexual organisms), and so on. Consequently, specific biological properties can be evaluated in terms of their contributions to these tasks. Biological improvements are just those properties that make a greater contribution to the performance of such tasks, and evolutionary progress is directional change toward such properties.

One consequence of this approach is that widely different organisms can be compared with respect to a particular biologically relevant property, with one emerging as clearly better than another with respect to that property. Barracuda are better in a straightforwardly factual sense for traveling quickly through water than are elephants, and this is so regardless of the fact that barracuda, but not elephants, have undergone selection for efficient movement through water. Judgments about evolutionary progress can be made in the same way. Creatures with complex nervous systems are generally better at responding in real time to rapidly changing environmental conditions than are creatures lacking such nervous systems. Even if it is true that trees would not obviously benefit from the presence of a nervous system (and indeed, would be hampered by it, given their needs), it is no less true that mammals are better at responding quickly to immediate changes in their environment than are trees. If complex nervous systems are a biologically relevant property, and organisms with such systems arise later in the history of life than organisms lacking such systems, then there has been improvement with respect to this property, despite the fact that not all organisms would benefit from having such a system.

Does Evolution Manifest Improvement?

Understood in this way, improvements are ubiquitous in the history of life. The most striking improvements are evolutionary innovations. Evolutionary innovations are adaptive breakthroughs that cross a functional threshold, and in so doing contribute to the solution of a general biological problem, thereby paving the way for additional developments (Nitecki 1990). Classic examples of evolutionary innovations include the bony skeleton of vertebrates, jaws of gnathostomes, the amniote egg, avian flight, continuously growing incisors of rodents, large brains of hominoids, the artiodactyl tarsus, the insect wing, rigid skeletons and complex spicules of sponges, and the insect pollination system of angiosperms (Cracraft 1990, pp. 21–22). Other examples of improvement in this sense might include the emergence of eukaryotic cells, sexual reproduction, multicellularity, symbioses, eusociality, language, and so on (Maynard Smith and Szathmáry 1995). Despite their striking character, however, innovations are merely especially obvious instances of evolutionary improvement. A slight increase in the running speed of a predator is no less an evolutionary improvement than is the development of wings for flight. Eyes have undergone gradual progressive evolution from light-sensitive patches to simple pinhole camera-type eyes to the complex eye of the hawk (Dawkins 1996). Wings have undergone progressive evolutionary improvement, from the proto-wings of *Archaeopteryx* to the highly specialized wings of modern birds. Arguably, the four-chambered heart of mammals is an improvement over the three-chambered heart of reptiles and birds (Walker and Liem 1994). Homeothermy is conceivably an improvement over poikilothermy. Intelligence (as evidenced by a rapid expansion of brain volume in the hominid lineage) is probably advantageous as well. Each of these evolutionary developments can rightly be regarded as biological improvement.[6]

Is There Long-Term Evolutionary Progress?

Innovations and less dramatic adaptive changes surely count as evolutionary improvements in one sense. With the appearance of each innovation came an increased ability to exploit some aspect of the world, and thereby to fulfill the biologically relevant function of survival and reproduction. If the ability to exploit new environments (e.g., the land, the air, etc.), or to exploit some environment more fully (e.g., through more acute sense perception, greater speed, etc.), improved during the history of life, then this would be evolutionary progress as defined above. But in order to count in favor of evolutionary progress, improvements must be linked to long-term directional evolutionary changes. Are there any candidates for evolutionary progress in this sense?

Complexity (again)

Ironically, although increasing complexity is perhaps the clearest example of a directional trend characterizing the history of life as a whole, and thus satisfies the first condition for long-term evolutionary progress (i.e., direction), it just as clearly fails to satisfy the second condition (i.e., improvement). There is no reason to consider more complex organisms better than simpler organisms in terms of any biologically relevant criterion of improvement. It is not obvious (and empirical studies have not shown) any positive correlation between degree of complexity and fitness. If it could be shown that complexity is favored by selection, then perhaps it could be inferred that complexity has increased because it is a good-making property. But in cases where more complex creatures are superior to their simpler ancestors, it could be that it is not complexity *per se* that provides this advantage but, rather, some other specific biological property – for example, a better functioning eye – that provides the advantage, and it just so happens that the better-functioning eye is more complex. So, even if complexity does increase in the history of life, it does not obviously constitute the criterion of biological improvement. There seems to be no *a priori* reason why adaptedness should give rise to, or even be linked to, complexity.

The Problem of Environmental Change and Stability

Are there any other candidates? Initially, the odds seemed stacked against finding any such candidates. The most fundamental problem facing proponents of long-term evolutionary progress is that environmental change appears to disrupt and negate any improvements that might arise, preventing such improvements from becoming cumulative. The sorts of characteristics identified above as constituting biological improvements might represent improvements in some contexts but not in others. In an evolving lineage subjected to selection pressures in a relatively unchanging environment, organisms later in the lineage may be better than those appearing earlier, in the sense that if they could be put into competition with their ancestors in that common environment, the descendants would beat the ancestors. If environments change, however, this sort of progress may be lost. What was an improvement may even become a liability. To the extent that adaptation is local or context-sensitive, it should not be cumulative (Sober 1984; Fisher 1986). Environmental change is thus the factor that threatens to undermine any proposed account of long term evolutionary improvement.

In light of the fact that the goodness of a biological property is determined by the environment(s) in which it appears, there seem to be just two closely related ways in which improvements could be cumulative and provide the basis for long-term improvement. First, there might be very general biological properties that are advantageous for any organism in any environment, regardless of changes taking place in the environment. An example of a context-independent property of this sort might be an organism's ability to obtain energy from the environment and convert this energy into copies of itself (van Valen 1978). Second, long-term progress via continual adaptive improvement could be supported by quite general features of the environment that do not change. If aspects of environments remain constant for long periods of time, then continued adaptation to these aspects could constitute long-term improvement. Are there any aspects like this? Factors like temperature, moisture, and the presence of enemies, and so on change quite often, and as they do the value of having certain organismal properties changes. Apparently, the one constant of all or most environments might be change itself. But if so, then *change* is precisely that aspect of environments that (when viewed from one level of abstraction up) remains constant.[7] The more environments change, the more accurate it is to say that all environments are characterized by change. So if organisms can adapt in such a way as to both survive in spite of, and even take advantage of, environmental changes, this would be an adaptive property that is in a sense immune to the vicissitudes of environmental change. Ability to survive and reproduce despite changing environmental conditions, therefore, would be a biological property the possession of which would always be an advantage. Recall that this was precisely why, at one stage of his career, Julian Huxley proposed as diagnostic of evolutionary progress features of organisms that, *by definition*, were not dependent on the particularities of local ever-changing environments.

Summary: Long-Term Evolutionary Progress

The question of whether evolution is progressive is just the question of whether there have been any long-term directional changes in the history of life that embody improvement relative to some standard. Evolution manifests *direction* just in case there has been an increase in some property in the history of life as a whole. The most promising kind of increase to consider is increase in the maximum of some property. Evolution manifests *progress* just in case there has been an increase in some biologically relevant good-making property in the history of life.

The most plausible candidate for this role is the ability to survive and reproduce despite changing environmental conditions, or what Huxley called independence from and control over the environment (Huxlean progress). A different approach would be to construct an array of biologically relevant standards of improvement, and then determine whether the maximum value of any biologically relevant properties has increased during the history of life (Simpsonian progress). A good case can be made that the history of life is characterized by both sorts of progress. By reasonable standards of justification, therefore, evolutionary progress is a fact.

Objections and Replies

So far I have introduced various distinctions that are essential for properly evaluating the idea of evolutionary progress, and have sketched alternative ways in which the thesis of evolutionary progress could be understood and justified. Huxlean evolutionary progress consists in improvements that characterize life as a whole, and permit it to continue advancing. Simpsonian evolutionary progress consists in improvements in particular biological functions. The history of life appears to embody progress in both senses. Nonetheless, in the interests of completeness, various objections must be acknowledged and addressed. I will try to show that the idea of evolutionary progress, as developed above, survives these objections.

No Theoretical Justification for Directionality

The great physicist Sir Arthur Eddington once offered the following advice: "Never accept a fact until it is verified by a theory!" Evidently some critics of evolutionary progress concur with this Eddingtonian injunction. They argue that *directionality* (the tendency for change to occur in a specific direction) does not follow from the theory of natural selection. Natural selection contains no mechanism biasing the evolution of lineages in any particular direction, but only insures that organisms will be relatively well adapted to whatever local environment their immediate ancestors occupied. As environments change, organisms will either track them or go extinct, but there is no reason to assume that this process has any inherent directionality to it. Therefore, the claim that evolution has a direction (or directions) is thought to be undermined.

This objection is vulnerable to two responses. First, many biologists (beginning with Darwin himself) have argued that natural selection *does* entail (or at least make likely) directionality. In the "Essay of 1844" Darwin notes that even though selection might cause some forms to become simpler, nonetheless "from the strong and general hereditary tendency we might expect to find some tendency to progressive complication in the successive production of new organic forms" (in F. Darwin 1909, p. 227). Similar remarks appear in the *Origin* (Darwin 1859, pp. 336–337; 345; 1959, pp. 222; 547). In addition to documenting increases in the maxima of size and complexity, Bonner (1988) also offers selectionist explanations of these trends. He suggests that larger size often provides a competitive advantage (i.e., in avoiding predation or in being a more successful predator). Greater organismal complexity may provide a selective advantage via division of labor and integration of different cell types. Saunders and Ho (1976) offer a different but complementary explanation. After noting that increases in complexity are what give a direction to evolution, they suggest that "a system which is not only organized but also capable of undergoing a continual process of self-organization which optimizes its structure with respect to some criterion [i.e., local fitness requirements] will tend to permit the addition of components more readily than their removal" (Saunders and Ho 1976, p. 376). In other words, once a well-organized system arises, and the components become tightly integrated, it will be more difficult to remove components and still have a well-functioning entity than it will be to add components.

Selection pressures will thus make evolution in one direction more likely than change in another. They identify this *asymmetry* as the principal cause of the observed increase in complexity in evolution.[8] Richard Dawkins has argued that coevolutionary arms races, in which predator and prey are locked into a contest to achieve any slight superiority over the other, each driving the other to greater adaptive refinement, have directionality built into them. As predators become more efficient at capturing prey, prey in turn evolve greater efficiency at escaping, which in turn places greater selective pressure on the predators, and so on in an upward spiral of adaptive improvement. As Dawkins notes, "The resulting positive feedback loop is a good explanation for driven progressive evolution, and the drive may be sustained for many successive generations" (Dawkins 1997, p. 1018). He regards such a process "as of the utmost importance because it is largely arms races that have injected such 'progressiveness' as there is in evolution" (Dawkins 1986, p. 178). Finally, as John Maynard Smith points out, "Although biologists might be reluctant to see [the stages through which life has passed] as representing 'advance,' they are progressive in one sense: the sequence in which they have occurred is not arbitrary, since each stage was a necessary precondition for the next" (Maynard Smith 1988, p. 219). The point of these references is not, of course, to establish that there *is* directionality in evolution, but rather merely to point out that the claim that Darwin's theory does not make directionality either inevitable or likely has hardly gone unchallenged.[9]

Second, even if *directionality* is not entailed by Darwin's theory, *directional trends* might nonetheless be consistent with the theory. A basic distinction within evolutionary biology is between pattern and process. The former refers to a sequence of events in the history of life or of a particular lineage, for example, as described in a phylogenetic tree. The latter refers to the events and causes responsible for generating this pattern. Establishing that a particular pattern has occurred is one thing; explaining why this pattern exists by identifying its causes is another.[10] Directional change describes a pattern of succession, but entails nothing about any processes that makes change in one direction more likely than change in others. Clearly, there can be directional change without directionality.[11]

Consequently, even if there is there is nothing about the operation of natural selection that would lead one to expect the value of any given property (e.g., size, complexity, intelligence, etc.) to increase, it is irrelevant to the question of evolutionary progress, which only requires directional change, not directionality.[12] Of course, if natural selection (or some other evolutionary mechanism) created a directional bias, then we would have a (potentially correct) explanation for any direction that is discovered. But the direction itself would still be an empirical fact that is logically distinct from its explanation.[13]

Directional Trends May Be Just the Result of Passive Forces

The foregoing response helps to dispose of a related objection to the claim of directional evolutionary change. Even if there are directional changes in evolution, they could be dismissed as simply the result of passive forces, on the model of an initially concentrated volume of gas diffusing throughout a container (Gould 1996). Any given particle is intrinsically as likely to move in one direction as another, but because movement is less restricted in one direction (i.e., toward regions of lower density), the gas eventually expands to fill the entire container. In such cases there has been a net directional change, even though no active force biased change in that direction. Applied to evolution, a passive trend is one in which no force (e.g., selection) is actively causing a lineage to evolve in a particular direction, but the lineage nonetheless evolves in a particular direction either because it is the only available evolutionary path or because evolution in one direction is less constrained than in other directions. For example, there might be a passive trend in the history of life toward increasing complexity, not because greater complexity is better, but just because if life begins in a simple form, there is simply more room to evolve in one direction than in the other (Maynard Smith 1970, 1988).

The response to this objection is similar to that offered above to the charge that there is no theoretical justification for directionality in evolution. First, passive forces can give rise to directional changes just as active forces can. Direction, not directionality, is the critical issue here. Second, active and passive trends may not be as distinct as they at first seem. Maynard Smith and Szathmáry (1995) identify eight major transitions in evolution. Sterelny and Griffiths (1999) suggest that at each one the left wall of complexity may have moved to the right as a result of selective (i.e., active) forces. Left-wall boundaries may appear as a result of selective forces, and then life may evolve rightward from each of these boundaries as a combination of active and passive forces (McShea 1993). Simply noting the presence of passive trends, therefore, does nothing to undermine the claim of directional evolutionary change.

Significant Biological Properties Have Not Increased Recently

One might object to the claim that directional change characterizes the evolutionary process by admitting that *of course* certain properties have increased in value since life began, but once life reached a significant stage of development such directional trends have been absent. For example, complexity must have increased in the history of life as a whole, because multicellular organisms (each consisting of millions of cells) have to be more complex than the first bacteria. But it is not obvious that complexity has continued to increase since the appearance of metazoans. Who can say whether a chimp is more complex than a trilobite? Therefore, the claim of directional evolutionary change is only trivially true.

This objection fails, however, for the following reason. In considering direction in evolution, either we are considering the *entire* history of life from the beginning until now or we are considering only *parts* of this history. If we consider the *entire* history of life, then it is clear that complexity (and many other biologically significant properties) has increased, and increased dramatically. If we are permitted to consider only *part* of the history of life, then the proponent of directional change, no less than the critic, is at liberty to consider any part of this history she likes, including the part(s) that show the most dramatic increases. *All* of hominid evolution is recent, in evolutionary terms, but is characterized by dramatic directional change, especially in cranial capacity. Although it is useful to partition the history of life and ask whether what is true of the whole is also true of its various parts, in considering the issue of evolutionary direction considering anything less than the entire history of life on earth seems arbitrary.

No Organism Is Better Overall

Progress requires improvement as well as directional change, and this is where the real problems begin. As we saw in Chapter 7, in his private notebooks, Darwin wrote: "It is absurd to talk of one animal being higher than another. – *We* consider those, where the cerebral structure/intellectual faculties most developed, as highest. – A bee doubtless would where the instincts were" (Darwin, B Notebook, p. 74; in Barrett et al. 1987, p. 189). D'Arcy Thompson, in his seminal book *On Growth and Form*, remarked "That things not only alter but improve is an article of faith, and the boldest of evolutionary conceptions. . . . I for one imagine that a pterodactyl flew no less well than does an albatross, and that Old Red Sandstone fishes swam as well and easily as the fishes of our own seas" (Thompson 1969, p. 201). Together these remarks embody two common objections to the thesis of evolutionary progress. First, it is impossible to compare different kinds of organisms whose ways of life are radically different, and to say which is better. Second, even with respect to organisms that are fundamentally similar in their ways of life (aerial, marine, etc.), it is impossible to say that there has been improvement between earlier and later kinds of organisms. Since it is impossible to judge one organism as better than another either as a whole or with respect to some particular function, the idea of evolutionary progress is mistaken as well.

Despite the confidence with which this objection is often delivered, it fails to undermine the core thesis of evolutionary progress. First, the claim that evolution embodies progress does not require that one organism be better *overall* than any other organism. It only requires that some biologically advantageous property increase during the history of life (in one of the senses defined earlier). Second, often it is possible to judge one thing as superior to another with respect to a particular function. We do not hesitate to judge one musician to be better than another, in terms of musicianship, despite the fact that the second musician is in fact a superior poker player. In like manner, there is no reason why progress must include improvement in every aspect of the organism under consideration. A given organism (e.g., *Homo sapiens*) may be superior to other organisms (e.g., bacteria) in some respects (e.g., intelligence), but inferior in others (e.g., in its ability to tolerate anaerobic conditions, etc.). Third, in principle it *is* possible to judge one organism as better than another overall. If we could list all the different sorts of biologically relevant functions there are, and determine that one kind of organism was better than all the others with respect to every one of these functions, then we would indeed have an organism that was best overall. Obviously, we should not expect there to be such an organism. Performing even a small number of biologically relevant functions maximally well may not be a compossible state of affairs. Every organism is a constellation of design tradeoffs, so an organism that is best at burrowing in the ground will probably not be the best at flying, and so on. But such an organism is not *impossible.*

The Idea of Progress Is "Value-Laden"

At a more fundamental level, critics argue that the idea of improvement is unacceptable because it introduces values into an otherwise objective, value-free intellectual enterprise. According to this view, science deals with strictly objective factors, eschewing all value judgments in favor of identifying facts (empirical data), establishing their interconnections (building models, identifying laws of nature), and then explaining such interconnections (theories). Assessments of good or bad are kept strictly at arm's length and are vigilantly prevented from intruding into the epistemic purity of scientific investigation. Since the idea of progress entails values, the idea has no place in serious science.

In reply, it can be pointed out that the idea of improvement is indeed value-laden, but then so is all good science. Some values are inherent in science as an intellectual enterprise. For example, scientists value theories that are simple, explanatory, predictive, of wide scope, empirically confirmed, and consistent with findings in other areas of science. Based on such standards, scientists routinely judge one theory as better than another (e.g., general relativity vs. Newtonian mechanics; descent with modification vs. special creation). A given scientific theory is deemed good relative to some set of standards that is intrinsic to the scientific enterprise, and some are judged to be better than, and an improvement over, others. Such assessments are thoroughly value-laden. Indeed, if all values were to be excluded from science, it would cease to function.[14]

"Good" in the context of assessing scientific theories is a term of epistemic appraisal. Additional nonepistemic values are part and parcel of evolutionary biology, although not necessarily a part of nonbiological sciences (Mayr 1988). A particular property of an organism might be described as good for achieving a certain result (e.g., capturing prey, evading a predator, sensing its environment). For any characteristic of an organism it makes sense to ask what it is good for (although there is no guarantee that every characteristic of every organism is good for something, nor that every good characteristic requires an evolutionary explanation). "Good" in this context typically refers to some kind of functional efficiency. A given characteristic might be described as good if it contributes to the solution of a problem facing the organism in its particular environment. Again, some characteristics might be better at solving a particular problem than others, and a transition to this better characteristic represents biological improvement. As Simpson noted, "[I]t is not true that value judgments are foreign to science. No scientific endeavor can be undertaken without some such judgment, and there must be a sense in which we can sometimes judge whether particular evolutionary changes are for the better" (Simpson 1974, p. 49). Consequently, the idea of improvement is not only respectable, it is also essential to science as an intellectual activity responsible to certain epistemic values, and to evolutionary biology as the pursuit of causal explanations for the functional characteristics of living things (Sober 1994).

The Idea of Progress Is Anthropocentric

Ironically, one of the chief reasons the idea of evolutionary progress is now so odious to biologists is precisely the same reason it was so desirable to naturalists of old: It seems tailor-made to justify our sense of occupying the highest rung of the ladder of life. Recall, for example, Julian Huxley's conclusion that *Homo sapiens* does indeed appear to be the most advanced species – assuming, that is, that we really have achieved the most independence from and control over the environment of any species. Even now popular depictions of evolution as an unbroken linear ascent starting from single-celled organisms, and progressing through mollusks, fish, amphibians, reptiles, birds, mammals, primates, and finally to *Homo sapiens* (perhaps stepping out of the earth's atmosphere into space, poised to conquer new worlds), are common. These depictions convey the impression that human beings represent the pinnacle of evolution, and the *raison d'être* for the entire process.

As artistic renderings of the human story, such depictions can be quite compelling. Contemporary biologists, however, are generally scornful of this view (to put it mildly), and are at pains to insist that the idea that human beings are "higher" than other organisms is utterly lacking in empirical support and, even worse, is probably scientifically incoherent as well. Huxlean progress is vulnerable to the charge that a characteristic of *Homo sapiens* has been elevated into a standard for all life. It is precisely because the notion of evolutionary progress has so often in the past been used to argue that *Homo sapiens* is the apogee of creation, and in virtue of this lofty position is warranted in exercising dominion over all other living things, that biologists are so wary of the notion. Add to this that notions of progress have in the past been linked with the noxious idea of racial superiority and its applied wing, eugenics, and it is easy to see why any right-thinking biologist would want to steer clear of the notion.

This objection fails for three reasons. First, it might be the case that as a matter of fact *Homo sapiens* is the most advanced species (either as a whole, or in some particular respect), however distasteful this might be for some to accept. Second, it is not obvious that *Homo sapiens* is the most advanced species for a range of biologically significant properties. Other kinds of living things can perform a multitude of functions better than we can. If evolvability is an important biological property, and if our science and technology allow us to take control of our genetic and evolutionary destiny, then we may indeed be the most evolvable species that has ever existed. By contrast, this honor might be better awarded to the very first living thing (if there was a first), because it gave rise to all other living things (including us), and therefore in a sense has already demonstrated its enormous capacity for further evolution. Finally, claims that evolution is not progressive are vulnerable to the same objection because they, too, can be said to reflect a merely human point of view (albeit a different one) As Simpson noted, "After all, it may be a fact that man does stand high or highest with respect to various sorts of progress in the history of life. To discount such a conclusion in advance, simply because we are ourselves involved, is certainly as anthropocentric and as unobjective as it would be to accept it simply because it is ego satisfying" (Simpson 1949, p. 242). The basic point is that the fact that *we* are the ones making judgments about evolutionary progress should cause us to consider the idea very carefully and skeptically before assenting to it, but it should *not* lead us to reject it simply because we are the ones making it.

Evolutionary Progress May Be Real, But It Is Boring

As we have seen, evolutionary progress may be understood in either Huxlean or Simpsonian terms. Huxlean progress, understood as advances that don't stand in the way of further advances, is open to the charge of being trivial. Simpsonian progress, by contrast, is vulnerable to the criticism that it is deflationary. Instead of progress being an objective fact about evolution, capturing what is essential about the history of life, it becomes merely a human construct. "Fine," critics might say, "if *that's* all you mean by progress. . . ." For example, Michael Ruse complains that, "progressionism may well exist in today's top-quality evolutionary thought but . . . it is of little moment because, *qua* cultural value, it has been effectively neutralized. Today one can define 'progress' in epistemic terms, measures of complexity and the like, and by doing so take all of the tension of out the issue" (Ruse 1996, p. 534). One almost gets the sense that such critics would prefer that the issue of evolutionary progress remain a problem. Some people are more easily bored than others, but I find nothing boring about any of the conceptions of evolutionary progress discussed here. A modest conception of evolutionary progress that might be correct is far more interesting than an extravagant conception of evolutionary progress that is almost certainly false. If evolution is indeed progressive in some sense, then understanding this simply enhances our understanding of the process that created all of us. How could that fail to be interesting?

Summary: Is Evolution Progressive?

The question of whether evolution manifests progress, and if so, in what sense, has divided biologists from Darwin to the present. On the one hand, one of the most striking features of the history of life on earth is the dramatic increase in characteristics such as size, complexity, sensory capacity, behavioral flexibility, and many other biological traits. On the other hand, the idea of evolutionary progress has seemed problematic (or even hopelessly "subjective") to merit any place in a rigorous scientific conception of nature. As I have argued in this and the preceding two chapters, however, the idea merits closer examination, with distinctions carefully drawn to separate justifiable from unjustifiable conceptions of evolutionary progress. Especially important in this regard are distinctions between succession, progression, and progress, as well as between directional change and directionality. When these distinctions are taken seriously, a good case can be made that evolution has manifested both Huxlean and Simpsonian progress, and that the various objections raised against the claim of evolutionary progress fail to demonstrate that such progress has no place in a Darwinian understanding of life on earth. Nonetheless, the question of evolutionary progress is likely to remain controversial in Darwinism, only rivaled in this respect by concerns over applications of Darwinian ideas to understanding human origins, human nature, and our destiny as a species. It is to these issues that we turn next.

10

Human Physical and Mental Evolution

What a chance it has been . . . that has made a man.
(Darwin, "E Notebook p. 68; in Barrett *et al.* 1987, p. 415)

Introduction

Recall once more Darwin's claim, the exploration of which is the central purpose of this book: "As natural selection works solely by and for the good of each being, all corporeal and mental endowments will tend to progress towards perfection" (Darwin 1859, p. 489; 1959, p. 758). In previous chapters we have examined the themes of "selection, perfection, and direction" as they pertain to living things in general, very few of which could be said to have impressive "mental endowments." Natural selection in conjunction with various chance elements operates on a range of causally interconnected biological entities, resulting in striking evolutionary trends and astounding (but ultimately imperfectly designed) living things. Such is Darwinian orthodoxy, at least concerning the physical evolution of nonhuman living things. But what about human beings and their most distinctive characteristic – intelligence? What does (or might) Darwinism say about *us*? In particular, how might the evolution of intelligence figure in a Darwinian understanding of life?

In this chapter, we will look at the past, present, and future of *Homo sapiens*, as viewed through Darwinian perspectives, and in so doing connect the themes of selection, perfection, and direction in evolution as applied to our own species. A range of additional questions present themselves for our consideration. How did Darwin treat the evolution of human "corporeal and mental endowments"? How did evolutionists who followed him view human mental and physical evolution? To what extent do our bodies and minds bear the telltale marks of our evolutionary history? What role has natural selection played in the evolution of our bodies and minds? Is human evolution different, or in some ways exempt, from the principles that govern the evolution of life in general? Was there anything predictable, or even inevitable, about the appearance of beings like ourselves? Finally, given our best contemporary understanding of the evolutionary process, what conclusions should we draw with regard to the future evolution of *Homo sapiens*? These are obviously far-reaching questions, and our discussion of them must necessarily be incomplete. But they are of too great interest to ignore. It will be enough if the broad sweep of Darwinian thinking on these issues is made clear. We will begin by examining the views of Darwin and Wallace, then briefly consider a number of twentieth-century evolutionary views, and finally explore current prospects for understanding our past, present, and future as a species.

Darwin and Wallace on Man

The *Origin of Species*, Darwin's greatest theoretical work, is filled with detailed observations, careful inductions, and bold speculations about the evolution of life on earth. Considered by Darwin to be a mere "abstract" of the much larger "Big Species Book" he had been writing, it was an instant bestseller, selling out its entire first printing of fifteen hundred copies the day it went on sale, and subsequently appearing in five new editions between 1860 and 1872. In some ways, the initial success of the *Origin* is surprising, since it offered up (in addition to some novel ideas about "natural selection" and the like) some fairly unexciting fare: domestic breeding (including much on pigeons), sterility and hybridism, classification, morphology, embryology, and so on. Educated Victorians were perhaps more interested in such topics than we tend to be, and it is true that the way had been prepared by other, more provocative books on evolution (e.g., Chambers 1844), thanks to which readers' curiosity had already been piqued. But such factors alone could hardly account for the book's sales, or the interest (and controversy) it engendered.

Darwin's Naturalism

Ironically, the success of the *Origin* lay as much in what it did *not* discuss as in what it did. The main ideas presented in the *Origin* as to how life had evolved were immediately clear to most readers. What people really wanted to know, however, was how the ideas developed in that book applied to *us*. In a letter to Alfred Russel Wallace, in response to the latter's question as to whether he would discuss human beings in his "species book," Darwin wrote: "I think I shall avoid the whole subject, as so surrounded by prejudice, though I fully admit that it is the highest and most interesting problem for the naturalist" (Wallace 1916, vol. 1, p. 133). True to his word, in the *Origin* Darwin had hardly anything to say on the matter, simply remarking in the concluding chapter that "Light will be thrown on the origin of man and his history" (Darwin 1859, p. 488).

Hoping that the central arguments of the book would be evaluated on their own merits (yet also knowing full well that readers would draw their own conclusions about the implications of his theory for human beings), Darwin chose to omit any discussion of human evolution in the *Origin*. Yet we know from his private notebooks that from the very beginning of his speculations about "transmutation" the question of human evolution had never been far from his mind. Long before he had worked out the implications of the theory of natural selection for human evolution, his notebook entries are distinguished by a consistently *naturalistic* approach to understanding human nature that could easily accommodate natural selection as its central explanatory concept. For example, in the "C Notebook" (composed between February and July 1838), Darwin recorded his firm conviction: "I will never allow that because there is a chasm between Man . . . and animals that man has different origin" (Darwin, C Notebook, p. 223; in Barrett et al. 1987, p. 310). He mused that perhaps even belief in God is simply a product of our brains, and then chided himself for his emerging materialism: "[L]ove of the deity effect of organisation – oh, you Materialist! . . . Why is thought being a secretion of brain, more wonderful than gravity a property of matter? It [is] our arrogance, our admiration of ourselves . . ." (Darwin, C Notebook, p. 166; in Barrett et al. 1987, p. 291). Already Darwin was speculating about how a naturalistic account of human origins would bear on our self-understanding. Yet he reserved such remarks for his own private contemplation, and occasionally for discussion with sympathetic friends. He would not publish his own views in any detail until *The Descent of Man* (1871), by which time others had rushed in to fill the conspicuous lacuna in his published work. The most important of these, as it turned out, was his friend and co-discoverer of the principle of natural selection, Alfred Russel Wallace.

Wallace on "The Antiquity of Man"

As we saw earlier (in Chapter 4), despite the many views they held in common, Darwin and Wallace came to disagree about the power of natural selection in forging adaptations. Whereas Wallace viewed (virtually) every feature of organisms as adaptations forged by natural selection, Darwin was reluctant to go that far, and acknowledged that there are many characteristics of organisms that have no discernible adaptive significance. Their disagreement over the power of selection and the scope of adaptations had direct implications for their understandings of human evolution, especially of human intellectual and moral faculties.

Initially, they were in complete agreement. In his paper "The Origin of Human Races and the Antiquity of Man Deduced from the Theory of 'Natural Selection'" (1864), Wallace proposed a completely naturalistic, selectionist explanation for human physical and mental evolution. According to Wallace, in the earliest stages of human evolution natural selection acted upon both the body and the mind, adapting each to the exigencies of the struggle for existence. But once the human intellectual and moral faculties had reached a sufficiently developed state, the mind became the means by which humans adapted to their changing environments, often by actively changing those environments to suit their needs, with natural selection then ceasing to work on the body. However, although selection had ceased to work on the human frame, it continued, and in the future would continue, to work on human intellectual and moral capacities. Optimistically, Wallace maintained that "the power of 'natural selection,' still acting on his mental organisation, must lead to the more perfect adaptation of man's higher faculties to the conditions of surrounding nature, and to the exigencies of the social state" (Wallace 1864, p. clxix.). Upon reading this paper, Darwin immediately wrote to Wallace to congratulate him on his splendid essay (Darwin and Seward 1903, vol. 2, p. 33), which he later declared to be "the best paper that ever appeared in the *Anthropological Review*" (Wallace 1916, vol. 1, p. 251). The two men, it seemed, were in complete harmony on this issue.

Wallace's Apostasy

Alas, Darwin and Wallace's fundamental agreement about human evolution was short-lived. In the five years after the publication of his paper, Wallace's view had undergone a thorough metamorphosis, putting the two men at odds. The first announcement that Wallace's view had undergone a dramatic reversal appeared in a review of some new editions of Charles Lyell's geological works (Lyell 1867/1868). At the end of his review, Wallace remarked that:

> Neither natural selection or the more general theory of evolution can give any account whatever of the origin of sensational or conscious life ... But the moral and higher intellectual nature of man is as unique a phenomenon as was conscious life on its first appearance in the world, and the one is almost as difficult to conceive as originating by any law of evolution as the other. (Wallace 1869b, p. 391)

This claim was developed further in a full-length essay, aptly entitled "The Limits of Natural Selection as Applied to Man" (1870). In a section entitled "What Natural Selection Can *Not* Do," Wallace reminded his readers of the "first principle of natural selection," namely, that "all changes of form or structure, all increase in the size of an organ or in its complexity, all greater specialisation or physiological division of labour, can only be brought about, in as much as it is for the good of [the] being so modified" (Wallace 1870, p. 333). In this he could enlist the assistance of Darwin, who has taught us that:

> "Natural selection" has no power to produce absolute perfection but only relative perfection, no power to advance any being much beyond his fellow beings, but only just so much beyond them as to enable it to survive them in the struggle for existence. Still less has it any power to produce modifications which are in any degree injurious to its possessor ... (Wallace 1870, p. 334)

It follows, therefore, that if we find in man at present any characteristics such that upon their first appearance could not possibly have been produced by natural selection (because they were either useless, or worse yet, would have actually been injurious to him), we can be sure that such characteristics were not the result of natural selection. So, too, if we find in man any specially developed organ whose development is disproportionate to its actual survival value, we can know with certainty that it did not come about through the operation of natural selection.

This is, Wallace pointed out, precisely what we do find when we consider the most distinctive physical characteristic of human beings, viz., our large brain, along with its attendant capacities. Wallace asks his readers to consider, for example,

> the capacity to form ideal conceptions of space and time, of eternity and infinity – the capacity for intense artistic feelings of pleasure, in form, colour, and composition – and for those abstract notions of form and number which render geometry and arithmetic possible. How were all or any of these faculties first developed, when they could have been of no possible use to man in his early

> stages of barbarism? How could "natural selection," or survival of the fittest in the struggle for existence, at all favour the development of mental powers so entirely removed from the material necessities of savage men, and which even now, with our comparatively high civilization, are, in their farthest developments, in advance of the age, and appear to have relation rather to the future of the race than to its actual status? (Wallace 1870, pp. 351–52)

With the human brain/mind, therefore, we discover "a surplusage of power; . . . an instrument beyond the needs of its possessor" (Wallace 1870, p. 338). In a gracious acknowledgment that philosophers occupy the highest rung of the intellectual ladder, Wallace notes that "Natural Selection could only have endowed savage man with a brain a little superior to that of an ape, whereas he actually possesses one very little inferior to that of a philosopher" (Wallace 1870, p. 356). This is, of course, good news for the "savage," but bad news for the theory of natural selection, which then becomes impotent to account for the most distinctive human characteristics.

The implications of this "problem of overdesign" (as Cronin 1991 aptly terms it) for understanding human evolution are, Wallace thought, clear and unavoidable. "[W]e must therefore admit, that the large brain he actually possesses could never have been solely developed by any of those laws of evolution, whose essence is, that they lead to a degree of organization exactly proportionate to the wants of each species, never beyond those wants" (Wallace 1870, p. 343). Nor can it be the case that natural selection somehow initially built into *Homo sapiens* characteristics it would need later, once it had evolved to a sufficient degree of sophistication. Natural selection has no power to foresee the future and plan accordingly. Some other sort of explanation must therefore be sought for such characteristics.

Wallace was quite clear about what was at stake for "Mr. Darwin's theory." In a letter to Lyell describing some of the unique features of human beings, for example, the ability of primitive people to quickly learn difficult musical skills, Wallace had issued an explicit challenge to Darwin:

> Unless Darwin can shew me how this rudimentary or latent musical faculty in the lowest races can have been developed by survival of the fittest – can have been of *use* to the individual or the race, so as to cause those who possessed it in a fractionally greater degree than others to win in the *struggle for life*, I must believe that some other power caused that development, – and so on with every other especially human characteristic. (Wallace to Charles Lyell, 28 April 1869; quoted in Richards 1987, p. 183)

Rather than simply plead agnosticism at this point, Wallace believed that he knew something about the nature of the "other power" responsible for human characteristics. "The brain of pre-historic and of savage man seems to me to prove the existence of some power, distinct from that which has guided the development of the lower animals through their ever-varying forms of being" (Wallace 1870, p. 343). In the special case of man, Wallace concluded, "a superior intelligence has guided the development of man in a definite direction, and for a special purpose" (Wallace 1870, p. 359).[1]

Darwin on the Descent of Man

Given his naturalistic perspective, Darwin was understandably dismayed at Wallace's dramatic supernaturalist turn. Wallace had written to Darwin, warning him in advance that his friend would be surprised at the special limitation he had placed on natural selection. Darwin feared the worst, writing back a few days later "I hope that you have not murdered too completely your own and my child" (F. Darwin and Seward 1903, vol. 2, p. 39). Upon reading Wallace's 1869 essay, Darwin concluded Wallace had, indeed, committed infanticide. In a letter he responded tersely to Wallace's apostasy, writing "I can see no necessity for calling in an additional and proximate cause in regard to Man" (Wallace 1916, p. 199).

Simply recording disappointment at Wallace's unorthodox position was not enough. Wallace had issued a challenge to Darwin that had to be taken up if the theory of natural selection was not to be run aground on the case of human beings. To adequately meet this challenge, Darwin had to provide an alternative evolutionary (and ideally, selectionist) account of the development of human physical and mental characteristics. To a large extent Darwin delegated the task to his friend and scientific advocate Thomas Henry Huxley (1825–95), who was only too happy to oblige. In his essay, "Mr. Darwin's Critics" (1871), Huxley noted that according to Wallace's own writings, the life of primitive people actually required extraordinary mental feats, including knowledge of a vast territory, reading signs of game or enemies, discovery of the properties of plants and the habits of animals, and so on. "In complexity and difficulty," Huxley estimated, "the intellectual labour of a 'good hunter or warrior' considerably exceeds that of an ordinary Englishman" (Huxley 1871, p. 471). The brain power actually needed by primitive peoples for survival was considerably greater than Wallace supposed, and not at all in excess of what could be forged by natural selection.

Possessed of enormous rhetorical skills, Huxley was by any standard a valuable ally, and Darwin was grateful for Huxley's assistance. But eventually Darwin himself had to address Wallace's challenge. Just as the writing of the *Origin of Species* was prompted by the arrival in the post of a letter from Wallace outlining a theory of evolution very much like his own, so too the *Descent of Man* (1871) was prompted by Wallace. But this time it was Darwin's desire to counter the claims of Wallace that provided the needed spur.

Darwin's aim in this book was to elaborate a thoroughly naturalistic account of human characteristics, physical and mental. His strategy was to focus, first, on homologous anatomical structures in humans and lower animals, and then on homologous mental structures, demonstrating the intellectual commonalties between humans and nonhumans. Elemental emotions like curiosity, fear, courage, affection, and shame are, Darwin showed, shared amongst humans and nonhumans alike. So, too, the essential elements of those supposedly distinctive human characteristics of tool use, a sense of beauty, language, and even religious sentiments can be found amongst nonhuman animals. If there was no need to appeal to some "higher power" for the appearance of such characteristics in the lower animals, and if there is a smooth continuum between the lower animals and ourselves, then there is no need to appeal to some higher power to account for the appearance of these characteristics in ourselves, either.

Such an argument, however, could only go so far. It is one thing to claim that something *must* be possible; it is another to show *how it is* possible. The challenge for Darwin was to show, not just that there is a continuum between nonhuman animals and humans but also to show that either (i) natural selection could have originally produced the distinctive characteristics we associate with humans, or (ii) such characteristics, although not the products of natural selection, could nonetheless have come about through purely natural processes. In principle, Darwin, not being a strict adaptationist, had a way out not available to Wallace. Unlike Wallace, Darwin could easily concede that a number of human physical and mental characteristics had never possessed any selective value in the struggle for existence, and therefore had not evolved directly by natural selection. In fact, this forms part of his response to Wallace:

> No doubt man, as well as every other animal, presents structures, which as far as we can judge with our little knowledge, are not now of any service to him, nor have been so during any former period of his existence, either in relation to his general conditions of life, or of one sex to the other. Such structures cannot be accounted for by any form of selection, or by the inherited effects of the use and disuse of parts. (Darwin 1871, vol. 2, p. 387)

Such a position helped to blunt the force of Wallace's challenge. By itself, however, such a response would be insufficient to counter the arguments of Wallace and others who doubted the sufficiency of natural selection to account for highly developed human mental capacities, because it would be implausible to suppose characteristics so distinctive of human beings would be selectively neutral. Darwin therefore had to try to show that *natural selection* could be sufficient to account for the origin of man's higher intellectual and moral faculties.

He broached this problem in Chapter V – "On the Development of the Intellectual and Moral Faculties During Primeval and Civilised Times." After praising Wallace's 1864 paper, Darwin noted that the intellectual and moral faculties of man are variable, and probably heritable. "Therefore, if they were formerly of high importance to primeval man and to his ape-like progenitors, they would have been perfected or advanced through natural selection" (Darwin 1871, vol. 1, p. 159). That they would be of high importance is beyond doubt, because it is precisely these faculties that make humans so biologically successful:

> All that we know about savages . . . shew that from the remotest times successful tribes have supplanted other tribes. . . . At the present day civilised nations are everywhere supplanting barbarous nations . . . and they succeed mainly, though not exclusively, through their arts, which are the products of the intellect. It is, therefore, highly probable that with mankind the intellectual faculties have been gradually perfected through natural selection; and this conclusion is sufficient for our purpose. (Darwin 1871, vol. 1, p. 160)

Tribes characterized by individuals with high intellects enjoyed a selective advantage over more intellectually challenged tribes, leading to the gradual elevation of the average level of intelligence. What holds for the evolution of intelligence holds for advancement in the moral faculties as well: "At all times throughout the world, tribes have supplanted other tribes; and as morality is one element in their success, the standard of morality and the number of well-endowed men will thus everywhere tend to rise and increase" (Darwin 1871, vol. 1, p. 166). Darwin essentially used the arguments of Wallace's own 1864 paper to argue against the claims of Wallace's 1869 paper!

Darwin had little more to say about the subject in the *Descent*, and contented himself by speculating (implausibly, it might be thought) that in civilised society perhaps those of superior intellect tend to rear a greater number of children, hence producing "some tendency to an increase in both the number and in the standard of the intellectually able" (Darwin 1871, vol. 1, p. 171). He optimistically concluded that: "Judging from all we know of man and the lower animals, there has always been sufficient variability in the intellectual and moral faculties, for their steady improvement through natural selection" (Darwin 1871, vol. 1, p. 180).

Darwinism and Human Nature

Wallace had posed the critical challenge to Darwin (and to fellow Darwinians): How could unaided natural selection account for the advanced intellectual and moral characteristics of human beings? In the *Descent of Man*, Darwin had sketched a solution reminiscent of Wallace's own earlier position: Tribes characterized by individuals with high intellects and upstanding moral character enjoyed a selective advantage over more intellectually and morally challenged tribes, leading to the gradual elevation of the average level of both intelligence and moral character. In this way society itself, thanks to natural selection, would become ever more perfect. Such an optimistic vision of societal progress fit perfectly the prevailing *Zeitgeist* of the day, and was therefore bound to be appealing to many. But it was another question whether such a sunny conclusion was in fact warranted by the theory of natural selection. Some doubted it.

Lending Evolution a Helping Hand

One of those responding to Wallace's 1864 paper (the one in which he argued *for* a selectionist account of distinctive human characteristics) was William Rathbone Greg, a Scots moralist and political writer. In his essay "On the Failure of 'Natural Selection' in the Case of Man" (1868), Greg agreed with Wallace that a struggle amongst nations and races had promoted those groups having superior mental abilities. But he dissented from Wallace's claim that natural selection continues to elevate the minds and morals characterizing larger social groups. According to Greg, the highly developed moral sympathies diagnostic of the more advanced societies would protect the physically, mentally, and morally unfit within them from the culling hand of natural selection. Because the mentally and morally unfit have an unfortunate tendency to reproduce more prolifically than the intellectually and morally superior, eventually even the most advanced societies would be awash with dullards and degenerates, subverting any further advance in intellectual and moral qualities. In a memorable passage (quoted by Darwin in the *Descent of Man*, vol. 1, p. 174), Greg declared:

> The careless, squalid, unaspiring Irishman, fed on potatoes, living in a pig-sty, doting on superstition, multiplies like rabbits or ephemera: – the frugal, foreseeing, self-respecting, ambitious Scot, stern in his morality, spiritual in his faith, sagacious and disciplined in his intelligence, passes his best years in struggle and celibacy, marries late, and leaves few behind him. . . . In the eternal "struggle for existence," it would be the inferior and less favored race that had prevailed – and prevailed by virtue not of its good qualities but of its faults. (Greg 1868, p. 361)

Like the steam governor on a locomotive that operates as a feedback mechanism regulating the engine's output, so, too, natural selection acting on the moral nature of man limits and damps the ongoing moral progress of humanity. The very element favored by natural selection at one stage in human evolution would serve at a later stage to retard any further advance in that direction. Since natural selection would have acted thus as soon as man's slightly higher intellectual and moral qualities had appeared, it was difficult to conceive how these qualities could have reached their present highly advanced state solely under the influence of natural selection. The unassisted operation of natural selection on the minds and morals of man, it seems, contains within itself the seeds of its own demise.

A range of responses to this problem were available. One was to insist that natural selection is sufficient to account for the advanced intellectual and moral qualities of man. Darwinians, for example, could follow Darwin himself by appealing to "community selection" to argue that while selection operating within a group might benefit the unfit, this might well be overcome by selection operating among groups, which would favor those groups displaying the social instincts to a greater degree. Another response was to appeal to some supernatural agency to account for human evolution (the solution of Wallace, Asa Gray, and many others). Yet another response was to agree that natural selection was indeed insufficient to guarantee the continued progress of man's intellectual and moral progress, but to point out that it is these very qualities themselves that permit us to take active control of our own evolution, directing it as we wish. By encouraging those with superior intellectual and moral qualities to interbreed, and to do so prolifically, an active role could be taken in stemming the rising tide of intellectual and moral inferiority, and in promoting those higher qualities associated with the best features of mankind. Just as selective breeding had been successfully applied to agricultural varieties and livestock, so too it could be applied by humans to humans.

The "eugenics" movement of the early decades of the twentieth century that stemmed from this response took a number of different forms, some benign, some sinister. One particularly depressing result was the Immigration Restriction Act of 1924, which was designed to prevent the dilution of "the great American type of citizenship" (i.e., those of "Nordic" descent) by those of inferior ethnic stock (i.e., from southern and eastern Europe) (Richards 1987, p. 514). As one eugenicist put it, "Society must protect itself, as it claims the right to deprive the murderer of his life, so also it may annihilate the hideous serpent of hopelessly vicious protoplasm" (Davenport 1910, p. 129). By the 1930s, however, almost all American geneticists of distinction had abandoned the eugenicist movement, realizing that it faltered on unsophisticated genetics and a grossly oversimplified picture of the relationship between heredity and environment.

As eugenics was cooling down in America, it was heating up in Germany, long fertile ground for loosely tethered speculations of a broadly Darwinian sort. At the turn of the century, Ernst Haeckel (1834–1919) was arguably the most enthusiastic German proponent of Darwinism, although his "Darwinism" contained generous admixtures of Lamarckism and extended into metaphysical speculations that seem almost Hegelian by contemporary standards. According to some scholars (e.g., Gasman 1971), Haeckel's speculations "created an intellectual environment congenial to the growth of Nazi pseudoscience," which located the "Aryan race" as embodied in the German people at the pinnacle of human civilization (Richards 1987, p. 533). It is hardly surprising, in the shadow of the implementation of Nazi policies to "purify" their society, exemplified in the death camps of Auschwitz and Bergen-Belsen, that biological (including Darwinian) accounts of human nature would come to seem to many in the postwar period irredeemably and forever tainted with the blood of over six million individuals unfortunate enough to be victimized by the ideological highjacking of poorly understood and grossly misapplied biological ideas.

The Sociobiology Controversy

Unsurprisingly, at mid-century, applications of evolutionary perspectives to humans were at a low ebb. By the 1960s, however, Darwinism as the foundation for explanations of human nature had begun to enjoy a renaissance of sorts as a spate of books appeared which attempted to explain human behavior in evolutionary terms. In books with titles like *The Territorial Imperative* (Ardrey 1966), *On Aggression* (Lorenz 1966), and *The Naked Ape* (Morris 1967), evolutionary ideas were pressed into service in order to account for one or another distinctive human characteristic. While appealing to a broad, nonspecialist readership, such books suffered from a serious theoretical defect, viz., a fuzzy, out-of-date conception of how natural selection actually operates. Arguments that one or another behavior had evolved "for the good of the species" were common. However, these books did place before the public serious (albeit flawed) attempts to consider human beings in a naturalistic Darwinian framework.

What was needed was a way to understand the basis of social behavior that did not rely upon discredited evolutionary ideas. A number of subsequent developments seemed to provide exactly what was needed. William D. Hamilton (1963, 1964a,b) showed that genes for altruistic behavior could be favored by natural selection if they were associated with behaviors that conferred benefits on other individuals possessing those same genes ("kin selection"). George C. Williams (1966) complemented this insight by launching an attack on accounts that uncritically assumed that selection operates at the level of groups to maximize group (rather than individual) fitness. Finally, Robert Trivers (1971) developed the idea of "reciprocal altruism," according to which an altruistic trait could be favored if it led to conferring benefits on individuals who were likely to return the favor in the future.

Such insights were among the most important theoretical foundations for the newly christened discipline of sociobiology, defined by Edward O. Wilson in his massive compendium *Sociobiology: The New Synthesis* (1975) as "the systematic study of the biological basis of all social behavior." The first chapter ("The Morality of the Gene") was unapologetic in its reductionistic approach to sociobiological explanation: "In a Darwinist sense the organism does not live for itself. Its primary function is not even to reproduce other organisms; it reproduces genes, and it serves as their temporary carrier. . . . [T]he organism is only DNA's way of making more DNA" (Wilson 1975, p. 3). In twenty-seven lushly illustrated, sweeping chapters, Wilson explored social behavior from genes (p. 7) to esthetics (p. 564), along the way summarizing the latest findings in social evolution. In the last chapter, "Man: From Sociology to Sociobiology," he broached the issue of human social behavior directly, inviting his readers to "consider man in the free spirit of natural history, as though we were zoologists from another planet completing a catalog of social species on Earth" (Wilson 1975, p. 547). While emphasizing the plasticity of human social organization (pp. 548–551), he nonetheless went on to suggest that the essential elements of culture, religion, and ethics have a strong genetic basis which makes them possible yet also constrains their potential for unlimited development. They exist as part of "a hierarchical system of environmental tracking devices" (p. 560) that have evolved for their fitness-enhancing effects. To account for the rapidity of human social evolution, Wilson speculatively postulated an autocatalytic "multiplier effect" whereby "[a] small evolutionary change in the behavior patterns of individuals is amplified into a major social effect by the expanding upward distribution of the effect into multiple facets of social life" (pp. 11–13; 569–72). Altered conditions of social life, in turn, changed the parameters of which behaviors would be selectively advantageous. The positive feedback loop that resulted, Wilson argued, helps to explain the distinctiveness of human culture, as it would be observed by one of his imagined extraterrestrial zoologists. Still, our biology sets strict limits to the range of cultures that are possible. Our genes keep culture on a leash – an elastic leash, to be sure, but a leash nonetheless.

As applied to nonhumans, sociobiology stirred little controversy. Applied to humans, it became the exemplar *par excellence* of a vociferous academic controversy (Segerstråle 2000). Moral and methodological criticisms were sometimes distinguished, but often not. Lewontin, Kamin, and Rose (1984), for example, launched a two-pronged attack on Wilson's sociobiology, arguing that it is both morally objectionable and methodologically flawed. They argued that because the central assertion of sociobiology is that all aspects of human culture and behavior are coded in the genes and have been molded by natural selection, sociobiology should be shunned as "a reductionist, biological determinist explanation of human existence" (p. 236). According to them, the academic and popular appeal of sociobiology flows directly from its simple reductionistic program and its claim that human society as we know it is both inevitable and the result of an adaptive process. Sociobiology is attractive to many insofar as it appears to lend scientific support to a conservative social agenda, indeed, to its "legitimation of the status quo" (p. 236). If present social arrangements are the ineluctable consequences of the human genotype, then nothing of any significance can be changed.

Other critics zeroed in on what they saw as sociobiology's methodological flaws. For example, Kitcher (1985) charged that "pop sociobiology" as it is casually applied to explain every conceivable human characteristic (e.g., by Wilson 1978) suffers from the scientific sins of anthropomorphism (explaining human behaviors on the model of animal behaviors requires classifying a specific animal behavior as "rape" or "murder," a move that requires substantive and controversial assumptions), reductionism (sociobiological explanations frequently assume that the behavior of the group reflects the propensities of its constituents), perfectionism (sociobiological explanations assume that organisms are perfectly adapted to their environments, thus ignoring cost/benefit tradeoffs), and adaptationism (sociobiological explanations frequently assume without proof that every trait is adaptive). A main thrust of such criticisms was that human sociobiology often failed to live up to the stringent methodological standards characteristic of the best nonhuman sociobiological studies. Whereas conclusions in the latter were typically based on carefully interpreted investigations, applications of sociobiological principles to humans were by comparison criticized as casual and cavalier.

Undeterred by what was often seen as politically motivated attacks, such critiques forced enthusiasts for Darwinian explanations of human nature to be more circumspect in the way they framed their explanations without, however, significantly abandoning their fundamental approach. The sociobiology controversy persisted well into the 1980s, and to some extent continues to simmer just beneath the surface of much work on the biological basis of human behavior. In recent years, biologists taking a sociobiological approach have quietly gone about their business under the less emotionally charged name "behavioral ecology," a blander designation calculated to attract less unwanted attention.

Darwinian Medicine

In *The Descent of Man*, Darwin summed up his survey of the status of *Homo sapiens* as products of the evolutionary process by declaring "that man with all his noble qualities . . . with his god-like intellect which has penetrated into the movements and constitution of the solar system – with all these exalted powers – Man still bears in his bodily frame the indelible stamp of his lowly origin" (Darwin 1871, vol. 2, p. 405). If (early) sociobiology erred in seeing all or most human *behaviors* as adaptive, those concerned with providing Darwinian analyses of the human *body* could hardly be charged with the same error. Indeed, a characteristic feature of the new science of "Darwinian Medicine" (and its subdiscipline, Darwinian Psychiatry) is the insistence that in many ways the human body falls short of optimal design, and therein lies the explanation for a range of human ailments (Nesse and Williams 1994). Darwinian evolutionary theory provides the theoretical foundation. Because our bodies are the products of evolution, we should expect that their specific functions (and malfunctions) can be better understood when viewed in an evolutionary perspective. Traditional medical science focuses on understanding the *proximate* causes of illness, that is, those causes operating on and in presently existing individuals. Darwinian medicine, by contrast, seeks to understand the *ultimate* causes of an organism's characteristics in light of its evolutionary history. As such, Darwinian medicine explicitly adopts the stance of "methodological adaptationism" (discussed in Chapter 6). When trying to explain some widespread biological characteristic, methodological adaptationism directs one to ask what adaptive function this characteristic has, or might have had in the past, and to then seek a plausible selectionist explanation of this function, while recognizing that there is no guarantee beforehand that every characteristic is adaptive, or that a plausible selectionist explanation will be forthcoming. Chance, historicity, and constraints combine with natural selection to shape organisms, and any given characteristic to be explained is likely to be a function of all of these.

Taking this view, human ailments can be classified according to a small set of categories. "Evolved Defenses" include "ailments" that are in fact beneficial adaptations; e.g., pain as an indication of potential or actual injury; coughing, vomiting, and diarrhea as evolved defenses designed to expel dangerous materials from the body; fever, which creates an unfavorable environment for pathogens; etc. Additional ailments arise from "Conflicts with Other Organisms" (having different genetic agendas that don't entail one's welfare). External predators pose an obvious threat, as do internal predators (i.e., pathogens). A third class of ailments arise from the problems associated with "Coping with Novelty" (i.e., cases where our adaptations are out of sync with rapid change of environments). Our bodies evolved in an "Environment of Evolutionary Adaptedness" (EEA) – the hypothetical set of environments in which 99.9 percent of human evolution took place. Common causes of death in our EEA included accidents, starvation, predation, and disease. Common causes of death now include heart attacks, strokes, other complications of atherosclerosis, cancer, and health hazards associated with obesity. The craving for fat, salt, and sugar which served us well in the Pleistocene now causes serious health problems. A fourth class of ailments arises from "Tradeoffs" (design compromises increasing net fitness). Lower back pain is one of the costs of bipedalism.[2] Some tradeoffs are genetic. Vulnerability to sickle-cell anemia is the cost of enjoying enhanced resistance to malaria.

Cancer (the unchecked growth of cells in the body) may be a risk incurred in organisms whose very existence depends on processes leading to multicellularity and tissue differentiation. Finally, some ailments are simply "Design Flaws." Selection is opportunistic, lacking foresight, and thus cobbles together structures in ways that enhance immediate fitness but that may not be optimal. The vertebrate eye is surely an adaptation, but the blind spot resulting from the fact that it is wired backward is an artifact of contingent historical conditions. The same is true of the ever-present possibility of choking on food, thanks to intersection of respiratory and food passages at the esophagus. Darwinian Medicine is still a young science, but it represents a way in which a critical understanding of Darwinian evolution can find practical application in diagnosing and treating human ailments.

Evolutionary Psychology

Darwinian Medicine concerns how our bodies work and why they too often fail. According to evolutionary psychologists, Darwinism is equally fruitful for understanding the workings of the human mind (Barkow, Cosmides, and Tooby 1992; Wright 1994; Buss 1999). They contrast two philosophical views of knowledge, one deriving from John Locke, which denies that there is any innate knowledge, insisting instead that all knowledge derives from the experiences of each individual. In Locke's famous phrase, the mind at birth is a "*tabula rasa*" (a "blank slate") to be written upon by experience. According to the other, constructivist view (associated with Immanuel Kant), the human mind uses experience to construct its image of reality, but it does so by using its innate "deep structure" to organize and make sense out of that experience. Despite this important historical anticipation of their views, this is where the indebtedness to Kant ends. For Kant the question of *why* the human mind organizes experience the way it does was an impenetrable mystery. Evolutionary psychologists believe that Darwin provided the solution. Evolution by natural selection is the source of the mind's innate structures.[3]

Accordingly, the starting point in applying evolutionary theory to understand the workings of the human psyche is to realize that our modern skulls house a stone-age mind. Rather than treating the human mind/brain as an all-purpose general computing machine (as some researchers in Artificial Intelligence tend to do), the mind/brain is treated as a set of information-processing "modules" (i.e., neural circuits) designed by natural selection over a ten-million-year period to solve the specific adaptive problems faced by our hunter-gatherer ancestors. These modules were designed to generate behavior that was appropriate to the environmental circumstances in which they evolved (the EEA), only some of which remain current. Nonetheless, these circuits continue to function, organizing the way we interpret our experiences, injecting certain recurrent concepts and motivations into our mental life, and providing universal frames of meaning that allow us to understand the actions and intentions of others. As Darwinian medicine teaches about our bodies, so too evolutionary psychology teaches about our minds. They are alike the legacies of our evolutionary past, and can only be understood as such.

In the *Origin of Species*, Darwin speculated on the implications of his theory for understanding in an entirely new way the workings of the human mind: "In the distant future I see open fields for far more important researches. Psychology will be based on a new foundation, that of the necessary acquirement of each mental power and capacity by gradation" (Darwin 1859, p. 488). According to some critics of evolutionary psychology, Darwin's prediction has yet to be realized, as enthusiasts for evolutionary approaches to the mind have rushed in to build a grand theoretical edifice without first laying a solid foundation in what is already known about the biological bases of cognition. Some of the criticisms of evolutionary psychology are reminiscient of the methodological complaints lodged decades before against sociobiology – for example, that it is guilty of uncritical adapationism (Gould 1977b). Critics charge as well that there is scant evidence demonstrating the existence of specialized neural modules associated with specific psychological strategies. What special-purpose neural circuits there are (e.g., those governing various emotions) evolved well before the EEA for humans (the Pleistocene), are located in the subcortical systems of the brain (rather than in the more recently evolved neocortex), and are shared by all mammals. Recent human evolution, by contrast, may have provided the context for a very general and flexible form of multipurpose intelligence that operates quite differently from the more primitive mammalian brain structures (Panksepp and Panksepp 2000). But, if so, then the search for specialized cognitive adaptations is misguided.

Such critiques have inspired revisionists from within the ranks of evolutionary psychologists to try to build upon rather than discard the insights to be gleaned from adopting a Darwinian perspective on the mind. They argue that the human mind is not like a Swiss Army knife consisting of specialized cognitive tools for solving specific problems, but instead a living "brain/mind construction system" that exploits pliable brain tissue which changes with each novel experience (La Cerra and Bingham 2002). Thanks to our common evolutionary ancestry, all humans share a common neural architecture. Yet personal experience determines which of the trillions of different ways of organizing the neural connections of the brain will be realized in any given person as they develop. Consequently, environmental factors such as socioeconomic conditions are imbedded in, rather than an alternative to, an evolutionary account of the mind. Such an approach goes a long way toward assuaging the concerns of those critics of earlier evolutionary psychology who charged that a Darwinian perspective on the mind was in danger of ignoring the significant influence of social context in explaining the nature of human nature.

Were We Inevitable?

> Man in his arrogance thinks himself a great work, worthy the interposition of a deity; more humble & I believe true to consider him created from animals.
>
> (Darwin, C Notebook, pp. 196–97; in Barrett et al. 1987, p. 300)

A common characteristic of many creation stories is that human existence is treated as anything but accidental. In some sense we were *meant* to be here; we are special, and our existence was, in some deep sense, inevitable. Darwinism seems to suggest a very different conclusion, viz., that our existence, like that of every other species that has ever existed, is the result of innumerable contingencies such that had any of them been different, we would not exist at all. Our very existence as a species is spectacularly improbable. This is the clear lesson of Darwinism.

Or is it? In fact, Darwinians of various stripes have maintained that, although chance factors have assuredly played a role in our evolution, our appearance is not entirely the result of forces operating randomly. They hold that we (or something like us) were in some sense either inevitable or at least highly probable, given the way in which evolution by natural selection operates. More recently, some Darwinians have argued that there are (nonspiritual) directional forces built into the evolutionary process that might have permitted a keen observer of the first replicating molecule, armed with the basic principles of Darwinian evolutionary biology, to predict that eventually we (or something like us in specified respects) would appear on the scene, and ponder its own existence. The issues here are difficult and the theories speculative, but they are too critical to our self-understanding to ignore. In a moment we will look at the arguments for and against the claim of human inevitability. First, however, it is important to distinguish between two claims that are easily conflated, viz., the claim that we are *special*, and the claim that we are *inevitable*.

Highness and Inevitability

No less of a Darwinian than Alfred Russel Wallace wrote without a shred of discomfort or apology that "To us, the whole purpose, the only *raison d'être* of the living world – with all its complexities of physical structure, with its grand kingdoms, and the ultimate appearance of man – was the development of the human spirit in association with the human body" (Wallace 1889, p. 47). We have already seen that earlier Wallace had abandoned his conviction that the human mind was entirely the product of natural selection in favor of the belief that a superior intelligence had "guided the development of man in a definite direction, purpose" (Wallace 1870, p. 359). Many biologists in the middle decades of the twentieth century followed Wallace in seeing humans as the pinnacle, the highest achievement, of the evolutionary process. George Gaylord Simpson insisted that nothing in our knowledge of the evolutionary process warrants the belief that it had human beings as its goal, or that the human evolutionary lineage is the central line of evolution. Nonetheless, when the range of different criteria for evolutionary progress are considered, "A majority of them do, however, show that man is among the highest products of evolution and a balance of them warrants the conclusion that man is, on the whole but not in every respect, the pinnacle so far of evolutionary progress" (Simpson 1949, p. 262; see also p. 285). Similarly, Theodosius Dobzhansky maintained that "Judged by any reasonable criteria, man represents the highest, most progressive, and most successful product of organic evolution. The really strange thing is that so obvious an appraisal has been over and over again challenged by some biologists" (Dobzhansky 1956b, p. 86). Finally, according to Ernst Mayr, not only is evolutionary progress a reality, but the evolutionary progress culminates in beings with precisely our characteristics:

> [W]ho can deny that overall there is an advance from the prokaryotes that dominated the living world more than three billion years ago to the eukaryotes with their well organized nucleus and chromosomes as well as cytoplasmic organelles; from the single-celled eukaryotes to metaphytes and metazoans with a strict division of labor among their highly specialized organ systems; within the metazoans from ectoderms that are at the mercy of the climate to the warm-blooded

endotherms, and within the endotherms from types with a small brain and low social organization to those with a very large central nervous system, highly developed parental care, and the capacity to transmit information from generation to generation? (Mayr 1988, pp. 251–52)

Such statements leave little doubt that some of the most influential biologists of the twentieth century had few qualms about assigning a privileged evolutionary status to *Homo sapiens.*

Although there are hints in the writings of these biologists that they viewed *Homo sapiens* as not only special, but in some sense inevitable (on one occasion Dobzhansky maintained that "the evolutionary line that produced man [is] the 'privileged axis' of the evolutionary process" (Dobzhansky 1967, p. 117), the claim of inevitability is conceptually distinct, and represents a different, much stronger, conclusion. Julian Huxley at times *sounds* as if he embraced both doctrines. Recall (from Chapter 8) the two characteristics that Huxley came to identify as distinguishing higher from lower creatures; namely, increased control over and independence from the environment. On both criteria, he believed, human beings, come out on top, and are thus "highest." In addition, he sometimes writes as if he thought that the appearance of *Homo sapiens* was inevitable. For example, in *Evolution: The Modern Synthesis* (1942), Huxley wrote: "One somewhat curious fact emerges from a survey of biological progress as culminating in the dominance of *Homo sapiens.* It could apparently have pursued no other general course than that which it has historically followed" (Huxley 1942, p. 569). Taken by itself, this certainly looks like a claim for inevitability. Yet it is clear from his remarks in a book from the previous year that he did not mean that the evolutionary process had to produce creatures like us, *tout court.* He was identifying necessary, rather than sufficient, conditions for creatures like ourselves to arise. In other words, his claim reflects a *conditional:* If creatures with our distinctive characteristics arise at all, then our appearance would have to be preceded by evolutionary developments just like those that did, in fact, precede our own appearance:

The essential character of man as a dominant organism is conceptual thought. And conceptual thought could have arisen only in a multicellular animal, an animal with bilateral symmetry, heart and blood system, a vertebrate as against a mollusc or an arthropod, a land vertebrate among vertebrates, a mammal among land vertebrates. Finally, it could have arisen only in a mammalian line which was gregarious, which produced one young at birth instead of several, and which had recently become terrestrial after a long period of arboreal life. (Huxley 1941, p. 15)

Huxley (1941, pp. 1–22) attempted to justify the necessity of each step in this sequence for the eventual emergence of conceptual thought. The exact details matter less than the central point at issue. A characteristic like conceptual thought requires a large number of biological antecedents in order for it to appear at all. Rather than being anathema to contemporary biologists, such a view is simply taken for granted (e.g., Maynard Smith and Szathmáry 1995). Contrary to first appearances, therefore, Huxley did not suggest that there was anything inevitable about the emergence of intelligent beings like ourselves. Despite the current hegemony of intelligence, we owe our existence to a series of contingent events. Being "highest" and being inevitable are two different issues. Perhaps there has to be a winner for every game, but it's not inevitable that a given individual specified beforehand *be* that winner (Dennett 1995).

Evolutionary Contingency

The theme of contingency regarding evolution as a whole, and regarding *Homo sapiens* in particular, has been a recurring theme in the writings of Stephen Jay Gould. We have already examined (in Chapter 8) in some detail his arguments against evolutionary progress in his book *Full House* (1996). In an earlier book, *Wonderful Life* (1989), he encouraged readers to think of the history of life on earth as thoroughly imbued with contingency. The occasion for advancing this thesis was reflection on fossils discovered in the Burgess Shale of British Columbia, which contain a bonanza of long extinct creatures – not just extinct species but extinct *kinds* of animals (i.e., phyla), the likes of which have not roamed the earth for half a billion years. These were creatures that dominated the seas forty million years after the "Cambrian Explosion" of life 570 million years ago, when multicellular organisms first appeared on the scene. They thrived in the ancient seas, yet virtually all of them came to an unfortunate end shortly thereafter. The few that remained gave rise to all the animals that would later populate the earth. Why did most of these forms vanish? Gould (1989, p. 47) emphasizes that in his view this was a genuine *decimation* in the sense that those that survived were a *random* sample of those previously existing. It was pure luck that allowed *Pikaia* to squeak by to give rise to later chordates (ourselves included), whereas *Anomalocaris, Opabinia*, and their myriad "weird wonder" cohorts vanished forever. An adaptationist, Gould points out, would interpret this pruning of the tree of life as yet another example of natural selection in operation, perhaps arguing that: "[A]ll but a small percentage of Burgess possibilities succumbed, but the losers were chaff, and predictably doomed. Survivors won for cause – and cause includes a crucial edge in anatomical complexity and competitive ability" (Gould 1989, p. 48). Gould, however, will have none of this. Those that survived, he insists, were just lucky:

> [T]he Burgess pattern of elimination ... suggests a truly radical alternative.... Suppose that winners had not prevailed for cause in the usual sense. Perhaps the grim reaper of anatomical designs is only Lady Luck in disguise. Or perhaps the actual reasons for survival do not support conventional ideas of cause as complexity, improvement, or anything moving at all humanward.... Groups may prevail or die for reasons that bear no relationship to the Darwinian basis of success in normal times. Even if fish hone their adaptations to peaks of aquatic perfection, they will all die if the ponds dry up. But grubby old Buster the Lungfish ... and his kin may prevail because a feature evolved long ago for a different use has fortuitously permitted survival during a sudden and unpredictable change in the rules. And if we are Buster's legacy, and the result of a thousand other similarly happy accidents, how can we possibly view our mentality as inevitable, or even probable? (Gould 1989, p. 48)

Using a fertile metaphor, Gould suggests that we engage in a thought-experiment. Picture the history of life on earth as a videotape – as "life's tape." Hit "rewind," taking us back to any time and place in the past, in the process erasing all that has actually transpired, then let the tape run again. Perform this exercise as many times as one likes. Is there any reason to suppose that the same biota would ever evolve again? Suppose that every run of the tape yields a strikingly different set of organisms from those that constitute the actual history of life. What could we then say about the inevitability of self-conscious intelligence? Gould is not optimistic that it would ever appear a second time. Multicellular life began with a range of different anatomical types, only a small set of which survived to give rise to all later animals. "If the human mind is a product of only one such set, then we may not be randomly evolved in the sense of coin flipping, but our origin is the product of massive historical contingency, and we would probably never arise again even if life's tape could be replayed a thousand times" (Gould 1989, pp. 233–34). Play the "tape of life" over again, and we are sure to get a (very) different set of organisms inhabiting the earth. The organisms that now inhabit the earth represent just a tiny fraction of the possible kinds of organisms that *could* exist, and might otherwise *have existed*, had various chance events in the distant past been just slightly different than they in fact were. Gould's book is a sober antidote for anyone who is tempted to see the present biota of the earth (including, of course, *Homo sapiens*) as in some way necessary or inevitable.

Competition and Convergence

According to Gould's "Evolutionary Contingency Thesis," were we able to replay life's tape again, altered ever so little at the outset, evolution would yield a sensible but entirely different biota than that which presently exists. Not everyone agrees. Simon Conway Morris, for example, one of the principal researchers of the Burgess Shale fauna, has argued at length that the basic premise of Gould's contingency thesis "is based on a fundamental misapprehension of both the nature of the Burgess Shale fauna and the processes of organic evolution" (Conway Morris 1998, p. 201). First, new findings since Gould wrote his book suggest very strongly that the animals fossilized in the Burgess Shale were part of a complex ecology in which predation (e.g., by the fearsome predator *Anomalocaris* – the terror of trilobites everywhere) played a critical role. Rather than representing some sort of "suspension" of familiar evolutionary mechanisms (e.g., natural selection), the animals of the Burgess Shale lived (and died) according to the same principles that have governed life before and ever since. In other words, being *well adapted* then, as now, gives one a "leg up," so to speak, in the struggle for survival – even if *legs, per se,* would not make an appearance for many millions of years. There is therefore little basis for concluding that chance factors, rather than natural selection, determined the fate of the Burgess Shale creatures.[4]

Second, emphasizing as well the pervasive importance of evolutionary convergence in the history of life, Conway Morris maintains that "Although there may be a billion potential pathways for evolution to follow from the Cambrian explosion, in fact the real range of possibilities and hence the expected end results appear to be much more restricted.... [Indeed] within certain limits the outcome of evolutionary processes might be rather predictable" (Conway Morris 1998, p. 202). Although it is true that *any given species* (e.g., *Trichechus manatus,* the Florida manatee) is an utterly contingent product of the evolutionary process, it does not follow that *every phenotypic characteristic* is equally contingent. The ubiquity of *convergent evolution,* in particular, seems to suggest that natural selection operating on random variations will again and again hit upon the same sorts of solutions to biological problems. Similar environments tend to favor similar adaptations, even in distantly related organisms. Ichthyosaurs, fish, and cetaceans all sport (or sported) streamlined bodies for efficient locomotion through a viscous medium (i.e., water). Pterosaurs, birds, and bats all evolved forelimbs with a large surface area in order to reap the advantages of an airborne lifestyle.

Marsupial saber-tooth cats look remarkably similar to their placental doppelgängers. In each case, organisms evolved to exploit a previously un- or underexploited niche in the economy of nature. Some phenotypic characteristics (e.g., eyes) are so advantageous in a side variety of environments that they have evolved many times (Salwini-Plawen Mayr 1977; Dawkins 1996). Rewind the tape of life to the first replicating molecules, and streamlined bodies, wings, and eyes would very likely eventually evolve once again. Conway Morris's summary of the implications of convergent evolution for Gould's contingency thesis is worth quoting in full:

> [A]t the heart of *Wonderful Life* are Gould's deliberations on the roles of contingencies in evolution. Rather than denying their operation – and that would be futile – it is more important to decide whether a myriad of possible evolutionary pathways, all dogged by the twists and turns of historical circumstances, will end up with wildly different alternative worlds. In fact the constraints we see on evolution suggest that underlying the apparent riot of forms there is an interesting predictability. This suggests that the role of contingency in individual history has little bearing on the likelihood of the emergence of a particular biological property. (Conway Morris 1998, p. 139)

Streamlined bodies, wings, and eyes are (under a wide array of conditions) beneficial (i.e., adaptive) biological properties. So, too, is intelligence. It could therefore be argued that once life got started, intelligence was bound to arise eventually because sooner or later the "smart niche" would be exploited – if not by *Homo sapiens*, then by some other intelligent species. Some nonbiologists have simply assumed that "other things being equal, it is better to be smart than to be stupid" (Sagan 1995, p. 2). Unfortunately, a survey of life on Earth seems to refute this claim, since the overwhelming majority of living things failed to evolve smartness. Unlike streamlined bodies, wings, and eyes, intelligence of the sort that characterizes *Homo sapiens* is hardly a widespread biological trait, having arisen (so far as we know) just *once* in all the fifty billion species that have existed since the origin of life.[5] It is not hard to see why this might be. Intelligence is not an unqualified good in relation to Darwinian fitness, as it entails both significant benefits *and* substantial costs. Thus, as one commentator correctly notes, "Natural selection does only one thing: it produces organisms better adapted to the local environment. It contains no built-in 'self-perfecting' principle that guarantees a particular outcome, such as intelligence" (Olson 1985, p. 6).

On the positive side of the ledger, however, even if natural selection doesn't guarantee increasing intelligence, there might nevertheless be significant selective advantages to greater intelligence in a range of environments that in fact occur. Greater intelligence is a way of being the ultimate generalist. When an ecological situation changes, intelligence permits learning, behavioral flexibility, and innovation, providing alternative strategies for the generalist to fall back on. A specialist, by contrast, while superbly adapted to its present environment, becomes vulnerable to changes in that environment which may render its specialized survival strategies obsolete. The long-term prospects of ecological generalists are better than those of specialists.

Back on the negative side of the ledger, large brains are high-maintenance machines. Our brains consume approximately 20 percent of the energy used by the body. In order to earn its keep, the human brain must be responsible for increasing energy intake by at least a fifth. Clearly the benefits of greater intelligence are sometimes enough to compensate for the increased energy demands a brain requires, as we are here, and it is unlikely that our intelligence was unrelated to our biological success as a species. But whether the benefits would *usually* outweigh the costs is a different matter. Some intelligence is often beneficial. But beyond a certain point greater intelligence may actually compromise an organism's reproductive capacity. High intelligence is just one way of making a living in the world; there are many others: superior speed, hearing, vision, smell, dodging skill, claws, teeth, venom, armor, and concealment; greater fecundity; changes in size; hunting cooperation; aggregation of prey species into groups to increase sensory coverage; nocturnal hunting; etc. Viewed from this perspective, there seems to be nothing inevitable about the evolution of intelligence of the sort possessed by humans. Certainly it is a minority strategy amongst living things, and even amongst mammals. As paleontologist Jack Sepkowski bluntly remarks, "I see intelligence as just one of a variety of adaptations among tetrapods for survival. Running fast in a herd while being as dumb as shit, I think, is a very good adaptation for survival" (quoted in Ruse 1996, p. 486).

Granted that there are many ways of making a living in the world, and that intelligence seems to be a distinctly minority strategy, might there not still be a bias in the evolutionary process toward greater intelligence, *once intelligence of a certain level appears in the first place*? Some biologists have thought so. First, there may be a genetically biased "ratchet effect" in relation to brain expansion. The same quantity of brain growth (e.g., 1 gm) requires more improbable genetic mutations for a small-brained creature than it does for an already large-brained creature, thus facilitating (although not, of course, causing) accelerated brain growth. If an increase in intelligence above a certain threshold provided an adaptive advantage, this could engage a "positive feedback loop" making additional intelligence even more advantageous. While the poor (in biological computing power) may not get any poorer, there may be a tendency for the rich to get richer. Second, many biological attributes (e.g., speed) generally plateau at a certain level because of the associated rapidly increasing energy costs, but this may not always be true of brain size. Beyond a certain level of intelligence, there may be "positive feedback" between increases in brain size, and the growing range of behavioral options opened up by the burgeoning culture which accompanies high intelligence. Thus, once higher intelligence evolves, it may accelerate its own evolution. (Cf. E. O. Wilson's "multiplier effect," discussed earlier.)

"The Ultimate Predator"

One more counterargument is available to the proponent of intelligence inevitability. Even if intelligence of the sort possessed by humans is rare, and therefore not an evolutionary convergence, it could be argued that it was nonetheless virtually inevitable due to evolutionary arms races. Recall Darwin's observations in the *Origin,* simultaneously chilling and optimistic: "[T]he structure of every organic being is related, in the most essential manner, to that of all other organic beings, with which it comes into competition . . . or from which it has to escape, or on which it preys. . . . Thus, from the war of nature . . . the most exalted object which we are capable of conceiving, namely, the production of the higher animals, directly follows" (Darwin 1859, pp. 77, 490). Gould focuses on adaptation to randomly changing environments, and argues that the evolutionary history of organisms should be effectively random as well. But part of the environment of organisms is other organisms with which it is in competition, and increased capacity for outsmarting, outmaneuvering, and so on, them will introduce nonrandom factors in evolution. Theoretically, arms races will propel the evolutionary process forward, giving it directionality (Vermeij 1987). Jerison (1973, p. 315) amassed data about the temporal succession of relative brain sizes ("encephalization quotients") for carnivores and herbivores, suggesting a coevolutionary arms race. As Gould explains,

Both herbivores and carnivores displayed continual increase in brain size during their evolution, but at each stage, the carnivores were always ahead. Animals that make a living by catching rapidly moving prey seem to need bigger brains

> than plant eaters. And, as the brains of herbivores grew larger (presumably under intense selective pressure imposed by their carnivorous predators), the carnivores also evolved bigger brains to maintain the differential. (Gould, 1977, p. 190)[6]

Could such a process account for the evolution of intelligence and other distinctive human characteristics? Some biologists have thought so (Levy 1999). Humans possess a suite of "weapons systems" that together provide a competitive edge over most (or all) other organisms. Intelligence permits the acquisition of knowledge about prey – their habits, movements, and vulnerabilities, making possible both cooperative hunting and transfer of information to the next generation. In conjunction with versatile pentadactyl forelimbs (i.e., hands), it also permits the construction of tools (including hunting weapons) to more efficiently capture, kill, and butcher prey. Language permitted efficient, versatile, and rapid communication of information. Bipedalism with swinging arms, an upright posture, "locking" knees, and modified leg musculature permitted efficient walking, effective structural support against gravity, and stamina in tracking prey over large distances. It is probably impossible to say at this juncture whether this suite of weapons systems found its primary use in relation to various nonhominid competitors, or whether it drove human evolution through competition between conspecifics. As Darwin noted, competition will usually be most intense between members of the same species since they vie for the same resources. Nonetheless, such characteristics may merit identifying *Homo sapiens* as "the ultimate predator" (Levy 1999) – a dubious honor in light of the invention of weapons of mass destruction that threaten not only our own survival but that of all living things.

The Evolutionary Destiny of *Homo Sapiens*

> What circumstances may have been necessary to have made man!
> (Darwin, C Notebook, p. 78; in Barrett et al. 1987, p. 263)

We have considered Darwinian explanations both for our origins and for the present state of our species. Whether or not one concludes that there was anything inevitable, or even likely, in the emergence of human beings, it is clear that "we" have come a long way from our hominid and pre-hominid ancestors. What does the future hold for *Homo sapiens*? What, if anything, might a Darwinian understanding of life predict about our destiny as a species? Has biological evolution has come to an end with

Homo sapiens – *not* because we are the predestined *telos* of the evolutionary process, but because our peculiar characteristics, especially the unprecedented ability to modify the planet to our own desires and thereby to control evolution to a significant degree, effectively thwarts the continued evolution of other species? It is risky to use the history of life as a firm basis to speculate about the future of life. (Sixty-five million years ago as reptiles dominated the earth, who would have supposed that their tenure would so abruptly come to an end, their suddenly vacant niches to be usurped by mammals, a hitherto marginal class of organisms?) It is even more risky to speculate about the future of a particular species, especially if the species in question is in many respects *atypical* of most life forms that have ever existed. Risky but also irresistible. Biologists (and others) from Darwin to the present have attempted to use our knowledge of the history of life (so far) as a basis for speculating about the future. Consideration of a few of these will round out our discussion of "the evolution of Darwinism."

Darwin and Wallace on the Future of Humankind

Speculation about the evolutionary future of humankind was present from the start. Once again, noting the similarities and differences between Darwin and Wallace is instructive, inasmuch as they represented two alternative visions of the role and future of human beings in the evolutionary drama. As we have seen, Darwin's naturalism, evident in his private notebook entries straight through his later works, stands in sharp contrast to Wallace's supernaturalist turn in the late 1860s. Along with their differing assessments of the power of selection and the scope of adaptation, these fundamental metaphysical differences resulted in two different accounts of the origin and development of distinctive human characteristics. Both men did not hesitate to describe human mental and moral characteristics as "perfect," but disagreed about whether a naturalistic, selectionist account could adequately explain this perfection. Nonetheless, in their own ways, Darwin and Wallace each envisioned a rosy future for *Homo sapiens*. For Wallace, a superior intelligence had already "guided the development of man in a definite direction, and for a special purpose" (Wallace 1870, p. 359), and would undoubtedly continue to do so in the future. Combining his Spencerian and his new-found spiritualist beliefs, he maintained that: "Progressive evolution of the intellectual and moral nature is the destiny of individuals" (Wallace 1874, p. 56) – a destiny vouchsafed not by natural selection but by higher intelligences that controlled human evolution.

Darwin took a more cautious but no less optimistic view of humanity's future. Although Darwin was a "progressionist" (in the senses already discussed) concerning the evolutionary process, he rarely played the role of soothsayer regarding the future course of evolution. At the end of the *Origin*, however, Darwin permitted himself to briefly speculate about the future of life on earth by considering life's past:

> Judging from the past, we may safely infer that not one living species will transmit its unaltered likeness to a distant futurity. And of the species now living very few will transmit progeny of any kind to a far distant futurity; for the manner in which all organic beings are grouped, shows that the greater number of species of each genus, and all the species of many genera, have left no descendants, but have become utterly extinct. (Darwin 1859, p. 489)

On this basis, one would not expect a bright future for any particular species, including *Homo sapiens.* Yet he tempered such pessimism with (perhaps) a scrap of hope for mankind: "We can so far take a prophetic glance into futurity as to foretell that it will be the common and widely-spread species, belonging to the larger and dominant groups, which will ultimately prevail and procreate new and dominant species . . ." (Darwin 1859, p. 489). According to some metrics (e.g., diversity of habitats inhabited) we are a widely spread species indeed. According to others (e.g., sheer numbers, total biomass, etc.), we are rather insignificant. Is *Homo sapiens* to be included amongst the common, widely spread, larger and dominant groups? Will *Homo sapiens* continue to exist into the distant future, or will we speciate into successor-species? Darwin didn't say. The *Origin* says as little explicitly about the future of man as it does on his history.

Darwin's remarks in the *Descent of Man,* a book explicitly devoted to human evolution, are hardly more informative. Although he emphasized that "progress is no invariable rule" (Darwin 1871, vol. 1, p. 177), it was still true that "man" had risen from humble beginnings to eventually occupy the pinnacle of the living world, and could probably look forward to attaining even greater heights in the future. In the final words of the *Descent of Man,* he gave poetic expression to this qualified optimism in words reminiscent of the famous last paragraph of the *Origin*: "Man may be excused for feeling some pride at having risen, though not through his own exertions, to the very summit of the organic scale; and the fact of his having thus risen, instead of having been aboriginally placed there, may give him hopes for a still higher destiny in the distant future" (Darwin 1871, vol. 2, p. 405). Rather than elaborate further, Darwin immediately qualified this statement: "But we are not here concerned with hopes or fears, only with the truth as far as our reason allows us to discover it" (Darwin 1871, vol. 2, p. 405). Unfortunately, reason can't tell us much about such matters – not that that would stop others after Darwin from speculating about what the future might hold.

The View from the Synthesis

Some later biologists were a little less reticent to speculate about the future. Julian Huxley, for example, was never one to shy away from considering the large-scale scope, direction, and significance of the evolutionary process. As we have seen (Chapter 8), Huxley was the preeminent apologist in the twentieth century for the reality of evolutionary progress. According to Huxley, progress in the history of life is a fact, and humans represent its highest achievement. Nonetheless, there has not been a simple, linear progression from microbe to man. It has been a rocky road indeed, strewn with false starts, dead-ends (literally), and brilliant but flawed creatures that couldn't quite cut it for the long haul. In addition, evolution has been "appallingly slow and appallingly wasteful," indicating clearly that human beings were not the predestined *telos* of the evolutionary process (Huxley 1941, p. 297). To put it mildly, in producing *Homo sapiens* evolution has taken a leisurely approach. Beginning with a generalized early type, various lines radiated out, each able to exploit their environments in various ways. Some of these lines (e.g., echinoderms – sea-urchins, starfish, brittle-stars, sea-lilies, sea-cucumbers, and other types now extinct) settled into a way of life eons ago, and have not advanced in any significant way for perhaps a hundred million years. Nor have they given rise to other major types. Although successful in their own way, they turned out to be evolutionary blind alleys. They were simply too specialized to permit further advance. Other lines, however, gave rise to genuinely new types – for example, certain reptilian lines, which evolved into birds and mammals. But such innovations have been the exception. The majority of lines sooner or later reached the limit of their development, whereas a few others radiated further into new types. In retrospect, Huxley argued, we see that all lines but one were closed off to further progress. "If we now look back upon the past history of life, we shall see that the avenues of progress have been steadily reduced in number, until by the Pleistocene period, or even earlier, only one was left" (Huxley 1941, p. 10). This was the line leading to the evolution of humans.

According to Huxley, the emergence of hominids fundamentally altered the nature of evolution. "[P]rogress has hitherto been a rare and fitful by-product of evolution. Man has the possibility of making it the main feature of his own future evolution, and of guiding its course in relation to a deliberate aim" (Huxley 1941, p. 32). The key biological adaptation making this possible, of course, was the emergence of human intelligence, which permits learning and transfer of accumulated knowledge from one generation to the next. "Experience could now be handed down from generation to generation; deliberate purpose could be substituted for the blind sifting of evolution. In man evolution could become conscious" (Huxley 1941, p. 297). With this special role comes special responsibility. In our hands alone lies the possibility to continue the creative process that began four billion years ago. "[T]he destiny of man on earth has been made clear by evolutionary biology. . . . man can now see himself as the sole agent of further evolutionary advance on this planet, and one of the few possible instruments of progress in the universe at large" (Huxley 1953, p. 31). Indeed, for Huxley this was a fact of truly cosmic importance: "As a result of a thousand million years of evolution, the universe is becoming conscious of itself" (Huxley 1957b, p. 13).

Huxley's friend and sometime intellectual sparring-partner George Gaylord Simpson made a similar, abeit more restrained, assessment of the situation. Befitting the last chapter of a book entitled *The Meaning of Evolution* (1949) Simpson allowed himself to speculate about "The Future of Man and of Life." Although evolution is likely to continue for many millions of years to come, it would be foolhardy, he warned, to attempt to predict its precise path. If humans were to be wiped out, either through their own doing or through natural disaster, it would be extremely improbable that anything like humans would ever evolve again, although given the selective advantage of intelligence, intelligence of the human sort *might* well evolve again. Although natural selection is the mechanism that produced human beings, future human evolution will not be determined by natural selection. "Man has too largely modified the impact of the sort of natural selection that produced him that desirable biological progression on this basis is not to be expected" (Simpson 1949, p. 334). Rather, "The only proper possibility of progress seems to be in voluntary, positive social selection to produce in offspring new and improved genetic systems and to balance differential reproduction in favor of those having desirable genes and systems" (Simpson 1949, pp. 334–35). Like Huxley, Simpson believed that future evolutionary progress would arise only through the deliberate conscious planning of human beings.

The Rise of "Non-Zero-Sumness"

The turn of a century, to say nothing of the dawn of a new millennium, inspires reflection and speculation on our past, present, and future as a species. Perhaps the most ambitious attempt in the twentieth century to defend a Darwinian account of directionality in human evolution appeared just as that century was ending. In *Nonzero: The Logic of Human Destiny* (2000), the science writer Robert Wright proposes a sweeping perspective that promises to integrate biological progress as well as cultural evolution. According to Wright, careful reflection on the path evolution has already taken provides a foundation (of sorts) for speculating about its future course. The key is realizing that human history and biological evolution alike are characterized by the rise of "non-zero-sumness": "From alpha to omega, from the first primordial chromosome on up to the first human beings, natural selection has smiled on the expansion of non-zero-sumness" (Wright 2000, p. 252). What exactly does that mean?

A *zero-sum* game is one in which gains for one participant entail equal losses for another. Playing poker for money, for example, would be a zero-sum game. Whatever money is lost by one player is won by another, while the total amount of cash being exchanged remains constant. A *non-zero-sum* interaction, by contrast, is one in which each participant gains in virtue of the interaction. Many commercial exchanges are of this sort. I pay a plumber several hundred dollars to fix the pipes in my house; he uses the money to buy a new set of tires for his minivan. In the biological realm, whenever two organisms enhance each other's fitness (i.e., one another's prospects for survival and reproduction), they create a non-zero-sum situation. Large fish visiting a coral reef "cleaning station" where small fish dart into their open mouth to remove (and consume) parasites would be an example. Baboons taking turns grooming each other would be another. So, too, would the division of labor in a termite colony. Examples could be multiplied indefinitely.

How does this lend directionality to the evolutionary process, considered as a whole? Wright's thesis is that "biological evolution, like cultural evolution, can be viewed as the ongoing elaboration of non-zero-sum dynamics" (Wright 2000, p. 252). In other words, cooperation pays. The key is to appreciate the fact that an examination of the large-scale tendency of evolution makes clear that it has produced ever more complex organisms. Complex organisms require coordination of parts in which genetic replicators form coalitions for mutual benefit. "To say that more and more complex organisms have evolved over time – as they have – is to say that genes have over time gotten involved in more vast and elaborate non-zero-sum interactions" (Wright 2000, p. 253). Drawing inspiration from the work of John Maynard Smith and Eörs Szathmáry (1995), he argues that "[T]his logic organized genes into little primitive cells, little primitive cells into complex eukaryotic cells, cells into organisms, organisms into societies" (Wright 2000, p. 263). At each stage of organization, cooperation (and non-zero-sumness) replaces competition (zero-sumness) as the defining feature of life.

So much for the past. What about the future? Wright hints that whereas (thanks to the advent of culture) natural selection has ceased to be a potent force in human evolution, non-zero-sum dynamics will continue to drive cultural evolution toward the sort of "global brain," possibly embued with consciousness (of a sort), that we already see emerging in nascent form in the burgeoning World Wide Web (Wright 2000, p. 302). What's more, such a vision would provide a basis for an optimistic interpretation of the entire cosmic drama, in which *Homo sapiens* is a vital link between the blind, groping opportunism of natural selection-driven evolution and the creative expansion of global self-consciousness. Wright's aim in proposing such speculations, he reassures the reader, is not to argue that such a conception is *true*, but only that if he *were* to argue for its truth, he wouldn't necessarily be insane. The point is argued with such a solid grasp of facts, breadth of knowledge, and good humor, it is tempting to agree.

Summary: Human Physical and Mental Evolution

> Intelligent life on a planet comes of age when it first works out the reason for its own existence.... Living organisms had existed on earth, without ever knowing why, for over three thousand million years before the truth finally dawned on one of them. His name was Charles Darwin.
>
> (Dawkins 1989, p. 1)

From Darwin to the present the issues of "selection, perfection, and direction" have been at the heart of evolutionary speculations about the origin, nature, and destiny of humankind. In the *Origin*, Darwin had little to say about how the theory of natural selection applied to human beings, but his readers had little doubt that it was meant to explain our own origin no less than that of every other kind of living thing. Yet its application to

Homo sapiens was not (and is not) straightforward. In particular, Darwin and Wallace disagreed about whether natural selection was sufficient to account for the advanced intellectual and moral characteristics of human beings. Whereas Wallace's strict adaptationism (and supernaturalism) inclined him to doubt its sufficiency, Darwin's more pluralistic naturalism convinced him that whatever difficulties there might be could be ironed out.

Still, Wallace had asked precisely the right question. Could unaided natural selection be expected to produce the distinctive moral and intellectual characteristics associated with human beings? In the early decades of the twentieth century, some concluded that natural selection needed a helping hand, either in the form of enlightened eugenics or, tragically, through genocidal megalomania. Unsurprisingly, in the aftermath of such aberrations, applications of Darwinian ideas to human nature seemed irredeemably tainted. But by the 1960s the time was ripe for a resurgence of interest in such accounts. The question of adaptation occupied center stage in these events. Given the way in which natural selection operates, how well adapted, how "perfect" should we expect the human frame and mind to be? How perfect do we find them, in fact, to be? Whereas sociobiology tended to uncritically assume that all or most human characteristics serve some adaptive function, and are well (or even optimally) designed, this Panglossian vision gave way to more cautious views that seek to recognize design flaws as well. Darwinian medicine locates many human ailments in such design flaws, while evolutionary psychology attempts to explain many human behaviors and mental structures as adaptations to environments which no longer exist, yet still structure and inform our every thought and action. Darwin's prediction that "psychology will be based on a new foundation" is in the process of becoming fulfilled.

Darwinian medicine and evolutionary psychology are both premised on the fact that natural selection has not yet perfected our various "corporeal and mental endowments." But perhaps evolution is not finished with us yet. Is there reason to hope that we are making progress in that direction? Is there any reason to think that directionality or progress have anything to do with human evolution in the first place? Despite its current status (to some) as a Darwinian heresy, many prominent Darwinians have insisted that according to reasonable criteria, human beings really are the "highest" product of the evolutionary process. Whether we (or something like us) were the inevitable (or even *a priori* likely) products of the evolutionary process, however, is a different matter. Stephen Jay Gould has been indefatigable in arguing that we (like all other living things) are the incredibly fortunate beneficiaries of myriad chance occurrences that collectively make our existence about as contingent as anything could possibly be. We owe our existence to luck, chance, and fortuitous circumstances, not preordained destiny, necessity, or inevitability. In a phrase, as a species we are a "happy accident." Not everyone has been convinced. Various nonrandom forces (competition, arms races, etc.) seem to introduce some degree of directionality into the evolutionary process, such that the eventual emergence of intelligence, either in us or in some other creature, seems highly likely.

It is tempting to go even further, and to use our knowledge of the history of life on earth to speculate about where we might be heading as a species. Darwin and Wallace were both cautiously optimistic, and seemed to foresee a bright future for our species. In the mid-twentieth century, evolutionists like Huxley and Simpson placed their hopes, not in our continued physical evolution, but instead in our ability to deliberately choose and shape our destiny (and that of other species) through enlightened application of our knowledge of evolution. Such speculations seem tame compared to those of some others, who attempt to look beyond the narrow confines of human evolution *per se* to envision *Homo sapiens* as a component of a larger evolving system of consciousness spreading across the face of the globe, driven by the impetus of "non-zero-sum" dynamics. In such a vision, individuals become like the cells of a gigantic "superorganism," a true cyborg whose "brain" is distributed among the billions of humans, and computers, inhabiting the earth. Whether this vision is credible or not, it poses a new perspective on our collective and individual existence that ultimately owes its existence to the discoveries and insights of a nineteenth-century English gentleman-naturalist whose voyage around the world and subsequent work changed not only his own view of life on earth but forever changed the way that anyone after him could reasonably think about such issues.

Epilogue

> We may confidently assert that it is absurd . . . to hope that maybe another Newton may some day arise, to make intelligible to us even the genesis of a blade of grass from natural laws that no design has ordered.
>
> (Kant 1790, p. 54)

So much for prophesy. What the great German philosopher confidently declared to be "absurd" transpired less than a hundred years later. Darwin was precisely the "Newton of a blade of grass" that Kant predicted would never appear. The *Origin of Species* was the beginning of something big; but, as we have seen, it was far from the last word. In *The End of Science: Facing the Limits of Knowledge in the Twilight of the Scientific Age* (1996), the science writer John Horgan argues that evolutionary biology has effectively come to an end, not because it has failed miserably, but rather because it has succeeded so brilliantly. It has already solved all the major problems, and there's nothing left to do but simply tie up a few loose ends. Unlikely. A theme running through this book is that although Darwinism has indeed succeeded brilliantly in unraveling some of the deepest mysteries of nature, many issues concerning natural selection, adaptation, and directionality remain to be sorted out. Evolutionary biology is as vigorous as ever. The next century should be interesting.

Darwin's vision of life on earth differed from all those that preceded it. Unlike the clockwork design vision characteristic of the natural theological tradition, species are not static entities, and life really does have a *history*, an unfolding story of coming to be. Unlike the evolutionary speculations of Lamarck, there was nothing preprogrammed about the evolutionary process. What probably struck Darwin's contemporaries as most striking about his theory was the degree to which it depended on nothing more than natural processes operating on chance occurrences. Darwin's theory introduced irreducible uncertainty into the evolution of life. Nonetheless, as Ron Amundson notes, "Darwin saw adaptation, progression, and diversity as following from natural selection without the need for a separate law of progress" (Amundson 1996, p. 31). Darwin found "grandeur in this view of life," and so should we, for it provides a crucial piece of the puzzle of how we got here, who we are, and where we might be heading.

Appendix

What Did Darwin Really *Believe About Evolutionary Progress?*

The "Mainstream" Interpretation

In Chapter 7, I constructed an interpretation of Darwin's thoughts on evolutionary progress gleaned from writings (private notebooks, unpublished manuscripts, published works, correspondence) spanning some twenty years. I argued that, based on the available evidence, it must be concluded that Darwin was a committed progressionist, in the qualified sense there explained. And yet, the question of Darwin's *real* view of evolutionary progress has been a contentious issue amongst scholars. The problem arises for two reasons. First, at times Darwin does indeed sound like a committed progressionist, whereas at other times he sounds positively scornful of the very idea. Unless Darwin was flat-out inconsistent (or just plain confused), some explanation of this disparity must be offered. Second, because "progress" is a noxious concept for many contemporary biologists (and historians), they find it incredible that Darwin himself took the idea seriously, or perhaps even embraced it enthusiastically. In support they can offer the following argument: The principles of Darwin's theory provide no justification for the belief in evolutionary progress. Surely Darwin himself realized this. Therefore he could not *really* have believed in it, his occasional (apparent) endorsements of progress notwithstanding.

I attempted to resolve the first of these two problems by arguing that in interpreting Darwin's view it is important to take chronological considerations into account (Darwin's view changed as his ideas developed), as well as the range of distinctions Darwin was careful to make in working out his own view (e.g., contingent vs. necessary evolutionary progress). However, a perusal of the literature on this issue suggests that this is a minority position. According to what one commentator describes as the "mainstream" interpretation (Nitecki 1988, p. 10) or "consensus view" (Radick 2000, p. 479), the puzzle of Darwin's view can be resolved by embracing two theses: (1) Darwin knew that the theory of natural selection provided no scientific justification for the idea of evolutionary progress, and indeed seemed to undermine this view, and therefore often said so in his private writings; yet (2) because of various psychological and social factors, he felt obliged in his publications to frame his discussion of evolution in progressionist terms. Thus, according to Peter Bowler, "Progressionism owed very little to the Darwinian theory of evolution by natural selection. . . . Darwin's mechanism challenged the most fundamental values of the Victorian era by making natural development an essentially haphazard and undirectional process" (Bowler 1986, p. 41). Again: "Darwin's theory of evolution was not based on a progressive trend, and sought to explain the origin of each form in terms of the circumstances to which it had been forced to adapt in the course of its history. Darwin himself made a notable attempt to break away from the progressionist assumption . . ." (Bowler 1986, p. 150). So Darwin recognized that his theory provided no support for progressionism. Yet despite this, he configured the theory of natural selection in his publications to make sure that "his theory would appear to fit the prevailing faith in progress. Whatever his opinions as a biologist, when it came to exploring the social and moral implications of evolution, Darwin again fell in with the conventional attitudes of his time" (Bowler 1993, p. 14).[1]

Other scholars present similar interpretations. As David Hull notes, "The traditional view is that Darwin had his doubts about biological evolution being progressive" (Hull 1988b, p. 30). He considers Darwin's apparent sympathies with the idea of evolutionary progress in his publications to be merely "poetic" and at odds with his more "hardheaded" portrayal of himself. Likewise, according to Michael Ruse, Darwin saw "evolution as a directionless process, going nowhere rather slowly" (Ruse 1988, p. 97). Although the logic of his theory seemed to dictate nonprogressionism, he was also deeply immersed in a "social current" that was thoroughly progressionist. Eventually, by the third edition of the *Origin* (1861), "Darwin gave in and went along with the progressionist current" (Ruse 1988, p. 103). Finally, according to Stephen Jay Gould, when Darwin apparently expresses progressionist sentiments, this should not be understood to represent his real views but, rather, as concessions to the then-prevailing *Zeitgeist* that had enshrined "progress" as an inevitable social law. Although Darwin in fact rejected any notion of evolutionary progress, he nonetheless sometimes "weakened" and included progressivist language in his writings so as not to upset the *status quo* of which he was such an indisputable beneficiary: "Darwin, the social conservative, could not undermine the defining principle of a culture . . . to which he felt such loyalty, and in which he dwelt with such comfort" (Gould 1996, p. 141).

Against the Mainstream Interpretation

According to a second view, however, Darwin's apparent sympathies with the idea of evolutionary progress should be taken at face value: He frames his discussions in progressionist terms because he was himself a progressionist. This is the interpretation I defended in Chapter 7. Some go even further. According to Robert J. Richards, Darwin not only accepted the reality of evolutionary progress, but a belief in progress was central to his biological theorizing: "[F]rom his earliest formulation of the idea of evolution through its mature form in the *Descent of Man,* Darwin conceived of the process as progressive (Richards 1988, p. 135). Likewise, Dos Ospovat takes the position, not only that Darwin believed in evolutionary progress, but that he recognized that natural selection leads "inevitably to the progressive development of life" (Ospovat 1981, p. 4), and that "progress was a necessary general consequence of organic change" (Ospovat 1981, p. 213). Additionally, whereas proponents of the "mainstream" interpretation see Darwin as a committed nonprogressionist who nonetheless cloaked his public statements about evolution in progressionist terms in order to accommodate the Victorian belief in progress, proponents of this second view see Darwin as a committed progressionist who, initially at least, intentionally *downplayed* his progressionism in order to make his theory more acceptable to his audience. Taking a view diametrically opposite to that of interpreters like Gould, Ospovat notes that "Darwin had nothing to gain and perhaps much to lose by exhibiting prominently his belief that progress is inevitable" (Ospovat 1981, p. 221).

What, then, did Darwin *really* believe about evolutionary progress? The only way to address the issue of Darwin's beliefs is to look at his writings – not just the published works, but also and perhaps especially those jottings – private notebooks, marginalia, letters, and unpublished manuscripts – in which he addresses the issue of progress either for himself or for his close confidants. It is, of course, possible that despite expressing progressionist beliefs in his writings, in the privacy of his own beliefs Darwin really disavowed evolutionary progress. But this is a position that has the weight of the evidence against it, and would need to be justified by good arguments for us to discount this evidence. What might such arguments look like?

Darwin as a Nonprogressionist

Gould provides the most detailed defense of the "mainstream" interpretation. He develops a three-pronged defense of the claim that, despite often sounding like a progressionist, Darwin was in fact a committed nonprogressionist. First, he argues that "Darwin was not shy in advertising his nonprogressivism" (Gould 1996, p. 137). As evidence he points out that in the margins of a book that did advocate progress in the history of life, Darwin jotted a note, reminding himself to "Never say higher or lower." Additionally, in a letter to the American paleontologist Alpheus Hyatt, who had proposed an evolutionary theory based on an internal progressive drive, Darwin confessed: "After long reflection, I cannot avoid the conviction that no innate tendency to progressive development exists" (Gould 1996, p. 137).[2] According to Gould, such remarks represent Darwin's *true* beliefs.

In assessing this evidence, consideration of the context is extremely important. We can simply observe in passing that a note in the margin of a book in one's personal library is hardly the best place to "advertise" one's views, and focus instead on the content and context of the remark itself. The note in question appears in the margins of Darwin's copy of Robert Chambers's *Vestiges of the Natural History of Creation* (1844), and the full (correct) quote reads as follows: "Never use the word higher & lower – use more complicated, as the fish type (& not a mere repetition of parts) where cartilaginous forms are higher for being nearer reptiles & consequently mammalia" (Di Gregorio 1990, vol. 1, p. 164). The additional remarks make clear that the self-imposed prohibition was not against the "higher" and "lower" terminology *per se* (because he goes on to use the term "higher" in the next line) but only a reminder to himself to carefully specify the *sense* in which one type of organism is higher than another. In this case, it is apparently "highness" in the sense of von Baer's "grade of development" that is at issue. Likewise, the letter to Alpheus Hyatt, cited by Gould, does not show that Darwin was opposed to evolutionary progress *per se,* but rather only that he rejected the existence of a particular explanation of evolutionary progress. Gould records Darwin's belief that "no innate tendency to progressive development exists," but he fails to include the remainder of Darwin's words where he is careful to specify that he rejects the idea of an innate tendency to progressive development "as is now held by so many able naturalists, and perhaps by yourself" (Darwin and Seward 1903, vol. 1, p. 344). In the letter to which Darwin was responding, Hyatt had proposed a "law of development" explaining "why the young of later-occurring animals are like the adult stages of those which preceded them in time" (Darwin and Seward 1903, vol. 1, pp. 340–41) – the famous "ontogeny recapitulates phylogeny" doctrine. Rejecting *this* idea and rejecting evolutionary progress *simpliciter* are clearly distinct issues. One can reject a "law of development" of the sort Hyatt suggested, while also holding that there might be processes which do, under the right circumstances, lead to progressive development. This was, of course, Darwin's view. As Ernst Mayr correctly notes, "Darwin, fully aware of the unpredictable and opportunistic aspects of evolution, merely denied the existence of a lawlike progression from less perfect to more perfect" (Mayr 1982, p. 531). The evidence Gould produces to show that Darwin was a nonprogressionist, when read in context, is insufficient to establish his claim.

As a second argument, Gould claims that Darwin must have been a nonprogressionist because the fundamental logic of his own theory dictated that he reject evolutionary progress. All adaptation is to *local* conditions, and what is locally superior almost certainly will not be globally so. Equally important, a superior set of traits in a local environment are unlikely to remain superior for long in the face of an ever-changing environment. There is simply no way that a series of randomly changing local environments can elicit progressive advance. In the *Origin,* Darwin treated the "struggle for existence" in a broad sense to include both struggle with other organisms (e.g., for food, territory, mates, etc.), as well as against the rigors of the physical environment (e.g., extremes of temperature, lack of water, etc.). Call the former sort of struggle *biotic competition* and the latter *abiotic competition.* According to Gould, although the logic of natural selection shows that abiotic competition cannot yield progress (because local environments change randomly, and hence adaptations to these shifting local environments will at most show random "backings and forthings"), biotic competition *can* yield progress. Struggle between organisms can lead to general biomechanical improvements (e.g., running faster, enduring longer, thinking better, etc.) transcending the particulars of any specific environment. Consequently, "a general trend to progress might be defended" only "if biotic competition is much more important than abiotic competition in the history of life" (Gould 1996, p. 142). This is, however, at most a necessary and not sufficient condition for progress. Biotic competition will lead to progressive evolution *only if* a certain additional specific condition obtains: "If environments are relatively empty – either because defeated forms can migrate somewhere else, or because losers can survive by switching to some other food or space in the same environment – then biomechanically inferior forms can continue to exist, and no ratchet to general progress will exist." By contrast, "if ecologies are always chock-full of species, and losers have no place to go, then the victors in biotic competition will truly eliminate the vanquished – and the buildup of these successive eliminations might produce a trend to general progress" (Gould 1996, pp. 142–43). So, Gould admits that under the right conditions, there can be progressive trends in evolution – that is, that such progress is theoretically *possible* – but maintains that the necessary conditions are unlikely to be satisfied. He therefore wonders why Darwin, after expelling progress from his account, would "bother to smuggle progress back in through the rear door of a complex and dubious ecological argument." He concludes that Darwin could not have taken this argument seriously. Consequently, "Natural selection can forge only local adaptations – wondrously intricate in some cases, but always local and not a step in a series of general progress or complexification" (Gould 1996, p. 140). According to Gould, Darwin took this theoretical argument for nonprogressivism very seriously and even "reveled in the radical character of this claim" (Gould 1996, p. 144).

Quite apart from the issue of whether Darwin ever advanced the argument Gould attributes to him (no citations to Darwin's work are provided), as a matter of fact Darwin seems to have believed that the conditions Gould identifies *do* obtain often enough to permit evolutionary progress to occur. Darwin treated the environment as constituted by other organisms as being the most significant factor upon which natural selection operates, thus making biotic competition more important than abiotic competition.[3] As Richards correctly notes, "He supposed that the environment against which organisms were most often selected would be the living environment of other creatures, so that reciprocal developmental responses would be invoked throughout the system" (Richards 1992, p. 86). As for whether Darwin believed that the second condition Gould specified is generally satisfied, we have ample evidence from his writings that he did believe it was satisfied. In a famous metaphor, Darwin compared the biotic world to a surface covered with wedges driven into it: "The face of Nature may be compared to a yielding surface, with ten thousand sharp wedges packed close together and driven inwards by incessant blows, sometimes one wedge being struck, and then another with greater force" (Darwin 1859, p. 67). Nature is here conceived as a "plenitude" in which each new species can find a place only by driving out another, precisely the sort of condition Gould deems necessary to bring about evolutionary progress. The fundamental problem with Gould's argument that Darwin rejected progress on theoretical grounds stems from reading Darwin through the lens of contemporary biology. The fact that *later* biologists challenged the claim that biotic competition is more important than abiotic competition, and that some spaces in nature remain unexploited, are reasons why *we* might doubt that evolution consistently manifests progress, but not reasons for concluding anything about what *Darwin* believed – which is the point at issue here.

This brings us to the third of Gould's claims. As the "wedging" metaphor and his other remarks show, Darwin believed that nature satisfies the basic conditions necessary for evolutionary progress to occur. Gould, however, assumes that whenever Darwin sounds as if he is endorsing evolutionary progress, he is really just accommodating his public presentation to satisfy the exigencies of his social situation:

> [Darwin's] strained and uncomfortable argument for progress arises from a conflict between two of his beings – the intellectual radical and the cultural conservative. The society that he loved, and that had brought him such rewards, had enshrined progress as its watchword and definition. . . . Darwin could not bear to fail his own world by denying its central premise. (Gould 1996, p. 144)

In other words, social and psychological factors were driving Darwin's public claims about progress, despite their biological implausibility. The explicit arguments Darwin proposed in support of evolutionary progress, rather than being a carefully thought out endorsement for a limited form of evolutionary progress, were instead merely a sop, generated by his own timidity, to the dominant cultural values of his day. Gould acknowledges that in the *Origin of Species* Darwin often *sounded* like a progressionist, and admits that such passages pose a problem for his claim that Darwin was a committed nonprogressionist. Although he is at first content to let great men think great (and sometimes inconsistent) thoughts, Gould argues that when Darwin apparently endorses progressionist sentiments, this should not be understood to represent his real views, but rather as concessions to the then-prevailing cultural values that had enshrined "progress" as an inevitable social law. Although Darwin rejected any notion of evolutionary progress, he nonetheless sometimes "weakened" and included progressivist language in his writings so as to not upset the *status quo* of which he was such an indisputable beneficiary: "Darwin, the intellectual radical, knew what his own theory entailed and implied; but Darwin, the social conservative, could not undermine the defining principle of a culture . . . to which he felt such loyalty, and in which he dwelt with such comfort" (Gould 1996, p. 141).

Gould assumes that when Darwin criticizes progressivism he is stating his true (scientifically defensible) view, but when he writes as a progressionist he is simply reflecting his Victorian culture, and hence such utterances must be explained by reference to social causes. But there is a fundamental problem with this interpretive strategy. If one is going to explain *some* of Darwin's statements about evolutionary processes in terms of his supposed need to validate the dominant cultural values of his day, why not also explain his occasional pronouncements *against* evolutionary progress in the same way? Why should some ostensibly reasonable scientific claims and arguments be interpreted as concessions to dominant cultural values, whereas others are best interpreted as being just what they appear to be, that is, generalizations based on empirically grounded observations? It is worth noting that this very same culture had enshrined even more deeply the belief that "man" (and indeed all living things) is specially created by God, and that Darwin's work threatened to undermine this cherished belief. Apparently Darwin was highly selective regarding which cultural premises he undermined and destroyed, and which ones he shored up. Darwin's decision to title his major work on human evolution *The Descent of Man* (rather than *The Ascent of Man*) symbolizes perfectly the fact that Darwin was unafraid to break with and indeed oppose the powerful cultural values of the time.

Fundamentally, however, speculative psychologizing (and sociologizing) is unnecessary for discerning Darwin's beliefs. Darwin found good theoretical reasons to deny progress in one sense, yet equally good reasons to affirm it in another. In general, appeal to social factors does not render theoretical factors otiose, or vice versa. It is possible for a position to be causally overdetermined such that theoretical and social factors may each be sufficient, by themselves, to generate the view being considered. Social factors may influence the adoption of a position directly, or they may operate by biasing the scientist toward certain epistemic values, principles, and standards, which in turn influence the arguments that are developed and deployed, and consequently the actual position taken (Shanahan 2001). It is indisputable that from the start Darwin was keenly aware of the beliefs and values of his readers, and carefully crafted his arguments accordingly (Manier 1978). Theoretical and social factors together contribute to a complete explanation of Darwin's thought.

Finally, before we leave Gould's interpretation behind, consider the famous closing words of the *Origin*: "There is grandeur in this view of life, with its several powers, having been originally breathed into a few forms or into one; and that, whilst this planet has gone cycling on according to the fixed law of gravity, from so simple a beginning endless forms most beautiful and most wonderful have been, and are being, evolved" (Darwin 1859, p. 490). These are also the very words Gould selects to close *Full House* (1996), his book-length attack on the notion of "higher" and "lower." Darwin's words seem to lend support to the book's central thesis, namely, that the history of life is primarily a history of diversification, the production of a "full house" of living things in which *variety* (not progress) is the fundamental fact. But Gould fails to include the sentence immediately preceding the words he quotes, in which Darwin writes: "Thus, from the war of nature, from famine and death, the most exalted object which we are capable of conceiving, namely, the production of the higher animals, directly follows" (Darwin 1859, p. 490).

It is worth pausing for a moment to reflect on the significance of these words.[4] The production of "higher animals," he says, follows directly from the struggle for survival. Far from privileging diversification over improvement, in the closing words of the *Origin* Darwin managed to remind his readers of the significance of *both.* This makes perfect sense in light of Darwin's belief that natural selection promotes both divergence of form and advancement of organization. Rather than being alternative visions of the evolutionary process, Darwin strove to show how both could be understood as the consequences of fundamental biological principles. Gould's selective reading (and quoting) of Darwin does little to justify his claim that Darwin was a committed nonprogressionist.

Conclusions: Darwin the Icon

If, as I have argued, Darwin was an evolutionary progressionist, why has there been such a strong tendency to view him as completely opposed to the idea of evolutionary progress? The answer must be sought by recognizing the special role Darwin plays in evolutionary biology. As John Horgan has noted, "No other field of science is as burdened by its past as is evolutionary biology. . . . The discipline of evolutionary biology can be defined to a large degree as the ongoing attempt of Darwin's intellectual descendants to come to terms with his overwhelming influence" (Horgan 1996, p. 114). One way of coming to terms with Darwin's influence is to appropriate him in support of one's own views. Richards's description of this phenomenon is right on the mark: "Among contemporary evolutionary theorists Darwin functions as an icon, an image against which theories receive approbation or reprobation. To select from the historical Darwin those features that best comport with one's own predilections in the contemporary scientific debate is to have those predilections sanctioned by the master" (Richards 1988, p. 146). Because the majority of contemporary evolutionists are extremely wary of the notion of "progress," they are predisposed to find in Darwin's writings the validation they seek. As Richards picturesquely puts it, "Most historians, philosophers, and biologists... regard attaching the idea of progress to Darwin's theory as comparable to stitching a Victorian bustle on the nylon running shorts of a woman marathoner, a cultural atavism disguising the slim grace supplying the real power" (Richards 1988, p. 129). This is perhaps why Gould's interpretation of Darwin as a nonprogressionist prefaces his all-out attack on evolutionary progress. But Darwin's view must be assessed in its own terms, not simply as a reflection of current beliefs. When we do so, a consistent (although not entirely unproblematic) picture emerges of Darwin as a committed but reflective advocate of qualified evolutionary progress.

Notes

Introduction

1. Editions of the *Origin of Species*, all published by John Murray of London, appeared in 1859, 1860, 1861, 1866, 1869, and 1872. For all editions after the first, references are to Peckham's variorum edition (Darwin 1959).
2. As Ernst Mayr (1991, pp. 93–96) notes, one of the many meanings given to "Darwinism" has been "anticreationism." After surveying, and rejecting as inadequate, that and a number of other meanings, he concludes that there are only two truly meaningful concepts of Darwinism, viz., "adaptive evolutionary change under the influence of natural selection, and variational instead of transformational evolution" (p. 107). We will examine both ideas in the pages that follow.
3. An historian of science once titled an essay: "Should Philosophers Be Allowed to Write History?" In a rejoinder, a philosopher asked: "Should *historians*?" (Hull 1979). Both questions are apt. Every historical account, no matter how "objective" its author might strive to be, necessarily reflects current concerns and perspectives, and embodies the historian's choice of methodology, whether it is made explicit or not. Even the attempt to avoid any value judgments about the events, ideas, or persons being discussed reflects certain value judgments about how history ought to be done, judgments that may not be universally shared. Nonetheless, historians and philosophers are likely to approach the same historical material in different ways. Historians of science are frequently at pains to exhibit a sequence of events and the causes of such events in great detail, exploring numerous factors (social, cultural, economic, etc.) that might have contributed to the events under study. The aim is to understand why the particular events in question happened in the way they did. The bearing of these events on the epistemic status of various theoretical positions is often left unexplored. The philosopher's interest in history of science is often quite different. He studies the history with specific questions in mind that he believes the historical material will provide evidence for or against. His primary concern is theoretical. For the

historian the history might well be sufficient. For the philosopher the history is treated as material for arriving at conclusions that transcend this particular parcel of history being considered. History is undeniably fascinating in its own right. But one very important reason we care about history is that it sheds light on the present. For genuine understanding, there is no substitute for history. This is also, of course, one of the chief lessons of evolutionary biology.

4. Another feature of this study is that (in the jargon of historians of science) it is largely "internalist" in character, which means that I will be concentrating on the scientific ideas themselves, paying relatively little attention to the "external" psychological, social, cultural, and economic factors that surely did influence the development and articulation of the ideas to be discussed. This is not to suggest that such factors are either insignificant or uninteresting. They are both significant and interesting, or at least can be. But my aim here is different. I want to understand the science itself, and to do this it is necessary to focus on the ideas, arguments, objections, responses, and the like. Hull (1979), Greene (1981), and Richards (1987) provide spirited defenses of the sort of methodology employed here.

Chapter 1

1. Upon his return from the *Beagle* voyage, Darwin began keeping a series of private notebooks in which he recorded his observations and conjectures concerning geology, transmutation of species, and metaphysical speculations (transcribed and edited by Barrett *et al.* 1987), dated as follows: Red Notebook (1836–37), A Notebook (1837–39), B Notebook (July 1837–February 1838), C Notebook (February–July 1838), D Notebook (15 July 1838–2 October 1838), E Notebook (October 1838–10 July 1839), M Notebook (15 July 1838–1 October 1838), N Notebook (2 October 1838–1 August 1839), Old and Useless Notes (1837–40).
2. Scholars have debated the reasons for Darwin's delay. Gould (1977, pp. 21–27) argues that Darwin delayed publishing because he feared the social consequences of the implied materialism of his account. Richards (1987, pp. 152–56) rejects Gould's account, showing that materialism of the sort implied by Darwin's theory was consistent with much theological orthodoxy, and emphasizes instead the combination of a range of scientific and personal reasons; for example, Darwin's realization that he lacked an adequate theory of heredity; problems in explaining the instincts of neuter insects; and Darwin's desire to firmly establish his scientific credentials before going public with his controversial theory. Richards (1983) reviews a variety of explanations for Darwin's delay.
3. Among contemporary biologists, "Lamarckism" has come to be virtually synonymous with biological error. But it is well to keep in mind that in many ways Lamarck was an outstanding biologist. (Incidentally, it was Lamarck who, in 1802, first coined the term "biologie.") As professor of "insects, worms, and microscopic animals" at the Museum of Natural History in Paris, he distinguished between vertebrates and invertebrates, and distinguished insects,

crustaceans, and arachnids as separate classes, all distinctions that we now take for granted. His *System of Invertebrate Animals* (1801) remains a major contribution in the historical development of zoology. For more on Lamarck, see Burkhardt 1977.

4. However, not all species will have instantiations at all times. Contrary to those who believed that the fossil record demonstrated unequivocally that certain species had gone extinct and had been replaced by new ones specially created by the Deity, Lamarck held that extinctions, in the strict sense, do not occur.
5. It is worth noting, however, that the issue is already, even at this early stage of discussion, conceptually problematic. Should the characteristic in question – "sterility" – be understood as a property of *individuals*, each of whom is sterile, or is it better characterized as a property of the community, which has a sterile *caste*?
6. This interpretation is repeated in his more recent work as well: "It must be emphasized that Darwin always thought of selection as working for the benefit of the individual rather than the group (Ruse 1980)" (Ruse 1996, p. 150). Malcolm Kottler, too, doesn't hesitate to assert without qualification that "Darwin rejected the possibility of group selection" (Kottler 1985, p. 388), although in an endnote he concedes that Darwin did make "exceptions" in explaining Hymenopteran sterility and the human moral sense!
7. Helena Cronin (1991, p. 305) argues, convincingly, that it would be a mistake to claim that Darwin was an "individual selectionist," or "kin selectionist," or "group selectionist" *simpliciter*, because Darwin's language is often confused, and no completely clear and unambiguous view emerges from his many statements. Why should we expect Darwin to have sorted out all the difficult issues that are the subject of considerable contemporary debate? Phillip R. Sloan (1981) also takes Ruse to task for classifying Darwin as an individual selectionist *simpliciter*.
8. The recognition that selection can, in principle, operate upon individuals at any level of the biological hierarchy has turned out to be critical for clarification of contemporary debates about the units of selection. Hull (1980) is the seminal work that first made this point explicit.
9. For example, in the *Descent of Man* Darwin wrote that "Sexual selection will ... be dominated by natural selection for the general welfare of the species" (Darwin 1871, vol. 1, p. 296).

Chapter 2

1. Although there might be a few others to vie with him, Wynne-Edwards might easily deserve the title of most ridiculed biologist of the last century. Amongst many evolutionary biologists his name has become virtually synonymous with deeply erroneous views about the operation of natural selection. For example, a section in Robert Trivers's book *Social Evolution* (1985) is entitled "The Wynne-Edwards Fallacy." Needless to say, having a fallacy named after one is not usually a welcome addition to one's curriculum vitae. At the other extreme, in her entertaining but partisan historical survey of group selectionist thinking, Helena Cronin (1991) devotes a mere two paragraphs to

Wynne-Edwards's ideas (pp. 282–83). Regardless of whether or not his theory was ultimately deemed correct, my aim in this chapter is to explore his ideas and the ensuing controversy they generated in considerably more detail than is typical, in an effort to understand, rather than merely dispose of, his crucial contributions to the development of Darwinian thinking.

2. Lack was educated at Magdalene College, Cambridge (M.A., 1936), and taught zoology in Devon from 1933 to 1938, when he joined an expedition to the Galápagos Islands. In 1945 he was appointed director of the Edward Grey Institute of Field Ornithology in Oxford. He was also a Fellow of Trinity College, Oxford, for the last ten years of his life.
3. In the third chapter of the *Origin* Darwin seems to suggest something along these lines: "But the real importance of a large number of eggs or seeds is to make up for much destruction at some period of life; and this period in the great majority of cases is an early one. If an animal can in any way protect its own eggs or young, a small number may be produced, and yet the average stock be fully kept up; but if many eggs or young are destroyed, many must be produced, or the species will become extinct" (Darwin 1859, p. 66).
4. Wynne-Edwards was Regius Professor of Natural History at the University of Aberdeen from 1945 to 1974.
5. Whereas Lack saw starvation as a common occurrence posing no special theoretical difficulty (Lack 1954, pp. 91–94, 143), Wynne-Edwards assumed that starvation is rare. This assumption is of the utmost theoretical importance. Without this assumption, Wynne-Edwards's critique of Lack's account collapses and there would be no need to postulate an alternative explanation of population regulation. Although the issue is clearly crucial to his argument, it is not explored at any length in his writings.
6. Hereafter simply *Animal Dispersion.* The stimulus for the book came from his reading of Lack's 1954 book, the final chapter of which was titled "Dispersion." Lack defined "dispersion" as "the non-random distribution of a species over the suitable habitats in its range" (Lack 1954, p. 264). Wynne-Edwards's definition added reference to causal processes: "Animal dispersion may be defined as comprising the placement of individuals and groups of individuals within the habitats they occupy, and the processes by which this is brought about" (Wynne-Edwards 1962, p. 1). Explaining why animal populations exemplify the patterns of dispersal they do is a central aim of the book. As the title of the book suggests, social behavior is the key explanatory resource. Wynne-Edwards advanced the same ideas in various subsequent publications (e.g., Wynne-Edwards 1963, 1964a, b, c, 1965).
7. Whether Wynne-Edwards had been thinking about group selection from the start is a difficult issue to resolve. In 1927, the year he graduated from Oxford, he bought a copy of Alexander M. Carr-Saunders's book *The Population Problem* (1922). Carr-Saunders (1886–1966) argued that the primitive human tribes that had survived into modern times had all, virtually without exception, practiced population control in one way or another. Sensitively attuned to the resources their territories could provide, they had maintained their populations close to the optimal density. By sharing and managing their resources

in this way, such peoples enjoyed good health. Those groups not practicing population control found themselves disadvantaged relative to those that did. "Those groups practising the most advantageous customs will have an advantage in the constant struggle between adjacent groups" (Carr-Saunders 1922, p. 223). While working on *Animal Dispersion,* Wynne-Edwards happened, he says, to take Carr-Saunders's book off the shelf and realized that he was applying to animal populations almost exactly the same ideas and principles that Carr-Saunders had applied to primitive peoples. It seems unlikely that he would have been unfamiliar with the ideas in Carr-Saunders's book, given the fact that he had the book in his possession for thirty-five years, and had bought the book on the recommendation of his Oxford tutor, who had himself been a close associate of Carr-Saunders.

8. The notion of groups of organisms as highly organized, integrated, adapted systems was a prevalent theme in works by ecologists at the University of Chicago from the 1930s through the 1950s (see Collins 1986; Mitman 1992). The *locus classicus* for this tradition is *Principles of Animal Ecology* (1949), by Warder Clyde Allee, Alfred E. Emerson, Orlando Park, Thomas Park, and Karl P. Schmidt. This 837-page tome, which became known as the Great AEPPS (pronounced "apes") Project because of the initials of its authors, was a veritable *Summa Ecologiae* of this school. The idea that populations are homeostatic entities is a leitmotif of the text. Characteristic of many of the explanations was the assumption that selection operates to insure the well-being of biological entities more inclusive than individual organisms. (For example, see pp. 692 and 728). Emerson later succinctly summed up the underlying principle of this approach: "There seems to be no reason to suppose that the unit of selection must be exclusively confined to a single system of organization, either at the individual, sexual, family, or social level of integration" (Emerson 1960, p. 319).
9. Of the fifteen reviews (or extended discussions) of *Animal Dispersion* appearing in the 1960s, only two could be considered generally positive (Nicholson 1962; King 1965). The remainder were moderately to highly critical (Anonymous 1962; Anonymous 1963; Braestrup 1963; Buechner 1963; Elton 1963; Christian 1964; Lack 1964, 1966; Maynard Smith 1964; Perrins 1964; Wiens 1966; Williams 1966). Brown (1969) efficiently sums up nearly a decade's worth of criticisms, as well as adding his own.
10. In the Preface to his book, Williams says that the writing dates from the summer of 1963 (i.e., just after the publication of *Animal Dispersion,* and during the height of the controversy about it that occupied the pages of *Nature*). However, the ideas in his book began years earlier as a visceral reaction to the views of influential ecologists at the University of Chicago, A. E. Emerson in particular. As a postdoctoral student in the mid-1950s Williams heard a lecture by Emerson in which the existence of higher-level adaptations was simply taken for granted. He left convinced that something had to be done about what seemed to him to be such misguided evolutionary thinking. Wynne-Edwards's book provided an additional catalyst to attack what he saw as faulty and pernicious theorizing in evolutionary biology (Williams, personal communication).

11. Williams did not entirely dismiss the possibility of selection for group-level adaptations. He conceded that if there are groups (e.g., herds) such that different individuals serve distinct functions, for example, as sentinels, as decoys, and so on, where such division of labor benefited the entire group, then "Such individual specialization in a collective function would justify recognizing the herd as an adaptively organized entity" requiring "something more than the natural selection of alternative alleles as an explanation" (Williams 1966, p. 17). Such specialization of labor is, of course, precisely what one finds in some of the social insects. On Williams's own criterion, therefore, groups of social insects could display group-level adaptations.
12. This case and its bearing on various different formulations of "group selectionism" is explored in considerably more detail in Shanahan (1990).
13. Like the one proposed by Wynne-Edwards, the mechanism proposed by Lewontin involves differential group extinction counteracting an individually advantageous (but group disadvantageous) trait – maximum individual reproduction. However, the evolution of myxoma avirulence does not involve social interactions between members of a randomly mating local population. Viruses are asexual, and thus do not form a *deme* in the strict sense. Viruses in a rabbit form a "group" because their fates are tied together in virtue of their common host, not because they form a social unit. Additionally, whereas the groups Wynne-Edwards considered can take over areas vacated by other groups that have gone extinct, the viruses in a rabbit cannot do so, because group extinction in this case includes the death of the host and consequently renders that "habitat" unsuitable for other would-be colonizers. Finally, although Lewontin's explanation involves a form of group selection, it does not postulate group-level *adaptations*.
14. Recall that trait groups need not split or send out propagules to form new groups; they need only disperse into the global population every so often, and be reassembled again (with different members) at some later time. "Reproduction" of trait groups (on the model of organismic reproduction) is absent because specific trait groups in one generation need bear no relation to any specific trait groups in subsequent generations. In addition, although trait groups might experience extinction, they need not in order for the process Wilson describes to operate. Indeed, if they do go extinct they fail to contribute their genes to the global gene pool during their dispersal phase.
15. Models of both types of processes have as their central assumption that populations are "viscous," that is, that individuals do not have equal chances of interacting with all other members of the population with respect to genetic similarity at one or more loci, and that such interactions affect an individual's realized (inclusive) fitness. Frequency-dependent selection is one way in which inclusive fitness values can become relevant (kin selection would be a special case of this in which interacting individuals are close genetic relatives). Maynard Smith argued that Wilson's model is insufficiently distinct from the ideas of frequency-dependent selection and inclusive fitness to justify classifying it as a different kind of model.

Chapter 3

1. As I will suggest later in this chapter, the metaphor of peeling an onion is apt, because in the end it becomes clear that there is no ultimate "core" level at which selection operates; there are only the various layers and their interrelations.
2. Then again, perhaps some behaviors that at first appear to be altruistic are in fact instances of the selfish manipulation of the behaviors of others for one's own benefit, and can be explained entirely at the level of individual advantage, without the need to invoke a gene-centered perspective. Recall again Darwin's observation: "[N]o instinct has been produced for the exclusive good of other animals, but . . . each animal takes advantage of the instincts of others" (Darwin 1859, p. 243). In a number of works, (Dawkins 1978; 1982b; 1989), Dawkins argues that much behavior that appears puzzling from the viewpoint of advantage for the behaving organism can best be understood by viewing it as benefiting some *other* organism which is manipulating the actor. Cuckoos who succeed in manipulating the hosts they have parasitized into caring for the offspring deposited in the latter's nests would be a classic example.
3. In truth, the situation is far more complicated (and interesting) than this. The entire issue of organism versus group selection can be reframed by "frame-shifting" down one level. From the gene's-eye point of view, individual organisms can be considered *groups* of identical genes that have banded together for their common benefit. As we have seen, genetic relatedness within groups (e.g., eusocial insects) is the key to understanding how they can form such cohesive units in which some individuals altruistically sacrifice themselves for the greater good. This is true in spades for organisms, in which each individual cell is genetically identical to every other cell in the body (i.e., they are all clones), and all are bound together with a common fate identified with that of the organism they compose (Buss 1987). Here, "self-sacrifice" reaches an extreme that makes the kamikaze behavior of sterile workers in a bee-hive pale in comparison. All the somatic cells in the body can be seen as relentlessly working toward one goal: to ensure that the germ cells (sperm and eggs) are passed on to the subsequent generation. In other words, from the genic perspective, organism selection is just a particularly prevalent and successful form of group selection. (Cancer is a *disease* from this point of view precisely because it involves cells that have begun to replicate without regard, and in a manner detrimental, to the welfare of all the other cells, whose fate is bound up with that of the individual organism they compose.)
4. Interestingly, the "bookkeeping argument" *against* the causal thesis of gene selectionism can also be seen as an argument *for* the representation thesis by admitting that gene selectionism provides a common currency for representing, comparing, and explaining evolutionary changes. Natural selection can always be viewed as selecting genes with certain phenotypic effects over other genes with different phenotypic effects. Or, if the genes in question have no phenotypic effects (e.g., in the case of so-called junk DNA that is never transcribed into RNA, and hence into proteins), then we can still track

evolutionary change in terms of changes in the fate of genes. Consequently, if we wish to have a quite general theory of evolutionary change, thinking in terms of selection's ultimate effect on genes seems to be a promising strategy.

5. Hull's later (1988) definitions are essentially the same as his earlier definitions, but more clearly emphasize the causal relationships between replicators and interactors. In his earlier work Hull (1981) also defined "evolvers" as the entities that actually change as a result of interaction and replication. In Hull (1988a) a *lineage* is defined as an entity that persists indefinitely through time either in the same or an altered state as a result of replication" (Hull 1988a, p. 409).
6. In Dawkins's terminology, "replicator survival" and "vehicle selection" together generate evolution: "Evolution results from the differential survival of replicators. Genes are replicators; organisms and groups of organisms are not replicators, they are vehicles in which replicators travel about. Vehicle selection is the process by which some vehicles are more successful than other vehicles in ensuring the survival of their replicators" (Dawkins 1982b, p. 46).
7. According to some biologists, even species can function as interactors since they can exhibit a variety of emergent species-level properties (e.g., population size, spatial and genetic separation between populations, etc.) that permit them to function as interactors. A discussion of "species selection" is beyond the scope of this book.
8. This follows in part from Hull's definition of a replicator as an entity that passes on its structure *directly* in replication. In virtue of the mechanisms of DNA replication, the copying process of genes is virtually direct. By contrast, the "replication" involved in most organismic reproduction is not. Replication takes place, in sexual species at least, *indirectly* through the processes of gametogenesis (formation of sex cells), fertilization, and ontogenesis (development of the offspring). In the case of colonies and higher-level entities, replication consists in the splitting of the entity into two or more parts. But splitting does not necessarily involve replication of structure. Structurally, what the sexually produced offspring and the new colony have in common with their progenitors are *genes*. So it seems that in most selection processes, the replicators are genes. Or so it is commonly thought. We will have occasion to question the exclusive identification of replicators and genes shortly.
9. At one point, however, Cronin does acknowledge the basic issue at stake here: "What light does all this throw on adaptations? Adaptations must be for the good of replicators, for the good of genes. But they are manifested in vehicles. Genes confer on vehicles properties that influence their own replication. So adaptations could, in principle, turn up at any level – at the level of organisms . . . at the level of groups and even higher" (Cronin 1991, p. 288).
10. It is worth noting that an instrumentalist interpretation of gene selectionism is suggested by the way its primary proponents sometimes characterize their doctrine. For example, Williams remarks that, "The formally disciplined use of the theory of genic selection for problems of adaptation . . . should foster progress and understanding regardless of the extent to which this theory

constitutes a true or adequate explanation" (Williams 1966, p. 270). At times Dawkins professes even less concern with the truth of gene selectionism: "I want to argue in favour of a particular way of looking at animals and plants, and a particular way of wondering why they do the things that they do. What I am advocating is not a new theory, not a hypothesis which can be verified or falsified, not a model which can be judged by its predictions.... What I am advocating is a point of view, a way of looking at familiar facts and ideas, and a way of asking new questions about them.... I am not trying to convince anyone of the truth of any factual proposition" (Dawkins 1982b, p. 1). Even so, it is clear that he, like Sterelny and Kitcher, wants to convince readers that his way of seeing things, in fact, is better (i.e., "truer"?) than any of the other available alternatives.

11. Rosenberg (1994, pp. 86ff.) takes Sterelny and Kitcher to task on similar grounds. Sterelny and Kitcher could respond by rejecting the presupposition upon which the above argument depends, namely, that there *is* a correct account of the causal structure of any selection process. In fact they seem to do just this when they write that, "There is no privileged way to segment the causal chain and isolate the (really) real causal story [about a selection process]" (Sterelny and Kitcher, 1988, p. 358). This sort of response will not do, however, because it is inconsistent with their articulation of Pluralism, which presupposes that the causal chain constituting a selection process can be segmented in a particular, privileged way (i.e., as including selection on genes, or on organisms, etc.) Hence the very articulation of their view presupposes that we *do* have an accurate understanding of the segments constituting the general causal structure of selection processes. Consequently, they cannot consistently reject the claim that there is a correct causal structure of any selection processes. The issues touched on briefly here are discussed in more detail in Shanahan (1996).
12. Amundson (2001) has some very enlightening things to say about the place of developmental biology within the Evolutionary Synthesis, including why it was largely excluded. In some ways what developmental systems theorists propose to do is to replace Dobzhansky's (1937) conception of evolution as a change in gene frequencies with one proposed by Leigh van Valen: Evolution is "the control of development by ecology" (van Valen 1974, p. 115). As Amundson notes, "The interposition of developmental processes between proximate and evolutionary processes, or between genotype and phenotype, shows that a fuller causal story can be told about how evolutionary change occurs" (Amundson 2001, p. 315).
13. Many analogies for this perspective come to mind. Consider the auto industry. The Ford Motor Company builds and sells vehicles such as the Escort, Taurus, and Probe. The sale of these vehicles is the end result of a long causal process, involving engineers who *design*, workers who *produce*, and dealers who *market* these vehicles in the hope of persuading consumers to purchase them. Information about customers' buying habits is used to improve (or at least change) the design of future models. The design features of very successful models are thereby more likely to appear in future models. Competition and selection can occur among agents at any level of the causal

process (design, construction, marketing), "driving it forward." However, the Ford Motor Company is just one player in the global auto industry. The Toyota Motor Company produces vehicles such as the Corolla, Camry, and Cressida. Within the global auto industry, Ford and Toyota are in competition. Because the final product of each manufacturer is the end result of a long causal chain involving design, production, and marketing, the relative success of rival auto companies can be seen as dependent on the entire causal chain resulting in their products. The entire Ford causal chain is in competition with the entire Toyota causal chain. If American and Japanese auto makers use somewhat different processes in the production of their vehicles (e.g., "top-down" management vs. "module-team" organization), then *methods* of producing autos are also in competition. Selection can operate among entities at any level of the entire causal process (among Ford designers; between Ford and Toyota designers; between Ford and Toyota production chains; and even between methods of producing autos). Selection operates *within* all levels of the "automobile hierarchy" as well as *between* the different causal processes constituting each hierarchy. Accounting for the characteristics of the vehicles we see around us requires understanding how selection operates on each and all of these levels.

Chapter 4

1. This insistence on the functional integration of organisms led Cuvier to classify animals into four "branches," or embranchements: Vertebrata, Articulata (arthropods and segmented worms), Mollusca (all other soft, bilaterally symmetrical invertebrates), and Radiata (cnidarians and echinoderms). For Cuvier, these embranchements were fundamentally different from each other. Any similarities between organisms were due to common functions, not to common ancestry.
2. Buffon's *Histoire Naturelle des Oiseaux* (1770–85) forms volumes 16 to 24 of his monumental forty-four-volume *Histoire Naturelle Générale et Particulière* (1749–1804). My thanks to Mary Beth Ingham, C.S.J., for translation from the French.
3. Strickland wrote Part I of this work (on "History and External Characters of the Dodo"), in which this quote appears. Part II (on the "Osteology of the Dodo") was written by A. G. Melville.
4. Ospovat calls this a doctrine of "limited perfection." Organisms are still believed to be "perfect," but only within the limits set by the laws that govern their existence. Even Paley, who eventually became the archetype of those who would argue from the perfection of living things to a divine Designer, acknowledged the principle that perfection has limits, limits set by the laws of matter (Ospovat 1981, pp. 36–37).
5. The remark appears in Darwin's notes on *Proofs and Illustrations of the Attributes of God* (1837), by John Macculloch (in Barrett et al. 1987, pp. 631–41).
6. However, it is hardly absent in his other works. The word "perfect" (or its cognates "perfection" and "perfected") appears fifty-nine times in *The Voyage of the Beagle* (1839); thirty times in *The Structure and Distribution of Coral Reefs* (1842);

ninety-two times in the monographs on barnacles (1854); ninety-two times in *The Descent of Man* (1871); and ten times in *The Expression of the Emotions in Man and Animals* (1872).

7. Prete (1990) suggests that Darwin's awareness of this fact (specifically, the difficulty of explaining such absolute perfection in terms of natural selection) was one factor in his long delay in publishing his views on evolution.
8. By the second edition of the *Descent* (1874), however, Darwin's view had apparently changed once again: "I am convinced, from the light gained during even the last few years, that very many structures which now appear to us useless, will hereafter be proved to be useful, and will therefore come within the range of natural selection" (Darwin 1874, p. 92; quoted in Cronin 1991, p. 87).
9. Kottler (1985, p. 410) notes that Wallace's early anti-adaptationism was so pronounced in his book *A Narrative of Travels on the Amazon and Rio Negro* (1853) that Darwin was led to comment, in a manuscript of the following year, that "Mr. Wallace . . . seems to doubt the strict adaptation even of very differently constructed birds; for he lays much stress on the fact of having repeatedly seen the ibis, spoon-bill, & heron feeding together on precisely the same food. . . . But until it can be shown that these birds feed throughout the year on exactly the same food, & are throughout their lives from the nest upwards exposed to the same dangers . . . the fact of their feeding together for a time or even a whole year, seems to me to tell as nothing against the strictest adaptation of their whole structure to their conditions of existence." This passage is interesting because it reveals as much about Darwin's own adaptationism at this stage in his thinking as it does about Wallace's non-adaptationism at the same time.
10. By contrast, Darwin also was open to the idea that such characteristics *might* turn out to have adaptive functions after all. In a letter of 30 November 1878, to Karl Semper, Darwin wrote: "As our knowledge advances, very slight differences, considered by systematists as of no importance in structure, are continually found to be functionally important. . . . Therefore it seems to me rather rash to consider the slight differences between representative species . . . as of no functional importance, and as not in any way due to natural selection" (in F. Darwin 1888, vol. 3, p. 61; quoted in Wallace 1889, p. 142).
11. Wallace (1896, p. 492) was willing to concede that occasionally nature would produce a "sport" whose traits were quite different from those characterizing the species as a whole, but he dismissed such cases as of only minor significance, as they rarely (or never) play an important role in either species formation or in distinguishing one species from another.
12. As Darwin breathlessly exclaimed in the C Notebook (1838), "Once grant my theory, & the examination of species from distant countries may give thread to conduct to laws of change of organization!" (Darwin, C Notebook, p. 70; in Barrett *et al.* 1987, p. 261).
13. Significantly, Darwin devoted an entire book – *On the Various Contrivances by which British and Foreign Orchids are Fertilised by Insects* (1862; 2nd edition 1877) – to demonstrating that the supposedly divinely designed

"contrivances" so beloved of utilitarian-creationists were in fact pieced-together contraptions shaped by natural selection but constrained by, and thus showing the unmistakable marks of, history.

Chapter 5

1. The discussion that follows is strongly indebted to Bowler (1983), to whom the reader should turn for a more detailed discussion of the issues touched on only briefly here. As he emphasizes, non-Darwinian evolutionary theories did not always exist in "pure" form. Indeed, they often could and did blend into one another (e.g., "orthogenetic Lamarckism," etc.). What follows is a simplified account bypassing many of the qualifications Bowler introduces.
2. As Bowler (1983, p. 201) and Provine (1985, p. 840) point out, however, both DeVries and Morgan retained a role for natural selection. Whereas Darwin generally (but not invariably) assumed that selection operates at the level of individual organisms, DeVries postulated that selection operates at the level of varieties, with those varieties displaying fitter characteristics displacing less fit varieties. In a similar vein, Morgan rejected selection among individuals as a driving force of evolution, but embraced "selection among species" as a significant force.
3. Consider, for example, Dawkins's interpretation of Wright's achievement: "Wright was in fact showing how a subtle mixture of drift and selection can produce adaptations *superior* to the products of selection alone" (Dawkins 1982b, p. 33; emphasis in original).
4. Beatty (1992) notes another trend, already discussed earlier – the "constricting" of the synthesis: "The term 'synthesis' suggests the coming together of many theories. But in fact the evolutionary synthesis effectively repudiated a large variety of Lamarckian, orthogenetic, and other theories of evolution. . . . The synthesis thus reduced, rather than increased, the number of alternative modes of evolution that could be taken seriously" (Beatty 1992, pp. 181–82).
5. Despite the title of his essay, Cain recognized that adaptations, although ubiquitous, need not be perfect. Imperfections in design are primarily because of an inability of species to track rapidly changing environments, to developmental constraints, and to tradeoffs between various specializations: "Every animal is always the resultant of a balance of often conflicting selective requirements and can only be as good a compromise as possible. All the functions to be performed and all the environmental circumstances that influence the life-history must be known before one can understand the design of an animal" (Cain 1964, p. 57).
6. Gould and Lewontin also hint at, but do not develop in any detail, their preferred alternative to "the adaptationist programme." Instead of viewing organisms as suites of interchangeable characteristics, they argue that "organisms must be analyzed as integrated wholes, with *Baupläne* (fundamental body plans) so constrained by phyletic heritage, pathways of development, and general architecture that the constraints themselves become more interesting and more important in delimiting pathways of change than the selective

force that may mediate change when it occurs" (Gould and Lewontin 1979, p. 147).

Chapter 6

1. The quote that begins this chapter is spoken by Cleanthes, the advocate for *a posteriori* natural theological arguments, in Hume's posthumously published *Dialogues Concerning Natural Religion* (1779, p. 15). Since the character Philo, who says some very Humean things about the relationship between experience and belief, criticizes such arguments, it is commonly assumed that Philo must represent Hume's own views. This straightforward interpretation is confounded by the fact that at the end of the dialogue (p. 89) Hume has the minor character Pamphilus declare Cleanthes' arguments to be the stronger. In any case, the fact that Hume has Philo declare that the first cause of the universe bears some resemblance to "MIND or THOUGHT" (p. 80) also suggests that Hume himself ultimately believed the natural order to be the work of an intelligent agent.
2. See, for example, Arnold and Fristrup 1982; Burian 1983; Gould and Vrba 1982; Griffiths 1992; Orzack and Sober 1994; Sober 1984, 2000; Sterelny and Griffiths 1999; West-Eberhard 1992.
3. See, for example, Bock 1980; Dennett 1995; Fisher 1985; Reeve and Sherman 1993; Thornhill 1990.
4. In the older literature such characteristics were often referred to as "preadaptations." This term was rightly discarded because it suggests that the evolutionary process looks forward to what might be useful later, and somehow prepares organisms for future challenges. Such an idea is completely contrary to the evolutionary process as understood within Darwinism, for which there is no foresight, no looking ahead, no anticipation of what might prove useful later. Natural selection always operates in the here and now. If a trait is beneficial now, then, under the right conditions, it will be selectively favored and the trait may spread in the population, and come to characterize future descendents. A trait that might be useful in the future, but is not immediately beneficial, will be selected against. Richard Dawkins (1986) has captured this aspect of the evolutionary process by calling it the "blind watchmaker": a watchmaker, because natural selection produces objects with complex functional designs; blind, because natural selection cannot look into the future to anticipate future needs and shape organisms in the present accordingly. This is another sense in which the evolutionary process is described as "opportunistic." It works entirely with present conditions, as delivered from the past, favoring what is beneficial now in particular circumstances, rather than what might be beneficial in the future under different circumstances.
5. Ironically, given his later critique of adaptationism (Gould and Lewontin 1979), at one point Lewontin himself espoused such a view: "That is the one point which I think all evolutionists are agreed upon, that it is virtually impossible to do a better job than an organism is doing in its own environment" (Lewontin 1967, p. 79; quoted in Dawkins 1982b, p. 30).

6. Orzack and Sober (2001) define it slightly differently: Adaptationism is the claim that "natural selection is the only important cause of the evolution of most nonmolecular traits and that these traits are locally optimal" (Orzack and Sober 2001, p. 6).
7. For the sake of simplicity, I will continue to assume that the living things in question are organisms. Whatever problems arise for thinking about perfection in relation to individual organisms will also arise with regard to higher-level biological entities, for example, groups.
8. In this respect, contemporary Darwinism differs dramatically from that of Darwin and Wallace. As Helena Cronin notes, early Darwinians had a strong interest in drawing attention to the apparent imperfections of adaptations, because such imperfections provided evidence against the idea that organisms are the products of conscious design rather than the products of opportunistic natural selection. They were less concerned to demonstrate how, if such imperfections exist, natural selection reconciled such costs with the overall fitness of the organism in question: "An imperfect adaptation was seen as an adaptation that fell short of ideal, not one that incurred costs because of that imperfection" (Cronin 1991, p. 68).
9. For more on constraints on possible design, see McMahon and Bonner 1983. For further discussion and application of these ideas to theological issues, see Nelson 1996 and Shanahan 1997.
10. If the adaptive landscape is thought of as rigid and inflexible, travel from one peak to another does seem impossible. But since environments change, it is more realistic to think of the adaptive landscape as a trampoline, in constant motion. Dramatic changes in environment that don't succeed in driving a species to extinction can provide the opportunity for dramatic evolutionary development. The advent of the last ice age may have been just such an environmental change responsible, in part, for the rapid encephalization of hominids (Calvin 2002).
11. Recall Gould and Lewontin's complaint that a problem with "the adaptationist programme" is that it "atomizes" organisms into discrete traits, and constructs an adaptive story for each, whereas in fact organisms exist as suites of characteristics that must function well together if the organism itself is to survive. True, it is often difficult to individuate traits, and there are plenty of one-many/many-one connections between genes and phenotypic characteristics (pleiotropy and epistasis). But it is also possible that selection has already taken this into account in the design of organisms. It would be useful (from the point of view of evolvability) for organisms to be modularized. As Leigh (2001) notes, "[C]ompartmentalization, or modularity, of organisms, whereby most genes affect specific, limited characteristics, is an adaptation that allows the evolution (and the analysis) of other adaptations . . . selection favors physiological organizations that enhance adaptive evolution if they impose no disadvantages on individual organisms . . . modular organization facilitates adaptive evolution withouth imposing countervailing disadvantage on individual organisms" (Leigh 2001, p. 369). He goes on to note that "[M]odularity increases the chance that a mutation that improves one feature will not be compromised by adverse effects on others . . . modularity

could arise only if it were selected. . . . Selection is required in order to achieve this degree of modularity" (Leigh 2001, p. 370).

Chapter 7

1. Characteristically, Michael Ruse expresses this idea with typical panache: "Saying the word 'progress' in the company of serious evolutionary biologists is like saying 'fuck' at a vicar's tea party . . . " (quoted in Lewin 1994, p. 37).
2. Ever alert to treading on theoretical issues that could not be settled by appeal to empirical observations, Darwin added: "but I can see no way of testing this sort of progress" (Darwin 1859, p. 337). An examination of successive editions of the *Origin* documents Darwin's increasing confidence that the foregoing theoretical claim is borne out in the fossil record. In the third edition (1861) he added that although a "large majority of palæontologists" would agree that the geological evidence demonstrates progressive improvement, he could "concur only to a limited extent" (Darwin 1959, p. 548). In the fourth edition (1866), this conclusion becomes "highly probable" (Darwin 1959, p. 548). By the fifth edition (1869), he says that this "must be admitted as true" (Darwin 1959, p. 549).
3. That Darwin had held this view for some time is evident from two letters to Hooker in 1858. In a letter dated 24 December 1858, he wrote that "species inhabiting a very large area, and therefore existing in large numbers, and which have been subjected to the severest competition with many other forms, will have arrived, through natural selection, at a higher stage of perfection than the inhabitants of a small area" (Darwin 1985, vol., 7, p. 221). In a follow-up letter dated 31 December 1858, Darwin elaborated on his thinking: "On our theory of Natural Selection, if the organisms of any area belonging to the Eocene or Secondary periods were put into competition with those now existing in the same area (or probably in any part of the world) they (i.e., the old ones) would be beaten hollow and be exterminated; if the theory be true, this must be so. . . . I do not see how this 'competitive highness' can be tested in any way by us. . . . Not that I doubt a long course of 'competitive highness' will ultimately make the organisation higher in every sense of the word; but it seems most difficult to test it" (Darwin 1985, vol. 7, pp. 228–29). Darwin here makes an important distinction which we return to later (in Chapter 9), viz., between the *meaning* of progress and our ability to *measure* it.
4. Clearly, "complexity" is not synonymous with either "specialisation" or "division of physiological labour." An organism could become more complex by acquiring additional parts that are either redundant or serve no functional purpose whatever. For brevity's sake, however, I will continue to use the term "complexity" to represent what Darwin meant by "specialisation of parts supporting a division of physiological labour."
5. That Darwin was very much concerned to distinguish his theory from Lamarck's is further evidenced by the fact that for the sixth edition of the *Origin* he added a historical sketch in which he distances himself from Lamarck's "law of progressive development." The same sentiment appeared in an 11 January 1844 letter to Hooker in which Darwin wrote: "Forfend me from

Lamarck nonsense of a 'tendency to progression'! But the conclusions I am led to are not widely different from his; though the means of change are wholly so" (Darwin and Seward, eds. [1903], vol. I, p. 41).

6. As Ospovat points out, in a significant sense Darwin's conception of progress was just the *opposite* of Lamarck's: "For Lamarck adaptation was the cause of deviations from the primary upward movement of life. But from Darwin's perspective, in which adaptation was the central concern, progress was a secondary consequence of adaptive change" (Ospovat 1981, pp. 212–13).
7. The inclusion of the phrase "survival of the fittest" along with the denial that it necessarily includes progressive development is ironic, since Darwin borrowed the phrase "survival of the fittest" from Herbert Spencer, for whom "evolution" and "progress" were virtually synonomous. As Ruse notes, "For Spencer, evolution was progress and progress was evolution.... For Spencer, biological progress was not an accidental side effect but a necessary outcome of the very possibility of change" (Ruse 1996, p. 188).
8. Darwin mentions [Charles-Guillaume] Nägeli as someone who "believes in an innate tendency towards progressive and more perfect development," or "an innate tendency towards perfection or progressive development" (Darwin 1872, pp. 170, 175).
9. As Robert J. Richards notes, "What Darwin objected to ... was Lamarck's theory that organisms exhibit an *innate* drive toward complexity, toward greater perfection. But this objection to Lamarck does not mean that Darwin rejected the idea that evolution was generally progressive.... Progress was the result, not of an internal drive pushing organisms to perfection, but of an external dynamic pulling them to perfection" (Richards 1988, p. 138; cf. Richards 1992, p. 86). This is essentially correct, although I would want to distance Darwin's view from the teleological flavor of Richards's formulation. Although on Darwin's view natural selection might and perhaps frequently does lead to higher levels of complexity, it is not *inevitable* that it do so, and it is misleading to talk of natural selection as "drawing" or "pulling" organisms to higher levels of complexity and perfection. I also part company with Richards when he claims that "Darwin crafted natural selection as an instrument to manufacture biological progress and moral perfection" (Richards 1988, p. 131). Progress was surely a *consequence* of the evolutionary process as Darwin conceived it, but it is far from clear that the theory of natural selection was intentionally designed to generate this result.
10. This is another sense in which Darwin's theory differs from Lamarck's. If Lamarck's vision of evolution as a ladder were correct, then it would be possible to say of any two organisms (e.g., cuttle-fish and bees), which was "higher" on the ladder of life. But Darwin's metaphor was not that of the ladder but, rather, of a branching tree. Cuttle-fish and bees occupy different branches of the tree, and so it is impossible to directly compare them and to say which is "higher." The great unanswered question here, of course, is the delineation of "types." If there is a single common ancestor for all of life, then all living things occupy the same tree of life. A given organism will be relatively farther from some organisms than from others. Cuttle-fish and bees are sufficiently different from one another to be unproblematically

placed in different types. Organisms of the same species are clearly of the same type. But what about organisms in the same class, order, family, genus, but in different species? Can different species of bees be compared? Can different members of the primates (e.g., chimpanzees, gibbons, orang-utangs, gorillas, etc.) be compared in terms of "grade of organization"? It is not clear how Darwin resolved this issue.

11. Settling on a definition of "highness" is one thing. Using this definition to render definite judgments about the relative "highness" or "lowness" of a specific organism is another. In his *Monograph on the Sub-Class Cirripedia*, Darwin discusses whether the barnacles he is studying are "high" or "low," and eventually concludes that in some respects they are both: "On the whole, I look at a Cirripede as a being of a low type, which has undergone much morphological differentiation, and which has, in some few lines of structure, arrived at considerable perfection, – meaning, by the terms perfection and lowness, some vague resemblance to animals universally considered of a higher rank" (Darwin 1854, vol. 2, p. 20).

Chapter 8

1. See, for example, Huxley 1923, 1928, 1936, 1942, 1953, 1957a. Throughout these writings, Huxley does not disguise the fact that his interest in evolutionary progress is closely related to specific extra-scientific ideological commitments. Swetlitz accurately reflects the consensus view among scholars (especially historians) that "Huxley's idea of evolutionary progress was inseparable from his social values and philosophical beliefs. . . . His conception of evolutionary progress . . . [offered] a cosmic sanction for his social ideology" (Swetlitz 1995, pp. 184, 211–212), and that it was this ideology of evolutionary humanism that provided the *raison d'être* of his lifelong advocacy of evolutionary progress. That Huxley's devotion to the idea of evolutionary progress was motivated by his vision of a new humanism based on evolution is beyond doubt. It might therefore be tempting to just dismiss his arguments for evolutionary progress as so much pseudoscience having no other justification than their role in undergirding what was essentially an ideological commitment. But as Gascoigne (1991) emphasizes, Huxley *did* offer arguments for the reality of evolutionary progress, and believed that it was precisely because evolutionary progress is an objective *fact* about the history of life that the ideology he advocated was scientifically justified. If we wish to avoid committing the "genetic fallacy," we must examine these arguments themselves, put forth in their most plausible form, to see whether he did offer any good arguments for evolutionary progress. If he did offer such arguments, it will, of course, make it no less true that Huxley's main purpose in developing them was ideological, rather than for purely scientific reasons. Arguments might be either good or bad regardless of the goodness or badness of the uses to which these arguments are put. A number of other authors have discussed in detail the connection between Huxley's science and his ideological commitments, emphasizing the impact of the latter on the former (e.g., Divall 1992; Durant 1989, 1992; Greene 1981, 1990;

Provine 1988; Ruse 1996; Smocovitis 1996; Swetlitz 1991, 1995). There is no need to repeat these discussions here. My concern is only with the arguments themselves, and their soundness.

2. According to the "law of the unspecialized" (e.g., as formulated by Cope 1896, pp. 172–74), generalized forms give rise to more specialized forms, and with increasing specialization comes decreased potential for further evolutionary change. Huxley was thoroughly familiar with this "law" and endorsed it in a number of early writings (e.g., Haldane and Huxley 1927; Huxley 1928). He also offered both structural and selectionist explanations for why increased specialization limited further evolutionary change. On the one hand, structural constraints will eventually come into play. For example, aerodynamic laws limit the size of animals specialized for flying, the respiratory system of insects limits their body size, and so on. On the other hand, selectionist forces will limit a highly specialized organism from significant further evolutionary change. In virtue of being efficient swimmers, whales have lost the potential for becoming equally efficient runners or flyers; in virtue of being efficient runners and grazers, horses have lost the potential for becoming efficient predators. For more examples, see Huxley 1928, p. 331.
3. Interestingly, a similar view is stated, but not emphasized, in Huxley's first book, *The Individual in the Animal Kingdom* (1912), where it is taken to be one of the distinguishing characteristics of biological individuality: "When a glance is thrown over the various forms of animal life to which the name of Individual is conceded, it is seen that in spite of many side-ventures, they can be arranged in a single main series in which certain characters are manifested more clearly and more thoroughly at the top than at the bottom. One of these characters is independence of the outer world and all its influences – in other words, immunity from accidents" (Huxley 1912, pp. 3–4). These criteria also appear in the textbook he coauthored with J. B. S. Haldane (Haldane and Huxley 1928, p. 232).
4. Kai Hahlweg (1991) has more recently defended the idea of evolutionary progress on a similar basis, albeit framed in the language of nonequilibrium thermodynamics. According to Hahlweg, "What improves in evolution is the ability of living things to stay alive in increasingly heterogeneous environments" (p. 436).
5. Simpson was never convinced by Huxley's disavowals of anthropocentrism. As he wrote decades later, "Huxley saw that his view might be considered anthropomorphic, or what I have been calling ad hoc with respect to *Homo sapiens*, but he rejected the criticism and insisted that his definition of progress is objective and general. A typical counter-argument was that man *does* possess 'greater power over nature' and *does* live 'in greater independence of his environment than any monkey'" (Simpson 1974, p. 47). Nonetheless, Simpson went on to say, "Huxley's concept of progress really is ad hoc and anthropomorphic. To point out that man has more power over nature and also more independence from it merely emphasizes the point: man is a specialist in manipulating his environment (not always to his advantage) and in doing so is more human than monkeys are. That is a legitimate *human* concept of progress. It does not follow and the evidence does not support the proposition

that other organisms progresses, became 'higher,' because of and in proportion to their acquisition of characteristics most highly developed in man" (Simpson 1974, pp. 48–49).

6. In his mature work, Huxley also distinguishes three senses of "improvement." A very small improvement is designated a "special adaptation." The continued improvement of a lineage in relation to its particular way of life is a "specialization." Finally, advance in the general organization or in the operation of some major system is properly called an "advance": Most improvement in specialization – it is improvement merely in relation to some restricted way of life or habitat. Some improvements, however, merit the name of advance. That is so whenever the efficiency of any major function of life is increased, whenever a higher and more integrated organization is achieved, whenever any radically new piece of biological machinery is evolved. Most specializations and most advances eventually come to a stop; but occasionally improvement continues. So we can conveniently define biological progress as improvement which permits or facilitates further improvement; or, if you prefer, as a series of advances which do not stand in the way of further advances" (Huxley 1953, p. 86).
7. As his one-time coauthor J. B. S. Haldane put it, "Evolution has been, on the whole, progressive, because a single species gaining a new faculty such as flight or temperature regulation can become the ancestor of thousands of species which exploit this capacity in different ways" (lectures on Darwinism given in winter 1953; Box 1, Haldane papers, London; quoted in Ruse 1996, p. 312).
8. Surprisingly, this move reopens the way to use human-centered criteria of evolutionary progress. Such a criterion is "perfectly valid in application to evolution in general, *provided we know what we are doing*" (Simpson 1949, p. 242; emphasis added). On Simpson's view, the problem with Huxley's account was not that he used an anthropocentric criterion of progress, but that he didn't realize that that is what he was doing. On Simpson's view, being self-aware of the anthropocentric character of one's criterion eliminates the only problem with using such a criterion: "Approximation to human status is a reasonable *human* criterion of progress. . . . It is merely stupid for a man to apologize for being a man or to feel, as with a sense of original sin, that an anthropocentric viewpoint in science or in other fields of thought is automatically wrong. It is, however, even more stupid, and more common among mankind, to assume that this is the *only* criterion of progress and that it has a *general* validity in evolution and not merely a validity relative to one only among a multitude of possible points of reference" (Simpson 1949, p. 242).
9. The distinction between Huxlean and Simpsonian evolutionary progress is useful for classifying other biologists. G. Ledyard Stebbins (1969, 1982), with his criterion of increasing levels of organizational complexity, clearly falls into the Huxlean camp. So, too, might Ernst Mayr: "Whenever there is strong competition, specialization undoubtedly gives an advantage, and there is always the possibility that such specialization will lead to a new 'adaptive plateau,' with unsuspected evolutionary possibilities, such as were discovered by the mammalian and avian branches of reptiles while most other reptilian lines

came to an end owing to over-specialization along a less-promising line" (Mayr 1942, p. 294). Francisco Ayala (1988), on the other hand, falls into the Simpsonian camp: "A standard [of evolutionary progress] is valid if it enables us to say illuminating things about the evolution of life.... [T]here is no standard of progress that is 'best' in the abstract or for all purposes. The validity of any one criterion of progress depends on the particular context or purpose of the discussion" (Ayala 1988, pp. 84, 90).

10. Gould does not isolate each of these arguments, but I have done so, and given each a specific name, in order to present Gould's claims and arguments more clearly.
11. As Simpson noted, we may find that criteria *not* selected with human beings in mind *still* indicate that humans are located relatively highly on the scale of evolutionary progress. In this case it would be foolish to reject the obvious conclusion. "After all, it may be a fact that man does stand high or highest with respect to various sorts of progress in the history of life. To discount such a conclusion in advance, simply because we ourselves are involved, is certainly as anthropocentric and as unobjective as it would be to accept it simply because it is ego satisfying" (Simpson 1949, p. 242).
12. See also Vrba and Gould (1986): '[I]f a historical system begins with simple components (as ours presumably did), and if complexity requires hierarchy and the bonding of lower-level individuals into higher entities (with a partial suppression of their independence and an altered status as parts of a larger whole), then a structural ratchet will ordain increasing complexity – 'progress' if you will – as hierarchy builds historically' (Vrba and Gould 1986, p. 226).
13. As Dawkins points out, Maynard Smith said as much some twenty-six years earlier: 'The obvious and uninteresting explanation of the evolution of increasing complexity is that the first organisms were necessarily simple.... And if the first organisms were simple, evolutionary change could only be in the direction of complexity' (Maynard Smith 1970; quoted in Dawkins 1997, p. 1017). Others have made essentially the same observation. See, for example, Stanley (1973), and Simpson (1974), p. 41.
14. Surprisingly, despite the battery of arguments he levels against the idea of evolutionary progress, at one point Gould concedes that there has been progress of a sort in evolution: "As the main claim of this book, I do not deny the phenomenon of increased complexity in life's history – but I subject this conclusion to two restrictions that undermine its traditional hegemony as evolution's defining feature. First, the phenomenon exists only in the pitifully limited and restricted sense of a few species extending the small right tail of a bell curve with an ever-constant mode at bacterial complexity – and not as a pervasive feature in the history of most lineages. Second, this restricted phenomenon arises as an incidental consequence . . . of causes that include no mechanism for progress or increasing complexity in their main actions" (Gould 1996, p. 197).
15. Discussion of arms races appears in Huxley 1912, pp. 114–15; 1923, pp. 37–38; Haldane and Huxley 1927, p. 237; and Huxley 1942, p. 495. See also Ruse 1996, p. 311.

Chapter 9

1. Ayala (1988) makes a related but different distinction between uniform and net *progress*: "*Uniform progress* takes place whenever every later member of the sequence is better than every earlier member of the sequence according to a certain feature.... *Net progress*... requires only that later members of the sequence be better, *on the average*, than earlier members" (Ayala 1988 p. 79; emphasis in original).
2. In his treatment of large-scale evolutionary trends, McShea (1998) asks: "Is there some feature of organisms that we can expect to have changed directionally, *on average*, over the entire history of life as a whole, at the largest temporal and taxonomic scale?" (McShea 1998, p. 294; emphasis added). The issue of directional change in maxima is simply not considered.
3. John Maynard Smith and Eörs Szathmáry begin their book *The Major Transitions in Evolution* (1995) by noting that "Living organisms are highly complex, and are composed of parts that function to ensure the survival and reproduction of the whole.... The increase has been neither universal nor inevitable. Bacteria, for example, are probably no more complex today than their ancestors two thousand million years ago. The most that one can say is that some lineages have become more complex in the course of time. Complexity is hard to define or to measure, but there is surely some sense in which elephants and oak trees are more complex than bacteria, and bacteria than the first replicating molecules" (Maynard Smith and Szathmáry 1995, p. 3).
4. There have been many attempts to define complexity. Chaitin (1975) proposed that the complexity of a series of symbols is the minimum algorithm needed to generate the sequence. Wicken (1979) and Papentin (1980) modified this approach and defined complexity as the minimum algorithm needed to generate the *description* of something, rather than the length of the algorithm for generating the thing itself. In what itself appears to be an evolutionary development, Hinegardner and Engelberg (1983) take the foregoing as their point of departure and define complexity as "the size of the minimum description of an object" (Hinegardner and Engelberg 1983, p. 8). Dawkins (1992) adopts this approach as well. For an important analysis of related issues, see McShea 1991, 1992, 1993, 1994, 1996, 1998.
5. Some prominent biologists have doubted that there was *any* consistent notion of complexity that can be said to have increased from the Paleozoic to the present day (e.g., Williams 1966, pp. 42–43; Gould 1989). But in light of Bonner's work, this doubt seems unfounded.
6. Incidentally, it is entirely possible that there be evolutionary progress even though we are unable to (accurately) measure it. Suppose that evolutionary progress is defined as an increase in some property, **p**, of organisms. There might be an increase in **p** over some time interval, and hence evolutionary progress, even though it is not possible to know or demonstrate that there is. This might be true even though the standard of evolutionary progress being applied is our own invention. This possibility is not nearly as odd as it might seem. As a conceptual invention, a constructed standard is known as soon as it is formulated. But knowing what the standard requires and knowing

precisely how the standard applies in a given situation are two different issues. The constructed standard might be crystal clear, but its application in a particular instance completely murky. Consider an analogy: The nanometer as a unit of measurement is perfectly clear, but it might be impossible to determine how many nanometers in diameter a given object is because the object is too small, is too large, is constantly moving, or is hidden inside other structures that prevent accurate measurement. Likewise, someone might propose increasing complexity as a standard of evolutionary progress and devise a rigorous definition of complexity but be unable to accurately measure the complexity of a given biological entity. Consequently, even if a standard of evolutionary progress is constructed by us for our own purposes, there is no guarantee that we can know whether evolution is progressive or not according to this standard. This was a point about which Darwin himself, writing in the *Origin*, was quite clear: "I do not doubt that [a] process of improvement has affected in a marked and sensible manner the organisation of the more recent and victorious forms of life, in comparison with the ancient and beaten forms; but I can see no way of testing this sort of progress" (Darwin 1959, p. 337).

7. This consideration may be one reason why in his discussions of evolutionary progress Dawkins (e.g., 1997; Dawkins and Krebs 1979) focuses on arms races. The mutual need to counteract the improved capacities among predator and prey is one of the great constants of evolution (see also Vermeij 1987). McShea (1998) remarks (without necessarily endorsing the idea) that "adaptedness of individuals in a species should increase in absolute terms if environments deteriorate directionally, perhaps as a result of relentless improvement by other species" (McShea 1998, p. 305), and cites van Valen's "Red Queen Hypothesis" in support of this idea (van Valen 1973). But even this is doubtful, because adaptedness concerns the fitness of an organism in a particular environment, and if part of the environment (in the form of other species) changes, it becomes meaningless to talk of adaptedness increasing in absolute terms. Even if arms races could give a good explanation of evolutionary *progression*, they don't solve the problem of evolutionary *progress*, because here instead of adapting to ever-changing abiotic environments, organisms adapt to ever-changing biotic environments, that is, other organisms. A later organism that is better equipped to exploit some other organism, or to avoid being exploited by others, is at an advantage relative to these other organisms; but should those other organisms cease to exist, it is not clear that these adaptations make the first organism better in any sense.

8. Contrary to the criticisms of McCoy 1977, Saunders and Ho do *not* argue that increases in complexity are universal, or inevitable, or even inherently advantageous. Instead, they argue that selection pressures drive organisms to local optima of fitness (within the restrictions imposed by history and constraints), and then also prevent these well-organized systems from eliminating components, with the consequence that one direction of further change (toward addition of components and greater complexity) is more probable than the opposite. Thus selection in their view introduces an *asymmetry* and *bias* into evolutionary change which might be sufficient to account for increases

in complexity. Castrodeza (1978) offers more serious criticisms of their proposal. First, he notes that the assumption that adding rather than subtracting components is more likely to be fitness-enhancing is questionable. Components require energy to maintain, and eliminating components (especially redundant ones) may well render the organism more energy-efficient. Second, "since the random removal of a component is judged detrimental to the organisms on the grounds that it may involve a loss of function from an other optimal organization, by the same token a random addition may *interfere* with any well connected function" (Castrodeza 1978, p. 470). Castrodeza concludes that although evolution produces complexity all the time, it is not obvious that it does so in any cumulative manner.

9. This response may well give rise to a related objection: If there is indeed directionality in evolution, why isn't directional change more manifest than it is? But, clearly, there can be directionality without directional change. A heavy object suspended above the ground has a directionality (i.e., an inherent tendency, thanks to the pull of gravity, to move downward), even though the object may remain motionless, and thus have no actual directional change. To the extent this object is permitted to descend, to that extent its directionality will be manifested as directional change. Likewise, there may well be directionality in evolution that is only partially manifested as directional change. For example, if selection favors more complex organisms, one would expect complexity to continually increase over time – unless, of course, various environmental factors intervene to thwart this tendency. In this case the result might be some increases in complexity, some stasis, and some simplification (in other words, exactly what we find). The fact that not all organisms continually increase in complexity therefore does nothing to undermine the claim that complexity (or some other property) is favored by natural selection.
10. As Laudan notes, "For centuries scientists have recognized a difference between establishing the existence of a phenomenon and explaining that phenomenon in a lawlike way. Our ultimate goal, no doubt, is to do both.... Galileo and Newton took themselves to have established the existence of gravitational phenomena, long before anyone was able to give a causal or explanatory account of gravitation. Darwin took himself to have established the existence of natural selection almost a half-century before geneticists were able to lay out the laws of heredity on which natural selection depended" (Laudan 1982, p. 18).
11. For example, an object can be moved west without there being anything about the object that biases its movement in that direction. Or, to take a different example: Flip a coin one thousand times. On any given flip, heads are as likely as tails. But it is possible, just by chance, that heads comes up in 80 percent of the flips. The sequence of results would have a clear direction. But there might not be any directionality.
12. My approach here is just the opposite of that proposed by McShea (1998), who suggests that "distinguishing trends based on their dynamics, rather than on the patterns they produce, would reduce confusion and encourage theorists to address certain fundamental issues" (McShea 1998, p. 295).

The best way to reduce confusion, in my view, is to first determine empirically what trends in fact exist, and to then explore alternative explanations for these trends. The two approaches are not mutually exclusive, however, and achieving a "reflective equilibrium" between empirical and theoretical perspectives is the ideal.

13. Just as some writers conflate directional change and directionality, there is often a similar (but harder to identify) conflation between two different senses of "tendency." On the one hand, one could speak of a tendency for organisms to become more complex, meaning simply that when one examines the fossil record, more complex organisms typically (but not necessarily invariably) appear later than simpler organisms. This use of "tendency" is simply reporting on an overall *trend*. On the other hand, one could speak of a tendency for organisms to become more complex, in this case meaning that there is some sort of impetus or force that moves organisms in this direction. The first use of "tendency" is simply reporting a pattern. The second use is inferring that there is some sort of inherent cause for the pattern. For discussion, see Radick (2000).
14. Additionally, it is not obvious that claims that evolution manifests progress must be understood as *scientific* claims, or that only if such claims are scientific can they be rationally justified. To argue that only scientific claims about evolution can be rationally justified is self-defeating, because *this* claim is not, itself, a scientific claim, and thus on its own terms cannot be rationally justified. "Scientism" is self-referentially incoherent.

Chapter 10

1. Wallace's dramatic about-face was motivated by two very different sets of considerations. In addition to the reasoning just discussed, between 1864 and 1869 he had became convinced that various psychical phenomena were real, and thus demonstrated "the existence of forces and influences not yet recognised by science" (Wallace 1916, p. 200). This led him to reevaluate his stance on human evolution, finding flaws in a selectionist account that previously had seemed perfectly adequate to him. Wallace's conversion to Spiritualism, and its implications for his evolutionary views, have fascinated historians. For discussion, see Kottler 1974, 1985; Turner 1974; Durant 1979; Gould 1980; Richards 1987; and Cronin 1991.
2. According to one theory, bipedalism evolved to facilitate use of the hands for tool use, and/or for carrying food back to camp. The only problem with this theory (granted, it is a rather large problem) is that it is contradicted by the evidence which suggests that these behaviors evolved several million years *after* the first appearance of bipedalism (although they might have later contributed to its refinement). A more plausible theory is that, because of climate changes, forests shrank and hominids took to the open savannahs in search of food. Walking upright has several distinct advantages in such an environment: more efficient locomotion; ability to spot food sources and predators at a distance; abililty to wade into deeper water to capture prey or

escape predators; a more vertical body orientation minimizes surface area exposed to the sun, and maximizes surface area exposed to the cooling effect of the wind. The relative importance of each of these or other factors is still being debated. One thing, however, is abundantly clear: Thanks to our upright posture, back pain is for many of us a fact of life.

3. Darwin himself hinted at such a view in the "M Notebook" (dated 16 August 1838): "Plato says in *Phædo* that our '*necessary ideas*' arise from the preexistence of the soul, are not derivable from experience – read monkeys for experience" (Darwin, M Notebook, p. 128; in Barrett *et al.* 1987, p. 551). In the same work he allowed himself to ponder the philosophical implications of this conviction: "Origin of man now proved. Metaphysic[s] must flourish. He who understands baboon would do more toward metaphysics than Locke" (Darwin M Notebook, p. 84; in Barrett et al. 1987, p. 539).
4. Gould (1989, pp. 234–36) considers, then rejects, the possibility that those creatures that survived early decimations and thus subsequently gave rise to all later animals did so because of their superior anatomical designs, noting that such arguments (e.g., "These forms survived, therefore they *must* have been adaptively superior") run the risk of making Darwinian explanations vacuously circular. An argument stated in that form would be question-begging. But there could conceivably be good reasons for judging one creature as better adapted than another.
5. Although tangential to the questions we are focusing on here, the issues being discussed have a direct bearing on the prospects for SETI (Search for ExtraTerrestrial Intelligence) projects. Conclusions about the inevitability of intelligence evolving on a planet can be arrived at by constructing either an optimistic or a pessimistic induction, with diametrically opposed results. The Optimistic Induction: On *every* planet that we know of, life has arisen very soon after its formation, and has eventually reached the stage of higher intelligence. Therefore the evolution of higher intelligence is extremely likely. The Pessimistic Induction: Out of the *billions* of species that have existed on the only planet with life that we know of, only *one* has developed higher intelligence. Therefore, the evolution of higher intelligence is unlikely in the extreme. Whereas the Optimistic Induction is generally favored by physical scientists (e.g., Sagan 1995), the Pessimistic Induction is the wet blanket thrown on the festivities by evolutionary biologists (e.g., Mayr 1985). However, it could be argued that the fact that at present only one species exists on Earth with humanlike intelligence (viz., us) may simply be due to the fact that once a species with intelligence appears, it tends to eliminate any close rivals, thus virtually guaranteeing that if higher intelligence evolves at all, it will characterize at most one species. This principle might explain the unfortunate fate of the Neanderthals (*Homo sapiens neanderthalensis*), a distinct hominid group that coexisted for a time with *Homo sapiens sapiens*).
6. In later writings, Gould draws a very different take-home lesson from this data (Gould, 1988a, p. 329; 1988b, pp. 321–22). See Shanahan (2001) for discussion.

Appendix

1. It is not clear that Bowler himself consistently wishes to endorse this interpretation, because in the same context he writes that "despite his warnings against simple-minded progressionism, Darwin did nevertheless accept that natural selection would, in the long run, produce higher levels of organization" (Bowler 1993, p. 14). Elsewhere he notes that "The relationship between Darwinism and progressionism is a complex one. . . . Natural selection did not guarantee progress, but it did allow progress to occur as a frequent byproduct of the drive toward better functioning organisms" (Bowler 1988, p. 33). Again: "Darwin continued to believe that natural selection could give rise to a form of progress, but he had to concede that it was at best a slow and irregular by-product of the mechanism's chief function of adaptation" (Bowler 1989, p. 181). Such claims are closer to the interpretation I defended in Chapter 7.
2. Gould's remarks reprise a claim he made two decades earlier: "In a famous epigram, Darwin reminded himself never to say 'higher' or 'lower' in describing the structure of organisms – for if an amoeba is as well adapted to its environment as we are to ours, who is to say that we are higher creatures? . . . [T]he father of evolutionary theory stood almost alone in insisting that organic change led only to increasing adaptation between organisms and their environment and not to an abstract ideal of progress defined by structural complexity or increasing heterogeneity – never say higher or lower" (Gould 1977, pp. 36–37).
3. When Gould considers the evidence from biotic competition that, *prima facie*, seems to indicate that Darwin *did* believe in some form of evolutionary progress, he treats this as an aberration, as "noise" in an otherwise clearly nonprogressionist program. He attributes to Darwin the view that a general trend to progress can be defended *only if* biotic competition is much more important than abiotic competition, but fails to address two obvious questions: Why can't biotic competition be relatively *insignificant* compared to abiotic competition, yet produce a *bona fide* progressive trend nonetheless? Why must life "as a whole" show a progressive trend in order for us to identify *any* progress in evolution? Gould supplies no reason why this restriction ought to be accepted, nor has he shown that *Darwin* accepted this restricted view of evolutionary progress. Consider an analogy: Would anyone seriously argue that "transportation" has not progressed since the days of the Model T as *most* people (globally) still do most of their traveling by a method which has not shown any improvement in a long time, viz., walking? Traveling by personal automobile (or high-speed train, or Concorde) might still be relatively rare when viewed in a global context, but nothing whatsoever follows about whether there has been progress in modes of transportation.
4. Lest it be thought that these are just "throw-away" lines composed by Darwin in his rush to publish the *Origin* before Alfred Russel Wallace could steal his thunder, an earlier version of this passage appears in the "Sketch of 1842" and the "Essay of 1844," and it appears as well in all six editions of the *Origin*, from the first in 1859 to the last in 1872.

References

Abrams, P. (2001), "Adaptationism, Optimality Models, and Tests of Adaptive Scenarios," in S. H. Orzack and E. Sober (eds.), *Adaptationism and Optimality* (Cambridge: Cambridge University Press), pp. 273–302.

Ahouse, J. (1998), "The Tragedy of *a priori* Selectionism: Dennett and Gould on Adaptationism," *Biology and Philosophy* 13:359–91.

Allee, W. C., Emerson, A. E., Park, O., Park, T., and Schmidt, K. P. (1949), *Principles of Animal Ecology* (Philadelphia: W. B. Saunders).

Amundson, R. (1996), "Historical Development of the Concept of Adaptation," in M. R. Rose and G. V. Lauder (eds.), *Adaptation* (San Diego: Academic Press), pp. 11–53.

Amundson, R. (2001), "Adaptation and Development: On the Lack of a Common Ground," in S. H. Orzack and E. Sober (eds.), *Adaptationism and Optimality* (Cambridge: Cambridge University Press), pp. 303–334.

Anonymous (1962), "The Nature of Social Life," *Times Literary Supplement* 14 December, p. 967.

Anonymous (1963), "Review of Wynne-Edwards' *Animal Dispersion in Relation to Social Behaviour,*" *Scientific American* 209:164.

Ardrey, R. (1966), *The Territorial Imperative* (New York: Atheneum).

Argyll, G. D. C., Duke of (1868), *The Reign of Law*, 5th edition (London: Alexander Strahan).

Arnold, A. J., and Fristrup, K. (1982), "The Theory of Evolution by Natural Selection: A Hierarchical Expansion," *Paleobiology* 8:113–29.

Ayala, F. J. (1988), "Can 'Progress' Be Defined as a Biological Concept?" in M. Nitecki (ed.), *Evolutionary Progress* (Chicago: University of Chicago Press), pp. 75–96.

Ayala, F. J. (1999), "Adaptation and Novelty: Teleological Explanations in Evolutionary Biology," *History and Philosophy of the Life Sciences* 21:3–33.

Barkow, J. H., Cosmides, L, and Tooby, J. (1992), *The Adapted Mind: Evolutionary Psychology and the Generation of Culture* (Oxford: Oxford University Press).

Barrett, P. H., Gautrey P. J., Herbert, S., Kohn, D., and Smith, S. (eds.), (1987), *Charles Darwin's Notebooks, 1836–1844: Geology, Transmutation of Species, Metaphysical Enquiries* (Ithaca, NY: Cornell University Press).

Beatty, J. (1980), "Optimal-Design Models and the Strategy of Model Building in Evolutionary Biology," *Philosophy of Science* 47:532–561.

Beatty, J. (1987), "Natural Selection and the Null Hypothesis," in J. Dupré (ed.), *The Latest on the Best: Essays on Evolution and Optimality* (Cambridge, MA: MIT Press), pp. 53–75.

Beatty, J. (1992), "Julian Huxley and the Evolutionary Synthesis," in C. Kenneth Waters and Albert Van Helden (eds.), *Julian Huxley: Biologist and Statesman of Science* (Houston, TX: Rice University Press), pp. 181–89.

Bock, W. J. (1980), "The Definition and Recognition of Biological Adaptation," *American Zoologist* 20:217–227.

Bonner, J. T. (1988), *The Evolution of Complexity by Means of Natural Selection* (Princeton, NJ: Princeton University Press).

Bowler, P. J. (1983), *The Eclipse of Darwinism: Anti-Darwinian Evolution Theories in the Decades Around 1900* (Baltimore, MD: Johns Hopkins Press).

Bowler, P. J. (1986), *Theories of Human Evolution: A Century of Debate, 1844–1944* (Baltimore, MD: Johns Hopkins University Press).

Bowler, P. J. (1989), *Evolution: The History of an Idea*, revised edition (Berkeley and Los Angeles: University of California Press).

Bowler, P. J. (1993), *Biology and Social Thought, 1850–1914* (Berkeley: Office for History of Science and Technology).

Braestrup, F. W. (1963), "Special Review [of Wynne-Edwards' *Animal Dispersion in Relation to Social Behaviour*]," *Oikos* 14:113–20.

Brandon, R. N. (1985), "Adaptation Explanations: Are Adaptations for the Good of Replicators or Interactors?" in D. Depew and B. Weber (eds.), *Evolution at a Crossroads: The New Biology and the New Philosophy of Science* (Cambridge, MA: MIT Press/Bradford), pp. 81–96.

Brandon, R. N. and Burian, R. (eds.) (1984), *Genes, Organisms, and Populations: Controversies over the Units of Selection* (Cambridge, MA: MIT Press).

Brown, J. L. (1969), "Territorial Behavior and Population Regulation in Birds: A Review and Re-Evaluation," *Wilson Bulletin* 81:293–329.

Buechner, H. K. (1963), "Review of *Animal Dispersion in Relation to Social Behaviour*," *Auk* 80:208–9.

Burian, R. M. (1983), "Adaptation," in M. Grene (ed.), *Dimensions of Darwinism* (Cambridge: Cambridge University Press), pp. 287–314.

Burkhardt, R. W. (1977), *The Spirit of System: Lamarck and Evolutionary Biology* (Cambridge, MA: Harvard University Press).

Buss, D. M. (1999), *Evolutionary Psychology: The New Science of the Mind* (Boston: Allyn and Bacon).

Buss, L. (1987), *The Evolution of Individuality* (Princeton, NJ: Princeton University Press).

Cain, A. J. (1951a), "Non-Adaptive or Neutral Characters in Evolution," *Nature* 168:1049.

Cain, A. J. (1951b), "So-Called Non-Adaptive or Neutral Characters in Evolution," *Nature* 168:424.

Cain, A. J. (1964), "The Perfection of Animals," in J. D. Carthy and C. L. Darlington (eds.), *Viewpoints in Biology* (London: Butterworth), 3:36–63.
Cain, A. J., and Sheppard, P. M. (1950), "Selection in the Polymorphic Land Snail *Cepaea nemoralis*," *Heredity* 4:275–294.
Cain, A. J., and Sheppard, P. M. (1952), "The Effects of Natural Selection on Body Colour in the Land Snail *Cepaea nemoralis*," *Heredity* 6:217–231.
Cain, A. J., and Sheppard, P. M. (1954), "Natural Selection in *Cepaea*," *Genetics* 39:89–116.
Calvin, W. H. (2002), *A Brain for All Seasons: Human Evolution and Abrupt Climate Change* (Chicago: University of Chicago Press).
Carpenter, W. B. (1889), *Nature and Man: Essays Scientific and Philosophical* (New York: Appleton).
Carr-Saunders, A. M. (1922), *The Population Problem: A Study in Human Evolution* (Oxford: Clarendon).
Cartwright, N. (1981), "The Reality of Causes in a World of Instrumental Laws," in P. Asquith and R. Giere (eds.), *PSA1980* (East Lansing, MI.: Philosophy of Science Association), pp. 38–48.
Cassidy, J. (1981), "Ambiguities and Pragmatic Factors in the Units of Selection Controversy," *Philosophy of Science* 48:95–111.
Castrodeza, C. (1978), "Evolution, Complexity, and Fitness," *Journal of Theoretical Biology* 71:469–71.
Chaitin, G. J. (1975), "Randomness and Mathematical Proof," *Scientific American* 232:47–52.
Chambers, R. (1844), *Vestiges of the Natural History of Creation* (London: John Churchill). [Tenth edition, 1853.]
Christian, J. J. (1964), "Review of Wynne-Edwards' *Animal Dispersion in Relation to Social Behaviour*," *Quarterly Review of Biology* 39:83–84.
Collins, J. P. (1986), "Evolutionary Ecology and the Use of Natural Selection in Ecological Theory," *Journal of the History of Biology* 19:257–88.
Cope, E. D. (1871), "The Laws of Organic Development," *American Naturalist* 5:593–605.
Cope, E. D. (1873), "The Method of Creation of Organic Forms," *Proceedings of the American Philosophical Society* 12:229–65.
Cope, E. D. (1896), *The Primary Factors of Organic Evolution* (Chicago: Open Court Publishing Company).
Cracraft, J. (1990), "The Origin of Evolutionary Novelties: Pattern and Process at Different Hierarchical Levels," in M. Nitecki (ed.) (1990), *Evolutionary Innovations* (Chicago: University of Chicago Press), pp. 21–44.
Cronin, H. (1991), *The Ant and the Peacock: Altruism and Sexual Selection from Darwin to Today* (Cambridge: Cambridge University Press).
Cuvier, G. (1805), *Leçons d'anatomie comparée*, 5 volumes (Paris: Crochard, Fantin).
Cuvier, G. (1817), *Le Règne Animal*, vol. 1 (Paris: Deterville).
Darwin, C. (1839), *Journal of Researches into the Geology and Natural History of the Various Countries Visited by H.M.S. Beagle etc.* (London: Henry Colburn). [Widely known as *The Voyage of the Beagle*.]
Darwin, C. (1842), *The Structure and Distribution of Coral Reefs* (London: Smith, Elder, and Co.).

Darwin, C. (1854), *A Monograph of the Fossil Balanidae and Verrucidae of Great Britain* (London: Palaeontological Society).

Darwin, C. (1859), *On the Origin of Species*, 1st ed. (London: John Murray). Reprinted, with an introduction by Ernst Mayr, Cambridge, MA: Harvard University Press, 1964.

Darwin, C. (1862), *On the Various Contrivances by which British and Foreign Orchids are Fertilised by Insects* (London: John Murray; 2nd edition, 1877).

Darwin, C. (1868), *The Variation of Animals and Plants under Domestication* (London: John Murray).

Darwin, C. (1871), *The Descent of Man and Selection in Relation to Sex*, 2 volumes (London: John Murray; 2nd edition, 1874).

Darwin, C. (1872), *The Expression of the Emotions in Man and Animals* (London: John Murray).

Darwin, C. (1958), *The Autobiography of Charles Darwin*, edited by N. Barlow (New York: W. W. Norton, 1958).

Darwin, C. (1959), *The Origin of Species; A Variorum Text*, edited by M. Peckham (Philadelphia: University of Pennsylvania Press).

Darwin, F. (ed.) (1887), *The Life and Letters of Charles Darwin, including an autobiographical chapter*, 2 vols. (New York: Appleton).

Darwin, F. (ed.) (1909), *The Foundations of the Origin of Species: Two Essays Written in 1842 and 1844 by Charles Darwin* (Cambridge: Cambridge University Press).

Darwin, F. and A. C. Seward (eds.) (1903), *More Letters of Charles Darwin*, 2 vols. (London: John Murray).

Davenport, C. (1910), "Report of the Committee on Eugenics," *American Breeders' Magazine* 1:126–29.

Dawkins, R. (1978), "Replicator Selection and the Extended Phenotype," *Zeitschrift für Tierpsychologie* 47:61–76.

Dawkins, R. (1979), "Twelve Misunderstandings of Kin Selection," *Zeitschrift für Tierpsychologie* 51:184–200.

Dawkins, R. (1982a), "Replicators and Vehicles," in King's College Sociobiology Group (ed.), *Current Problems in Sociobiology* (Cambridge: Cambridge University Press), pp. 45–64.

Dawkins, R. (1982b), *The Extended Phenotype: The Gene as the Unit of Selection* (Oxford: W. H. Freeman). [Revised edition, 1999.]

Dawkins, R. (1986), *The Blind Watchmaker: Why the Evidence of Evolution Reveals a World Without Design* (New York: W. W. Norton).

Dawkins, R. (1989a), *The Selfish Gene*, revised edition (Oxford: Oxford University Press).

Dawkins, R. (1989b), "The Evolution of Evolvability," in C. Langton (ed.), *Artificial Life* (Santa Fe, NM: Addison-Wesley), pp. 201–220.

Dawkins, R. (1992), "Progress," in E. Fox Keller and E. Lloyd (eds.), *Keywords in Evolutionary Biology* (Cambridge, MA: Harvard University Press), pp. 263–272.

Dawkins, R. (1996), *Climbing Mount Improbable* (New York: W. W. Norton).

Dawkins, R. (1997), "Human Chauvinism" *[review of S. J. Gould's Full House], Evolution* 51:1015–20.

Dawkins, R., and J. R. Krebs (1979), "Arms Races Between and Within Species," *Proceedings of the Royal Society of London, B Biological Sciences* 205:489–511.

Dennett, D. C. (1996), *Darwin's Dangerous Idea: Evolution and the Meanings of Life* (New York: Simon & Schuster).

DeVries, H. (1906), *Species and Varieties: Their Origin by Mutation*, revised edition, edited by D. T. MacDougal (Chicago: Open Court).

DeVries, H. (1910), *The Mutation Theory: Experiments and Observations on the Origin of Species in the Vegetable Kingdom*, translated by J. B. Farmer and A. D. Darbyshire, 2 vols. (London: Kegan Paul).

Di Gregorio, M. A. (1990), *Charles Darwin's Marginalia* (London: Garland Publishing).

Divall, Colin (1992), "From a Victorian to a Modern: Julian Huxley and the English Intellectual Climate," in C. K. Waters and A. van Helden (eds.), *Julian Huxley: Biologist and Statesman of Science* (Houston: Rice University Press), pp. 31–44.

Diver, C. (1940), "The Problem of Closely Related Snails Living in the Same Area," in J. Huxley (ed.), *The New Systematics* (Oxford: Oxford University Press).

Dobzhansky, Th. (1937), *Genetics and the Origin of Species*, 1st edition (New York: Columbia University Press). [2nd edition, 1941. Third edition, 1951.]

Dobzhansky, Th. (1956), *The Biological Basis of Human Freedom* (New York: Columbia University Press).

Dobzhansky, Th. (1967), *The Biology of Ultimate Concern* (New York: New American Library).

Doolittle, W. F., and Sapienza, C. (1980), "Selfish Genes, the Phenotypic Paradigm and Genomic Evolution," *Nature* 284:601–3.

Dunford, C. (1977), "Kin Selection for Ground Squirrel Alarm Calls," *American Naturalist* 111:782–85.

Durant, J. (1979), "Scientific Naturalism and Social Reform in the Thought of Alfred Russel Wallace," *British Journal for the History of Science* 12:31–58.

Durant, J. R. (1989), "Julian Huxley and the Development of Evolutionary Studies," in M. Keynes and G. A. Harrison (eds.), *Evolutionary Studies: A Centenary Celebration of the Life of Julian Huxley* (London: Macmillan), pp. 26–40.

Durant, J. R. (1992), "The Tension at the Heart of Huxley's Evolutionary Ethology," in C. Kenneth Waters and A. van Helden (eds.), *Julian Huxley: Biologist and Statesman of Science* (Houston, Texas: Rice University Press), pp. 150–64.

Eimer, Th. (1898), *On Orthogenesis and the Impotence of Natural Selection in Species-Formation*, translated by T. J. McCormack (Chicago: Open Court).

Elton, C. S. (1963), "Self-Regulation of Animal Populations," *Nature* 197:634.

Emerson, A. E. (1960), "The Evolution of Adaptation in Population Systems," in S. Tax (ed.), *Evolution After Darwin*, vol. 1 (Chicago: University of Chicago Press), pp. 307–48.

Emlen, S. T. (1984), "Cooperative Breeding in Birds," in J. R. Krebs and N. B. Davies (eds.), *Behavioural Ecology: An Evolutionary Approach*, second edition (Oxford: Blackwell), pp. 305–39.

Fenner, F., and Marshall, I. D. (1965), *Myxomatosis* (Cambridge: Cambridge University Press).

Fisher, D. C. (1985), "Evolutionary Morphology: Beyond the Analogous, the Anecdotal and the Ad Hoc," *Paleobiology* 11:120–38.

Fisher, D. C. (1986), "Progress in Organismal Design," in D. M. Raup and D. Jablonski (eds.), *Patterns and Processes in the History of Life* (Berlin: Springer-Verlag), pp. 99–117.

Fisher, R. A. (1930), *The Genetical Theory of Natural Selection* (Oxford: Clarendon Press). [2nd edition, New York: Dover, 1950]

Fisher, R. A. (1936), "The Measurement of Selective Intensity," *Proceedings of the Royal Society of London*, Series B, 121:58–62.

Gascoigne, Robert M. (1991), "Julian Huxley and Biological Progress," *Journal of the History of Biology* 24:433–55.

Gasman, D. (1971), *The Scientific Origins of National Socialism* (New York: Science History Publications).

Godfrey-Smith. P. (1999), "Adaptationism and the Power of Selection," *Biology and Philosophy* 14:181–94.

Godfrey-Smith. P. (2001), "Three Kinds of Adaptationism," in S. H. Orzack and E. Sober (eds.), *Adaptationism and Optimality* (Cambridge: Cambridge University Press), pp. 335–57.

Gould, S. J. (1977), *Ever Since Darwin: Reflections in Natural History* (New York: W. W. Norton).

Gould, S. J. (1980), *The Panda's Thumb: More Reflections in Natural History* (New York: W. W. Norton).

Gould, S. J. (1983), "The Hardening of the Modern Synthesis," in M. Grene (ed.), *Dimensions of Darwinism* (Cambridge: Cambridge University Press), pp. 71–93.

Gould, S. J. (1988a), "On Replacing the Idea of Progress with an Operational Notion of Directionality," in M. Nitecki (ed.), *Evolutionary Progress* (Chicago: University of Chicago Press), pp. 319–38.

Gould, S. J. (1988b), "Trends as Changes in Variance: A New Slant on Progress and Directionality in Evolution," *Journal of Paleontology* 62:319–29.

Gould, S. J. (1989), *Wonderful Life: The Burgess Shale and the Nature of History* (New York: W. W. Norton).

Gould, S. J. (1996), *Full House: The Spread of Excellence from Plato to Darwin* (New York: Harmony Books). [Published in the U.K. as *Life's Grandeur.*]

Gould, S. J. (1997a), "Self-Help for a Hedgehog Stuck on a Molehill" [review of R. Dawkins, *Climbing Mount Improbable*], *Evolution* 51:1020–1023.

Gould, S. J. (1997b), "Evolution: The Pleasures of Pluralism," *The New York Review of Books*, June 26, 1997, pp. 47–52.

Gould, S. J., and Lewontin, R. C. (1979), "The Spandrels of San Marco and the Panglossian Paradigm: A Critique of the Adaptationist Programme," *Proceedings of the Royal Society of London Series B*. 205:581–98.

Gould, S. J., and Vrba, E. S. (1982), "Exaptation – A Missing Term in the Science of Form," *Paleobiology* 8:4–15.

Gray, A. (1876), *Darwiniana: Essays and Reviews Pertaining to Darwinism* (New York: D. Appleton).

Gray, R. D. (1992), "Death of the Gene: Developmental Systems Strike Back," in P. E. Griffiths (ed.), *Trees of Life: Essays in the Philosophy of Biology* (Dordrecht: Kluwer), pp. 165–209.

Gray, R. D. (2000), "Selfish Genes or Developmental Systems: Evolution without Replicators and Vehicles," in R. Singh, C. Krimbas, D. Paul, and J. Beatty

(eds.), *Thinking About Evolution: Historical, Philosophical, and Political Perspectives* (Cambridge: Cambridge University Press), pp. 184–207.
Greg, W. R. (1868), "On the Failure of 'Natural Selection' in the Case of Man," *Fraser's Magazine* 78:353–62.
Greene, J. C. (1981), *Science, Ideology, and Worldview* (Berkeley: University of California Press).
Greene, J. C. (1990), "The Interaction of Science and World View in Sir Julian Huxley's Evolutionary Biology," *Journal of the History of Biology* 23:39–55.
Griffiths, P. E. (1992), "Adaptive Explanation and the Concept of a Vestige," in P. E. Griffiths (ed.), *Trees of Life: Essays in the Philosophy of Biology* (Dordrecht: Kluwer), pp. 111–31.
Griffiths, P. E., and Gray, R. D. (1994), Developmental Systems and Evolutionary Explanation," *Journal of Philosophy* 91:277–304.
Griffiths, P. E., and Gray, R. D. (1997) "Replicator II – Judgment Day," *Biology and Philosophy* 12:471–92.
Hahlweg, K. (1991), "On the Notion of Evolutionary Progress," *Philosophy of Science* 58:436–51.
Haldane, J. B. S. (1932), *The Causes of Evolution* (New York: (Longmans, Green, and Co.).
Haldane, J. B. S., and Huxley, J. S. (1927), *Animal Biology* (Oxford: Clarendon).
Hamilton, W. D. (1963), "The Evolution of Altruistic Behavior," *American Naturalist* 97:31–33.
Hamilton, W. D. (1964), "The Genetical Theory of Social Behavior: I & II," *Journal of Theoretical Biology* 7:1–52.
Hinegardner, R., and Engelberg, J. (1983), "Biological Complexity," *Journal of Theoretical Biology* 104:7–20.
Holcomb, H. (1989), "Expecting Nature's Best: Optimality Models and Perfect Adaptation," *Philosophy in Science* 4:124–47.
Horgan, J. (1996), *The End of Science: Facing the Limits of Knowledge in the Twilight of the Scientific Age* (Reading, MA: Addison-Wesley Publishers).
Hull, D. L. (1979), "In Defense of Presentism," *History and Theory* 18:1–15. Reprinted in D. L. Hull, *The Metaphysics of Evolution* (Albany: SUNY Press), pp. 205–20.
Hull, D. L. (1980), "Individuality and Selection," *Annual Review of Ecology and Systematics* 11:311–32.
Hull, D. L. (1981), "Units of Evolution: A Metaphysical Essay," in R. Jensen and R. Harre (eds.), *The Philosophy of Evolution* (Brighton: Harvester), pp. 23–44.
Hull, D. L. (1988a), *Science as a Process: An Evolutionary Account of the Social and Conceptual Development of Science* (Chicago: University of Chicago Press).
Hull, D. L. (1988b), "Progress in Ideas of Progress", in M. Nitecki (ed.), *Evolutionary Progress* (Chicago: University of Chicago Press), pp. 27–48.
Hume, D. (1777), *An Enquiry Concerning Human Understanding,* edited by E. Steinberg (Indianapolis, IN: Hackett Publishing, 1977).
Hume, D. (1779), *Dialogues Concerning Natural Religion,* edited by R. H. Popkin (Indianapolis, IN: Hackett Publishing Company, 1998).
Huxley, J. S. (1912), *The Individual in the Animal Kingdom* (Cambridge: Cambridge University Press).

Huxley, J. S. (1923), "Progress, Biological and Other," in *Essays of a Biologist* (London: Chatto and Windus), pp. 1–64.

Huxley, J. S. (1928), "Progress Shown in Evolution," in Francis Mason (ed.), *Creation by Evolution* (New York: The Macmillan Company), pp. 327–339.

Huxley, J. S. (1936), "Natural Selection and Evolutionary Progress," *Proceedings of the British Association for the Advancement of Science* 106:81–100.

Huxley, J. S. (1941), *The Uniqueness of Man* (London: Chatto & Windus).

Huxley, J. S. (1942), *Evolution: The Modern Synthesis* (London: Allen & Unwin). [2nd edition, 1963.]

Huxley J. S. (1953), *Evolution in Action* (London: Chatto & Windus).

Huxley, J. S. (1954), "The Evolutionary Process," in J. S. Huxley, A. C. Hardy, and E. B. Ford (eds.), *Evolution as a Process* (London: George Allen & Unwin), pp. 1–23.

Huxley, J. S. (1957a), "Three Types of Evolutionary Process," *Nature* 180:454–55.

Huxley J. S. (1957b), *New Bottles for New Wine* (New York: Harper & Brothers).

Huxley, T. H. (1871), "Mr. Darwin's Critics," *Contemporary Review* 18:443–476. [Reprinted in T. H. Huxley, *Collected Essays* (New York: D. Appleton, 1896–1902), vol. 2, pp. 120–86.]

Jablonka, E., and Lamb, M. J. (1995), *Epigenetic Inheritance and Evolution: The Lamarckian Dimension* (Oxford: Oxford University Press).

Jacob, F. (1977), "Evolution and Tinkering," *Science* 196:1161–66.

Jerison, H. J. (1973), *Evolution of the Brain and Intelligence* (New York and London: Academic Press).

Kant, I. (1785), *Foundations of the Metaphysics of Morals*, translated by L. W. Beck (New York: Macmillan Publishing Company, 1990).

Kant, I. (1790), *Critique of Teleological Judgement*, translated by J. C. Meredith (Oxford: Oxford University Press, 1952).

Keller, E. F., and Ross, K. G. (1993), "Phenotypic Plasticity and Cultural Transmission in the Fire Ant, *Solenopsis invicta*," *Behavioural Ecology and Sociobiology* 33:121–29.

Kettlewell, H. B. D. (1955), "Selection Experiments on Industrial Melanism in the Lepidoptera," *Heredity* 9:323–42.

Kettlewell, H. B. D. (1956), "Further Selection Experiments on Industrial Melanism in the Lepidoptera," *Heredity* 10:287–301.

Kettlewell, H. B. D. (1961), "The Phenomenon of Industrial Melanism in the Lepidoptera," *Annual Review of Entomology* 6:245–62.

Kettlewell, H. B. D. (1973), *The Evolution of Melanism* (Oxford: Clarendon Press).

King, J. (1965), "Social Behavior and Population Homeostasis," *Ecology* 46: 210–11.

Kitcher, P. (1985), *Vaulting Ambition: Sociobiology and the Quest for Human Nature* (Cambridge, MA: MIT Press).

Kitcher, P. (1987), "Why Not the Best?" in J. Dupré (ed.), *The Latest on the Best* (Cambridge, MA: MIT Press), pp. 77–102.

Kottler, M. (1974), "Alfred Russel Wallace, the Origin of Man and Spiritualism," *Isis* 65:144–92.

Kottler, M. (1985), "Charles Darwin and Alfred Russel Wallace: Two Decades of Debate Over Natural Selection," in D. Kohn (ed.), *The Darwinian Heritage* (Princeton, NJ: Princeton University Press, 1985), pp. 367–432.

Kragh, H. (1987), *An Introduction to the Historiography of Science* (New York: Cambridge University Press).

La Cerra, P. and Bingham, R. (2002), *The Origin of Minds: Evolution, Uniqueness, and the New Science of the Self* (New York: Harmony Books).

Lack, D. (1945), "The Galapagos Finches (*Geospizinae*)," *Occasional Papers of the California Academy of Sciences*, no. 21.

Lack, D. (1947), *Darwin's Finches* (Cambridge: Cambridge University Press).

Lack, D. (1954), *The Natural Regulation of Animal Numbers* (Oxford: Clarendon Press).

Lack, D. (1960), *Darwin's Finches*, reprint with new preface (New York: Harper Torchbooks).

Lack, D. (1964), "Significance of Clutch-Size in Swift and Grouse," *Nature* 203:98–99.

Lack, D. (1965), "Evolutionary Ecology," *Journal of Animal Ecology* 34:223–31.

Lack, D. (1966), *Population Studies of Birds* (Oxford: Clarendon).

Lamarck, J. B. (1801), *Système des Animaux sans Vertèbres* (Paris: Chez Deterville).

Lamarck, J. B. (1815–1822), *Histoire Naturelle des Animaux sans Vertèbres* (Paris: Verdière).

Lamarck, J. B. (1809), *Zoological Philosophy*, translated by H. Elliot (New York: Hafner, 1963; reprinted Chicago: University of Chicago Press, 1984).

Laudan, L. (1982), "Commentary: Science at the Bar – Causes for Concern," *Science, Technology, and Human Values* 7:16–19.

Lauder, G. V. (1996), "The Argument from Design," in M. R. Rose and G. V. Lauder (eds.), *Adaptation* (San Diego: Academic Press), pp. 55–91.

Leigh, E. G., Jr. (2001), "Adaptation, Adaptationism, and Optimality," in S. H. Orzack and E. Sober (eds.), *Adaptationism and Optimality* (Cambridge: Cambridge University Press), pp. 358–87.

Lorenz, K. (1966), *On Aggression*, translated by M. K. Wilson (New York: Harcourt, Brace & World).

Leroi, A. M., Rose, M. R., and Lauder, G. V. (1994), "What Does the Comparative Method Reveal About Adaptation?" *American Naturalist* 143:381–402.

Levy, C. K. (1999), *Evolutionary Wars: A Three-Billion Year Arms Race* (New York: W. H. Freeman and Company).

Lewin, R. (1980), "Evolutionary Theory Under Fire," *Science* 210:883–87.

Lewin, R. (1994), "A Simple Matter of Complexity," *New Scientist* 141:37–40.

Lewontin, R. C. (1967), Spoken remark in P. D. Moorhead and M. Kaplan (eds.), *Mathematical Challenges to the Neo-Darwinian Interpretation of Evolution* (Wistar Institute Symposium Monograph) 5, p. 79.

Lewontin, R. C. (1970), "The Units of Selection," *Annual Review of Ecology and Systematics* 1:1–14.

Lewontin, R. C. (1979), "Sociobiology as an Adaptationist Program," *Behavioral Science* 24:5–14.

Lewontin, R. C., Rose, S., and Kamin, L. J. (1984), *Not in Our Genes: Biology, Ideology, and Human Nature* (New York: Pantheon).

Lloyd, E. (2000), "Units and Levels of Selection: An Anatomy of the Units of Selection Debates," in R. Singh, C. Krimbas, D. Paul, and J. Beatty (eds.), *Thinking About Evolution: Historical, Philosophical, and Political Perspectives* (Cambridge: Cambridge University Press), pp. 267–91.

Lyell, C. (1867/1868), *Principles of Geology*, 10th edition, 2 vols. (London).

Lyon, J., and Sloan, P. R. (1981), *From Natural History to the History of Nature: Readings from Buffon and His Critics* (Notre Dame, IN: University of Notre Dame Press, 1981).

Macculloch, J. (1837), *Proofs and Illustrations of the Attributes of God*, 3 vols. (London: J. Duncan).

Manier, E. (1978), *The Young Darwin and His Cultural Circle* (Dordrecht: D. Reidel).

Maynard Smith, J. (1964), "Group Selection and Kin Selection," *Nature* 201:1145–47.

Maynard Smith, J. (1970), "Time in the Evolutionary Process," *Studium Generale* 23:266–72.

Maynard Smith, J. (1976), "Group Selection," *Quarterly Review of Biology* 51:277–83.

Maynard Smith, J. (1982), "The Evolution of Social Behaviour – A Classification of Models," King's College Sociobiology Group (ed.), *Current Problems in Sociobiology* (Cambridge: Cambridge University Press, 1982), pp. 29–44.

Maynard Smith, J. (1987), "How to Model Evolution," in J. Dupré (ed.), *The Latest on the Best* (Cambridge, MA: MIT Press), pp. 119–31.

Maynard Smith, J. (1988), "Evolutionary Progress and the Levels of Selection," in M. Nitecki (ed.), *Evolutionary Progress* (Chicago: University of Chicago Press), pp. 219–30.

Maynard Smith, J., Szathmáry, E. (1995), *The Major Transitions in Evolution* (Oxford: Freeman/Spektrum).

Mayr, E. (1942), *Systematics and the Origin of Species* (New York: Columbia University Press).

Mayr, E. (1963), *Animal Species and Evolution* (Cambridge, MA: Harvard University Press).

Mayr, E. (1976), *Evolution and the Diversity of Life* (Cambridge, MA: Harvard University Press).

Mayr, E. (1982), *The Growth of Biological Thought: Diversity, Evolution, and Inheritance* (Cambridge, MA: Belknap Press).

Mayr, E. (1983), "How to Carry Out the Adaptationist Programme?" *American Naturalist* 121:324–34.

Mayr, E. (1988), *Towards a New Philosophy of Biology* (Cambridge, MA: Harvard University Press).

Mayr, E. (1991), *One Long Argument: Charles Darwin and the Genesis of Modern Evolutionary Thought* (Cambridge, MA: Harvard University Press).

Mayr, E. (1994), "The Resistance to Darwinism and the Misconceptions on Which It Was Based," in J. H. Campbell and J. W. Schopf (eds.), *Creative Evolution?!* (Boston and London: Jones and Bartlett Publishers), pp. 35–46.

Mayr, E. (1985), "The Probability of Extraterrestrial Intelligence," in E. Regis, Jr. (ed.), *Extraterrestrials: Science and Alien Intelligence* (Cambridge: Cambridge University Press), pp. 23–42.

McCoy, J. W. (1977), "Complexity in Organic Evolution," *Journal of Theoretical Biology* 68:457–58.
McMahon, T. A., and Bonner, J. T. (1983), *On Size and Life* (New York: Scientific American Books).
McMullin, E. (1984), "A Case for Scientific Realism," in J. Leplin (ed.), *Scientific Realism* (Berkeley: University of California Press), pp. 8–40.
McShea, D. W. (1991), "Complexity and Evolution: What Everybody Knows," *Biology & Philosophy* 6:303–24.
McShea, D. W. (1992), "A Metric for the Study of Evolutionary Trends in the Complexity of Serial Structures," *Biological Journal of the Linnean Society of London* 45:39–55.
McShea, D. W. (1993), "Evolutionary Change in the Morphological Complexity of the Mammalian Vertebral Column," *Evolution* 47:730–40.
McShea, D. W. (1994), "Mechanisms of Large-Scale Evolutionary Trends," *Evolution* 48:1747–63.
McShea, D. W. (1996), "Metazoan Complexity and Evolution: Is There a Trend?" *Evolution* 50:477–92.
McShea, D. W. (1998), "Possible Largest-Scale Trends in Organismal Evolution: Eight 'Live Hypotheses'," *Annual Review of Ecology and Systematics* 29:293–318.
Michod, R. E. (1980), "Evolution of Interactions in Family Structured Populations: Mixed Mating Models," *Genetics* 96:275–96.
Michod, R. E. (1982), "The Theory of Kin Selection," *Annual Review of Ecology and Systematics* 13:23–55.
Miller, K. (1999), *Finding Darwin's God: A Scientist's Search for Common Ground Between God and Evolution* (New York: HarperCollins).
Milne Edwards, H. (1827), "Organisation," in *Dictionnaire Classique d'Histoire Naturelle.*
Mitman, G. (1992), *The State of Nature: Ecology, Community, and American Social Thought, 1900–1950* (Chicago: University of Chicago Press).
Mivart, St. George (1871), *On the Genesis of Species* (London: Macmillan).
Morgan, T. H. (1903), *Evolution and Adaptation* (New York: Macmillan).
Morris, D. (1967), *The Naked Ape: A Zoologist's Study of the Human Animal* (New York: McGraw-Hill).
Nelson, P. A. (1996), "The Role of Theology in Current Evolutionary Reasoning," *Biology & Philosophy* 11:493–517.
Nesse, R. M., and Williams, G. C. (1994), *Why We Get Sick: The New Science of Darwinian Medicine* (New York: Vintage Books).
Nicholson, E. M. (1962), "Special Review [of Wynne-Edwards' *Animal Dispersion in Relation to Social Behaviour*]," *Ibis* 104: 570–71.
Nitecki, M. (ed.) (1988), *Evolutionary Progress* (Chicago: University of Chicago Press).
Nitecki, M. (ed.) (1990), *Evolutionary Innovations* (Chicago: University of Chicago Press).
Olson, E. C. (1985), "Intelligent Life in Space," *Astronomy Magazine* 13:6–22.
Orgel, L. E., and Crick, F. H. C. (1980), "Selfish DNA: The Ultimate Parasite," *Nature* 284:604–6.

Orzack, S., and Sober, E. (1994), "Optimality Models and the Test of Adaptationism," *American Naturalist* 143:361–80.

Orzack, S. H. and Sober E. (eds.) (2001), *Adaptationism and Optimality* (Cambridge: Cambridge University Press).

Ospovat, D. (1981), *The Development of Darwin's Theory: Natural History, Natural Theology, and Natural Selection, 1838–1859* (Cambridge: Cambridge University Press).

Ostrom, J. H. (1974), "Archaeopteryx and the Origin of Flight," *Quarterly Review of Biology* 49:27–47.

Ostrom, J. H. (1979), "Bird Flight: How Did it Happen?" *American Scientist* 67:46–56.

Owen, R. (1866), *Memoir on the Dodo*, with an Historical Introduction by the Late William John Broderip (London: Taylor and Francis).

Oyama, S. (2000), *The Ontogeny of Information*, revised edition (Cambridge: Cambridge University Press).

Oyama, S., Gray, R., and Griffiths, P. (eds.) (2001), *Cycles of Contingency: Developmental Systems and Evolution* (Cambridge, MA: MIT Press).

Paley, W. (1802), *Natural Theology, Or Evidences of the Existence and Attributes of the Deity, Collected from the Appearances of Nature* (London: Faulder; 13th edition, 1810).

Panksepp, J., and Panksepp, J. B. (2000), "The Seven Sins of Evolutionary Psychology," *Evolution and Cognition* 6:108–31.

Papentin, F. (1980), "On Order and Complexity. I. General Considerations," *Journal of Theoretical Biology* 87:421–56.

Perrins, C. (1964), "Survival of Young Swifts in Relation to Brood-Size," *Nature* 201:1147–48.

Prete, F. R. (1990), "The Conundrum of the Honey Bees: One Impediment to the Publication of Darwin's Theory," *Journal of the History of Biology* 23:271–90.

Provine, W. B. (1983), "The Development of Wright's Theory of Evolution: Systematics, Adaptation, and Drift," in M. Grene (ed.), *Dimensions of Darwinism* (Cambridge: Cambridge University Press), pp. 43–70.

Provine, W. B. (1985), "Adaptation and Mechanisms of Evolution After Darwin: A Study in Persistent Controversies," in D. Kohn (ed.), *The Darwinian Heritage* (Princeton, NJ: Princeton University Press, 1985), pp. 825–66.

Provine, W. B. (1986), *Sewall Wright and Evolutionary Biology* (Chicago: University of Chicago Press).

Provine, W. B. (1988), "Progress in Evolution and Meaning in Life," in M. Nitecki (ed.), *Evolutionary Progress* (Chicago: University of Chicago Press), pp. 49–74.

Radick, G. (2000), "Two Concepts of Evolutionary Progress," *Biology & Philosophy* 15:475–91.

Reeve, H. K., and Sherman, P. W. (1993), "Adaptation and the Goals of Evolutionary Research," *Quarterly Review of Biology* 68:1–31.

Resnik, D. (1997), "Adaptationism: Hypothesis or Heuristic?" *Biology & Philosophy* 12:39–50.

Richards, R. J. (1983), "Why Darwin Delayed, or Interesting Problems and Models in the History of Science," *Journal of the History of the Behavioral Sciences* 19:45–53.

Richards, R. J. (1987), *Darwin and the Emergence of Evolutionary Theories of Mind and Behavior* (Chicago: University of Chicago Press).

Richards, R. J. (1988), "The Moral Foundations of the Idea of Evolutionary Progress: Darwin, Spencer, and the Neo-Darwinians," in M. Nitecki (ed.), *Evolutionary Progress* (Chicago: University of Chicago Press), pp. 129–48.

Richards, R. J. (1992), *The Meaning of Evolution: The Morphological Construction and Ideological Reconstruction of Darwin's Theory* (Chicago: University of Chicago Press).

Richardson, R. C., and Kane, T. C. (1988), "'Orthogenesis and Evolution in the 19th Century: The Idea of Progress in American Neo-Lamarckism," in M. Nitecki (ed.), *Evolutionary Progress* (Chicago: University of Chicago Press), pp. 149–67.

Roger, J. (1997), *Buffon: A Life in Natural History* (Ithaca, NY: Cornell University Press).

Rose, M. R., and Lauder, G. V. (eds.) (1996), *Adaptation* (San Diego: Academic Press).

Rosenberg, A. (1994), *Instrumental Biology, or the Disunity of Science* (Chicago: University of Chicago Press).

Ruse, M. (1980), "Charles Darwin and Group Selection," *Annals of Science* 37:615–30.

Ruse, M. (1988), "Molecules to Man: Evolutionary Biology and Thoughts of Progress," in M. Nitecki (ed.), *Evolutionary Progress* (Chicago: University of Chicago Press), pp. 97–126.

Ruse, M. (1993), "Evolution and Progress," *Trends in Ecology and Evolution* 8:55–59.

Ruse, M. (1996), *Monad to Man: The Concept of Progress in Evolutionary Biology* (Cambridge, MA: Harvard University Press).

Ruse, M. (2000), *Can a Darwinian Be a Christian? The Relationship Between Science and Religion* (Cambridge: Cambridge University Press).

Sagan, C. (1995), "The Abundance of Life-Bearing Planets," *Bioastronomy News* 7:1–4.

Salwini-Plawen, L. V., and E. Mayr (1977), "The Evolution of Photoreceptors and Eyes," *Evolutionary Biology* 10:207–63.

Saunders, P. T., and Ho, W. H. (1976), "On the Increase in Complexity in Evolution," *Journal of Theoretical Biology* 63:375–84.

Scriven, M. (1959), "Explanation and Prediction in Evolutionary Theory," *Science* 130:477–82.

Segerstråle, U. (2000), *Defenders of the Truth: The Sociobiology Debate* (Oxford: Oxford University Press).

Seeley, T. D. (1989), "The Honey Bee Colony as a Superorganism," *American Scientist* 77:546–53.

Shanahan, T. (1990), "Group Selection and the Evolution of Myxomatosis," *Evolutionary Theory* 9:239–54.

Shanahan, T. (1996), "Realism and Antirealism in Evolutionary Biology," in R. S. Cohen, R. Hilpinen, and Q. Renzong (eds.), *Realism and Anti-Realism in the Philosophy of Science* (Dordrecht: Kluwer Academic Publishers), pp. 447–64.

Shanahan, T. (1997), "Darwinian Naturalism, Theism, and Biological Design," *Perspectives on Science and Christian Faith* 49:170–78.

Shanahan, T. (2001), "Methodological and Contextual Factors in the Dawkins/Gould Dispute Over Evolutionary Progress" *Studies in History and Philosophy of Biology and the Biomedical Sciences* 32:127–51.

Sherman, P. (1977), "Nepotism and the Evolution of Alarm Calls," *Science* 197:1246–53.

Simpson, G. G. (1944), *Tempo and Mode in Evolution* (New York: Columbia University Press).

Simpson, G. G. (1949), *The Meaning of Evolution: A Study of the History of Life and of Its Significance for Man* (New Haven, CT: Yale University Press).

Simpson, G. G. (1953), *The Major Features of Evolution* (New York: Columbia University Press).

Simpson, G. G. (1960), "The History of Life," in S. Tax (ed.), *Evolution After Darwin*, vol. 1 (Chicago: University of Chicago Press), pp. 117–80.

Simpson, G. G. (1974), "The Concept of Progress in Organic Evolution," *Social Research* 41:28–51.

Sloan, P. R. (1981), [Review of *The Darwinian Revolution*, by Michael Ruse], *Philosophy of Science* 48:623–27.

Smocovitis, V. B. (1996), *Unifying Biology: The Evolutionary Synthesis and Evolutionary Biology* (Princeton, NJ: Princeton University Press).

Sober, E. (1980), "Evolution, Population Thinking, and Essentialism," *Philosophy of Science* 47:350–83.

Sober, E. (1984), *The Nature of Selection: Evolutionary Theory in Philosophical Focus* (Chicago: University of Chicago Press).

Sober, E. (1985), "Darwin on Natural Selection: A Philosophical Perspective," in D. Kohn (ed.), *The Darwinian Heritage* (Princeton: Princeton University Press), pp. 867–99.

Sober, E. (1994), "Progress and Direction in Evolution," in J. H. Campbell and J. W. Schopf (eds.), *Creative Evolution?!* (Boston and London: Jones and Bartlett Publishers), pp. 19–33.

Sober, E. (2000), *Philosophy of Biology*, 2nd edition (Boulder, CO: Westview Press).

Sober, E., and Lewontin, R. C. (1982), "Artifact, Cause, and Genic Selection," *Philosophy of Science* 49:157–80.

Sober, E., and Wilson, D. S. (1994), "A Critical Review of Philosophical Work on the Units of Selection Problem," *Philosophy of Science* 61:534–55.

Sober, E, and Wilson, D. S. (1998), *Unto Others: The Evolution and Psychology of Unselfish Behavior* (Cambridge, MA: Harvard University Press).

Spencer. H. (1887), *The Factors of Organic Evolution* (London: Williams & Norgate).

Stanley, S. M. (1973), "An Explanation for Cope's Rule," *Evolution* 27:1–26.

Stauffer, R. C. (ed.) (1975), *Charles Darwin's Natural Selection: Being the Second Part of His Big Species Book Written from 1856–1858* (Cambridge: Cambridge University Press).

Stebbins, G. L. (1950), *Variation and Evolution in Plants* (New York: Columbia University Press).

Stebbins, G. L. (1969), *The Basis of Progressive Evolution* (Chapel Hill: University of North Carolina Press).

Stebbins, G. L. (1982), *Darwin to DNA, Molecules to Humanity* (San Francisco: W. H. Freeman).
Steele, E. J. (1981), *Somatic Selection and Adaptive Evolution: On the Inheritance of Acquired Characters*, 2nd edition (Chicago: University of Chicago Press).
Stenmark, M. (2001), *Scientism: Science, Ethics and Religion* (Aldershot: Ashgate).
Sterelny, K. (2001), *Dawkins vs. Gould: Survival of the Fittest* (Cambridge: Icon Books, Ltd.).
Sterelny, K., and Griffiths, P. (1999), *Sex and Death: An Introduction to Philosophy of Biology* (Chicago: University of Chicago Press).
Sterelny, K., and Kitcher, P. (1988), "The Return of the Gene," *Journal of Philosophy* 85:339–61.
Sterelny, K., Smith, K., and Dickson, M. (1996), "The Extended Replicator," *Biology & Philosophy* 11:377–403.
Strickland, H. E., and Melville, A. G. (1848), *The Dodo and Its Kindred, or, The History, Affinities, and Osteology of the Dodo, Solitaire, and Other Extinct Birds of the Islands Mauritius, Rodriguez, and Bourbon* (London: Reeve, Benham, and Reeve).
Swetlitz, M. (1991), "Julian Huxley, George Gaylord Simpson and the Idea of Progress in Twentieth Century Evolutionary Biology" (Ph.D. Dissertation, University of Chicago).
Swetlitz, M. (1995), "Julian Huxley and the End of Evolution," *Journal of the History of Biology* 28:181–217.
Thompson, D. W. (1969), *On Growth and Form* (Cambridge: Cambridge University Press).
Thornhill, R. (1990), "The Study of Adaptation," in M. Bekoff and D. Jamieson (eds.), *Interpretation and Explanation in the Study of Behavior*, vol. 2 (Boulder, CO: Westview Press), pp. 31–62.
Toulmin, S. (1961), *Foresight and Understanding: An Enquiry into the Aims of Science* (New York: Harper & Row).
Trivers, R. (1971), The Evolution of Reciprocal Altruism," *Quarterly Review of Biology* 46:35–57.
Trivers, R. (1985), *Social Evolution* (Menlo Park, CA: Benjamin/Cummings).
Turner, F. M. (1974), *Between Science and Religion: The Reaction to Scientific Naturalism in Late Victorian England* (New Haven, CT: Yale University Press).
Van Valen, L. M. (1973), "A New Evolutionary Law," *Evolutionary Theory* 1:1–30.
Van Valen, L. M. (1974), "A Natural Model for the Origin of Some Higher Taxa," *Journal of Herpetology* 8:109–21.
Van Valen, L. M. (1978), "Evolution as a Zero-Sum Game for Energy," *Evolutionary Theory* 4:289–300.
Vermeij, G. (1987), *Escalation and Evolution: An Ecological History of Life* (Princeton, NJ: Princeton University Press).
Vogel, S. (1988), *Life's Devices: The Physical World of Animals and Plants* (Princeton, NJ: Princeton University Press).
Von Baer, K. E. (1828–1837), *Ueber Entwickelungsgeschichte der Thiere*, 2 vols. (Königsberg: Borntrager).
Vrba, E. S., and Gould, S. J. (1986), "The Hierarchical Expansion of Sorting and Selection: Sorting and Selection Cannot Be Equated," *Paleobiology* 12: 217–28.

Wade, M. J. (1976), "Group Selection Among Laboratory Populations of *Tribolium*," *Proceedings of the National Academy of Science* 113:399–417.

Walker, W. F., and Liem, K. F. (1994), *Functional Anatomy of the Vertebrates* (New York: Harcourt Brace).

Wallace, A. R. (1856), "On the Habits of the Orang-utan of Borneo," *Annals and Magazine of Natural History, second series*, 18:26–32.

Wallace, A. R. (1858), "On the Tendency of Varieties to Depart Indefinitely from the Original Type," *Journal of the Proceedings of the Linnean Society: Zoology* 3:53–62.

Wallace, A. R. (1864), "The Origin of Human Races and the Antiquity of Man Deduced from the Theory of 'Natural Selection'," *Anthropological Review* 2:clviii–clxxxvii.

Wallace, A. R. (1867), "Mimicry and other Protective Resemblances among Animals," *Westminster and Foreign Quarterly Review* 32:1–43.

Wallace, A. R. (1869), "Review of *Principles of Geology* by Charles Lyell, 10th ed., 2 vols. (London, 1867, 1868) and of *Elements of Geology* by Charles Lyell, 6th ed. (London, 1865)," *Quarterly Review* 126:359–94.

Wallace, A. R. (1870), "The Limits of Natural Selection as Applied to Man," in *Contributions to the Theory of Natural Selection* (London: Macmillan), pp. 332–71. [Originally published in 1869.]

Wallace, A. R. (1874), *A Defense of Modern Spiritualism* (Boston: Colby and Rich).

Wallace, A. R. (1889), *Darwinism: An Exposition of Natural Selection with Some of Its Applications* (New York: Humboldt).

Wallace, A. R. (1896), "The Problem of Utility: Are Specific Characters Always or Generally Useful?" *Journal of the Linnean Society* (Zoology) 25:481–96.

Wallace, A. R. (1916), *Alfred Russel Wallace: Letters and Reminiscences*, edited by J. Marchant (New York: Harper).

Waters, K. (1991), "Tempered Realism about the Forces of Selection," *Philosophy of Science* 58:553–73.

Weismann, A. (1891–1892), *Essays upon Heredity and Kindred Biological Problems*, edited by E. B. Poulton, S. Schönland, and A. E. Shipley, 2 vols. (Oxford: Oxford University Press).

West-Eberhard, M. J. (1992), "Adaptation: Current Usages," in E. F. Keller and E. A. Lloyd (eds.), *Keywords in Evolutionary Biology* (Cambridge, MA: Harvard University Press), pp. 13–18.

Wiens, J. A. (1966), "On Group Selection and Wynne-Edwards' Hypothesis," *American Scientist* 54:273–87.

Williams, G. C. (1966), *Adaptation and Natural Selection: A Critique of Some Current Evolutionary Thought* (Princeton, NJ: Princeton University Press).

Williams, G. C. (1992), *Natural Selection: Domains, Levels, and Challenges* (New York: Oxford University Press).

Wilson, D. S. (1975), "A Theory of Group Selection," *Proceedings of the National Academy of Science (USA)* 72:143–46.

Wilson, D. S. (1979), "Structured Demes and Trait-Group Variation," *The American Naturalist* 113:606–10.

Wilson, D. S. (1980), *The Natural Selection of Populations and Communities* (Menlo Park, CA: Benjamin/Cummings).

Wilson, D. S. (1983), "The Group Selection Controversy: History and Current Status," *Annual Review of Ecology and Systematics* 14:159–88.

Wilson, E. O. (1975), *Sociobiology: The New Synthesis* (Cambridge, MA: Harvard University Press).

Wilson, E. O. (1978), *On Human Nature* (Cambridge, MA: Harvard University Press).

Wilson, E. O. (1992), *The Diversity of Life* (Cambridge, MA: Harvard University Press).

Wilson, L. G. (ed.) (1970), *Sir Charles Lyell's Scientific Journals on the Species Question* (New Haven, CT: Yale University Press).

Wimsatt, W. (1980), "Reductionistic Research Strategies and Their Biases in the Units of Selection Controversy," in T. Nickles (ed.), *Scientific Discovery* (Dordrecht: Reidel), pp. 213–59.

Wright, R. (1994), *The Moral Animal: Evolutionary Psychology and Everyday Life* (New York: Vintage Books).

Wright, R. (2000), *Nonzero: The Logic of Human Destiny* (New York: Pantheon).

Wright, S. (1931), "Evolution in Mendelian Populations," *Genetics* 16:97–159.

Wright, S. (1932), "The Roles of Mutation, Inbreeding, Crossbreeding and Selection in Evolution," *Proceedings of the Sixth International Congress of Genetics* 1:356–66.

Wright, S. (1945), "*Tempo and Mode in Evolution*: A Critical Review," *Ecology* 26:415–19.

Wynne-Edwards, V. C. (1955), "The Dynamics of Animal Populations," *Discovery* (October), pp. 433–36.

Wynne-Edwards, V. C. (1959), "The Control of Population-Density Through Social Behaviour: A Hypothesis," *Ibis* 101:436–41.

Wynne-Edwards, V. C. (1962), *Animal Dispersion in Relation to Social Behaviour* (Edinburgh: Oliver and Boyd).

Wynne-Edwards, V. C. (1963), "Intergroup Selection in the Evolution of Social Systems," *Nature* 200:623–26.

Wynne-Edwards, V. C. (1964a), "Group Selection and Kin Selection," *Nature* 201:1147.

Wynne-Edwards, V. C. (1964b), "Survival of Young Swifts in Relation to Brood-Size," *Nature* 201:1148–49.

Wynne-Edwards, V. C. (1964c), "Reply to Lack's 'Significance of Clutch-Size in Swift and Grouse'," *Nature* 203:99.

Wynne-Edwards, V. C. (1965), "Self-Regulating Systems in Populations of Animals," *Science* 147:1543–48.

Wynne-Edwards, V. C. (1968), "Population Control and Social Selection in Animals," in D. C. Glass (ed.), *Biology and Behavior: Genetics* (New York: Rockefeller University Press).

Wynne-Edwards, V. C. (1970), "Feedback from Food Resources to Population Regulation," in A. Watson (ed.), *Animal Populations in Relation to Their Food Resources* (Oxford: Blackwell).

Wynne-Edwards, V. C. (1971), "Space Use and the Social Community in Animals and Men," in A. H. Esser (ed.), *Behavior and Environment: The Use of Space by Animals and Men* (New York: Plenum Press), pp. 267–80.

Wynne-Edwards, V. C. (1977), "Society Versus the Individual in Animal Evolution," in B. Stonehouse and L. Perrins (eds.), *Evolutionary Ecology* (Baltimore: University Park Press), pp. 5–17.

Wynne-Edwards, V. C. (1978), "Intrinsic Population Control: An Introduction," in F. J. Ebling and D. M. Stoddart (eds.), *Population Control by Social Behaviour* (London: Institute of Biology), pp. 1–22.

Wynne-Edwards, V. C. (1986), *Evolution Through Group Selection* (Oxford: Blackwell).

Wynne-Edwards, V. C. (1989), "Upstage and Backstage with Animal Dispersion," in D. A. Dewsbury (ed.), *Studying Animal Behavior: Autobiographies of the Founders* (Chicago: University of Chicago Press), pp. 486–512.

Wynne-Edwards, V. C. (1991), "Ecology Denies Neo-Darwinism," *The Ecologist* 21:136–41.

Wynne-Edwards, V. C. (1993), "A Rationale for Group Selection," *Journal of Theoretical Biology* 162:1–22.

Young, D. (1992), *The Discovery of Evolution* (London: Natural History Museum Publications).

Index